NMR of Biological Macromolecules

NATO ASI Series

Advanced Science Institutes Series

A series presenting the results of activities sponsored by the NATO Science Committee, which aims at the dissemination of advanced scientific and technological knowledge, with a view to strengthening links between scientific communities.

The Series is published by an international board of publishers in conjunction with the NATO Scientific Affairs Division

A Life Sciences **B Physics**	Plenum Publishing Corporation London and New York
C Mathematical and Physical Sciences **D Behavioural and Social Sciences** **E Applied Sciences**	Kluwer Academic Publishers Dordrecht, Boston and London
F Computer and Systems Sciences **G Ecological Sciences** **H Cell Biology** **I Global Environmental Change**	Springer-Verlag Berlin Heidelberg New York London Paris Tokyo Hong Kong Barcelona Budapest

NATO-PCO DATABASE

The electronic index to the NATO ASI Series provides full bibliographical references (with keywords and/or abstracts) to more than 30000 contributions from international scientists published in all sections of the NATO ASI Series. Access to the NATO-PCO DATABASE compiled by the NATO Publication Coordination Office is possible in two ways:

- via online FILE 128 (NATO-PCO DATABASE) hosted by ESRIN, Via Galileo Galilei, I-00044 Frascati, Italy.
- via CD-ROM "NATO Science & Technology Disk" with user-friendly retrieval software in English, French and German (© WTV GmbH and DATAWARE Technologies Inc. 1992).

The CD-ROM can be ordered through any member of the Board of Publishers or through NATO-PCO, Overijse, Belgium.

Series H: Cell Biology, Vol. 87

NMR of Biological Macromolecules

Edited by

Chariklia Ioannidou Stassinopoulou

Institute of Biology
National Centre for Scientific
Research "Demokritos"
153 10 Ag. Paraskevi, Greece

Springer-Verlag
Berlin Heidelberg New York London Paris Tokyo
Hong Kong Barcelona Budapest
Published in cooperation with NATO Scientific Affairs Division

Proceedings of the NATO Advanced Study Institute on NMR of Biological Macromolecules, held at Kolympari, Crete, August 23–September 2, 1993

ISBN-13:978-3-642-79160-4 e-ISBN-13:978-3-642-79158-1
DOI: 10.1007/978-3-642-79158-1

CIP data applied for

Softcover reprint of the hardcover 1st edition 1994

Typesetting: Camera ready by authors
SPIN 10120195 31/3130 - 5 4 3 2 1 0 - Printed on acid-free paper

PREFACE

The purpose of the Advanced Summer Institute held in Crete from August 23 to September 2, 1993 was to inform scientists in the fields of biophysics, biochemistry and molecular biology about the recent advances and the latest techniques of NMR as applied to the study of proteins, carbohydrates and nucleic acids. The first chapters of this volume are devoted to an introduction on NMR theory and parametres related to molecular structure and molecular interactions. NMR experiments from basic 1D to 2D, 3D and 4D used in combination with isotopically labelled molecules are described and a general strategy is presented for the biomacromolecular structure determination. The following chapters are dealing with more advanced principles and techniques and their applications on structure and dynamic processes involving biomacromolecules in solution. Some very advanced results on peptide,protein, oligosaccharide and nucleic acid structure and recognition are presented. The need for sceptical use of the procedures leading from experimental data to molecular structures is stressed at many points.

The Advanced Summer Institute was financed mainly by NATO. It was further supported by IUBMB, the University of Crete, the European Peptide Society and the Greek General Secretariat for Energy and Technology. Generous support was also granted by Shell Hellas, Varian Ass. and Bruker AG.

The meeting was under the auspices of the National Center for Scientific Research "Demokritos" and the Hellenic Biochemical and Biophysical Society.

I wish to acknowledge the help of the members of the organizing committee, Drs. Ph.Dais, J.Feeney, I.Gerothanassis, G.Wagner and K.Wuthrich before, during and after the meeting. I wish also to thank once aigain Kate Apostolou for skilled and dedicated secretarial support.

April 1994 Chariklia Ioannidou Stassinopoulou

Athens

TABLE OF CONTENTS

BASIC PULSED NMR EXPERIMENTS

Charikllia Ioannidou Stassinopoulou
Institute of Biology
NCSR "Demokritos"
153 10 Aghia Paraskevi, Attiki
GREECE

NMR studies of biological macromolecules

Proteins, nucleic acids and polysaccharides are often called biopolymers because of their size and the repetitive nature of their structure. However, the structural complexity of biological polymers exceeds by far that of man-made polymers. The monomeric units of a biopolymer although they belong to the same class of chemical compounds exhibit a large variety:more than twenty aminoacids occur in proteins,the nucleic acids are composed of four purine and pyrimidine bases plus two types of sugars and,even in the simplest polysaccharides, a plethora of simple carbohydrate derivatives appear as monomers, interconnected by a variety of linkages. An additional level of complexity is introduced by the fact that biopolymers are also organized at higher orders of structure. One of the goals of molecular biology is to determine the structures of biological macromolecules in order to be able to study their biological functions.

Biopolymers can be examined at the level of primary, secondary or tertiary structure. Up to the middle seventies X-ray diffraction crystallography was the only method available for determining 3-dimensional structures of molecules . Attempts to use NMR to elucidate the structure of biological macromolecules started as early as 1957. The main problem in these early studies was the poor resolution of the available instrumentation. The advent of superconducting magnets, the use of Fourier

NATO ASI Series, Vol. H 87
NMR of Biological Macromolecules
Edited by C. I. Stassinopoulou

transform techniques and the introduction of computers for the control of the instrumentation opened the way for the application of NMR to the study of biopolymers. Even so, up to the end of the seventies the use of NMR was confined to the study of short oligomers or of some selected regions of larger molecules where the appropriate resonance lines in the spectra could sometimes be recognized against a complicated spectral background. In certain molecules containing paramagnetic centers or anisotropic shielding groups, as for example in metalloproteins or in heme-proteins, some resonances are shifted away from the mean envelope of signal,and a wealth of information concerning active sites can be retrieved.

Proton NMR was predominant in the early studies. In many cases fluorine substitution in crucial sites, the introduction of spin labels or substitution of existing diamagnetic metals with paramagnetic metals have all provided useful strategies for simplifying spectra and facilitating structural and kinetic studies. Nowadays incorporation of isotopic labels such as ^{13}C and ^{15}N play an increasingly important role in assisting NMR studies of biological macromolecules.

A very important breakthrough occurred when the techniques of 2D NMR were introduced for the study of biopolymers. Around 1977, Wuthrich's group in their pioneering work on proteins showed that two dimensional NMR techniques not only significantly increased the resolution but also provided a means of making sequential assignments for all the resonances in the molecule. Subsequently they showed how these methods could be used to derive the 3-dimensional structure of a small protein in solution. These methods were also rapidly applied to the two other classes of macromolecules. New 2D techniques were devised at an ever increasing rate and multidimensional NMR methods (3D and 4D) were introduced and have now become standard practice.Because of the complexity of modern NMR spectra computer assistance is frequently used for assisting peak assignments and is essential for obtaining three dimensional structures of macromolecules.

The one-pulse NMR experiment

The nuclear Zeeman splitting

There are two allowed energy states for a nucleus of quantum number $I = 1/2$ in the presence of a magnetic field B_0 : one corresponding to the nuclear magnetic moment, μ, aligned with the field, α, and one corresponding to μ opposed to the field, β (Fig.1). We will confine our attention to nuclei with spin 1/2 since most of the important nuclei involved in NMR of biomolecules namely ^{1}H, ^{13}C, ^{15}N and ^{31}P belong to this group.

In a real sample the population of nuclei will be distributed between the two states according to Boltzmann

$$N_\beta = N_\alpha \exp(-\Delta E/kT)$$

where ΔE is the difference in energy between the two states. There is only a slight excess of nuclei in the α state because ΔE is of the order of kT. This excess ΔN gives rise to a net macroscopic magnetization. We shall very briefly describe the behaviour of the magnetization during an NMR experiment using the vector model which is more easy to visualize than the quantum mechanical model and is adequate for an elementary description of several modern NMR experiments. A better understanding is provided by the product operators approach described in a following chapter.

Detection of the NMR signal

First we consider the vector M of the net nuclear magnetization in a frame of Cartesian coordinates. This is what we call the laboratory frame.

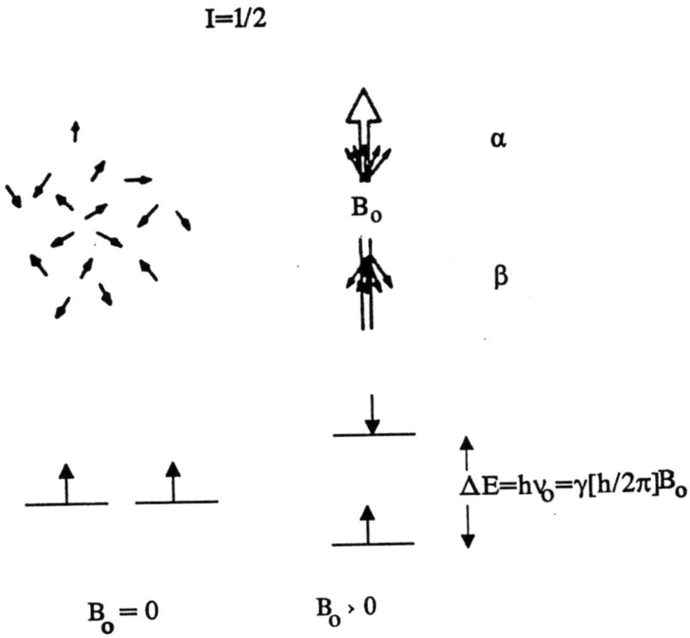

Figure 1. Orientation of the nuclear spins, I = 1/2, when $B_0 = 0$ (left) and when $B_0 > 0$ (right).

For simplicity B_0 is made to coincide with the z axis . At equilibrium M is aligned with B_0 along the z axis (Fig.2a). In order to obtain a signal we must perturb the system by switching on an oscillating field B_1 for a very short time (of the order of microseconds). This is referred to as a pulse of radiofrequency (because the energy given by the pulse is in the region of radiofreqencies as dictated by $\Delta E = h\nu$). B_1 is perpendicular to B_0 and may be along x or along y. Under the influence of the pulse the vector M is tipped away from the z axis. The tipping angle depends on B_1 and on the duration of the pulse. In a simple experiment it is equal to 90 (90 degree pulse). A 90_x pulse rotates M about the x axis and aligns it with the y axis. A 90_y pulse rotates M about y and aligns it with x.

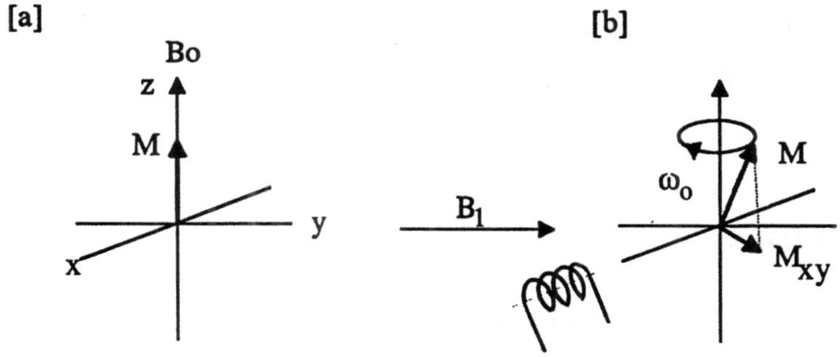

Figure 2. Effects of B_1 pulse on the magnetization vector M. At equilibrium [a] M is aligned with B_0 along the z axis. After the pulse [b] M is precessing about B_0 with a Larmor frequency ω_0 ; there is a component $M_{xy} = 0$ the precession of which induces an oscillating current in the receiver coil positioned along x.

After the pulse we are left with two vectors, M and B_0 interacting with each other. According to classical physics, B_0 exerts a torque on M. As a result M starts precessing about B_0 at an angular frequency ω_0 (Larmor frequency) related to B_0 by

$$\omega_0 = \gamma B_0$$

where $\omega_0 = 2\pi\nu_0$ and γ is the magnetogyric ratio characterizing the observed nucleus.

Thus, in the non equilibrium position, M has a component M_{xy} precessing in the plane perpendicular to B_0 and a component along the z axis. M_{xy} decays to zero as the equilibrium is restored whereas M_z builds up to become M, the equilibrium value of the total magnetization. For the time interval during which M_{xy} is non-zero its precession induces an oscillating current in a receiver coil placed at the proper position perpendicular to the static magnetic field (Fig.2b). This current after being properly amplified and mathematically treated produces the detected NMR signal. The signal before the mathematical manipulation is called a free induction decay (FID) signal. It is a function of time and it is also referred to as the signal in the time domain.

The rotating frame

Instead of using a fixed coordinate system as we have done in describing the NMR experiment it proves to be simpler to use a coordinate system in which the x and y axes are rotating about the z axis. We define a new set of coordinates x′, y′, z′ (Fig.3). The z′ axis is fixed and coincides with z of the laboratory frame and with B_0. The x′ and y′ axes rotate with an angular frequency ω' given by

$$\omega' = \gamma B_1$$

In the rotating frame we have the advantage of simplifying the observed motion of the M vector away from equilibrium (during the NMR experiment). Thus , if we choose $\omega' = \omega_0$ (condition of resonance) we are effectively "switching off" the Larmor precession. If one sits on the x′ axis and observes from "inside" the M_{xy} vector the system will appear to be stationary. We say that M_{xy} is fixed in the rotating frame. In fact, immediately after a 90 degree pulse and before the M_z component starts building up it is as if the B_0 were non existent.

The concept of spin packets (or spin isochromats)

In a real sample, the energy difference, ΔE between states α and β, depends not only on γ and B_0 but also on the magnetic environment of the nucleus observed:

$$\Delta E = (h/2\pi)\ \gamma\ (1-\sigma)\ B_0$$

where σ is the shielding cause by its environment. The effect of the static

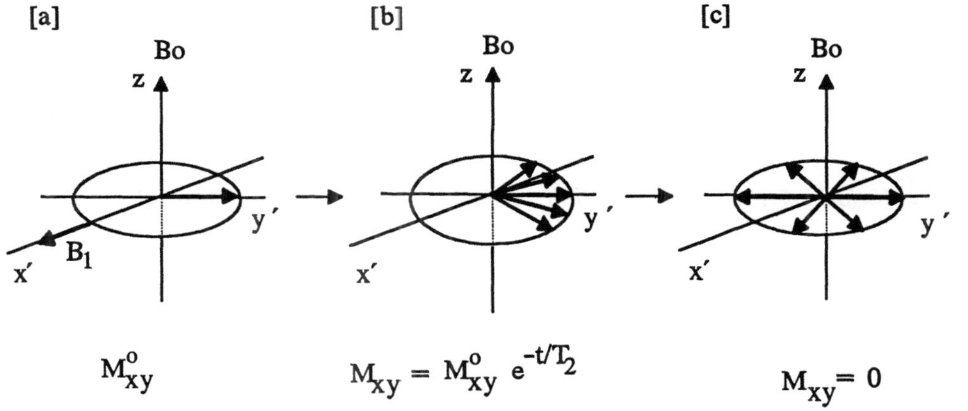

Figure 3. Evolution of the magnetization in the xy plane after a 90_x pulse. Immediately after the pulse M_{xy} is aligned with the y axis and has its maximum value, M^0_{xy}. At time t after the pulse [b] the spins have started fanning out. After $t = T_2$, where T_2 is the transverse relaxation time, the spins are distributed in the four quadrants of the xy plane [c] and $M_{xy} = 0$.

magnetic field is felt by the nucleus after the shielding imposed by the environment. The parameter σ is directly related to the chemical shift.

The net magnetization due to nuclei of one species (same γ) in a sample is made up from the magnetic moments of groups of magnetically equivalent nuclei such as the protons of a freely rotating methyl group. In each group the spins have a characteristic Larmor frequency. After a 90 degree pulse the individual groups start precessing about the z axis at their characteristic frequencies.

A spin packet can be defined as the average spin vector of a group of magnetically equivalent nuclei characterized by a certain σ so that its average Larmor frequency is equal to ω_0 (1-σ) . For the observer inside the rotating frame the spin packet will precess with an angular frequency, ω, given by the difference

$$\omega = \omega_0 (1-\sigma) - \omega'$$

Evolution of the magnetization after the pulse in the rotating frame of reference

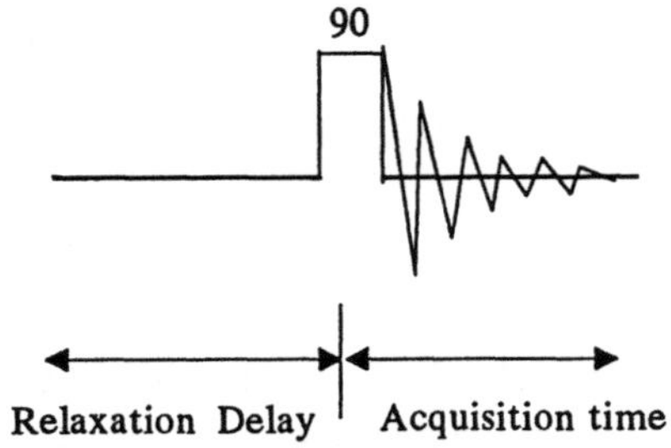

Scheme I

The radiofrequency pulse has the effect of abolishing the differences in ω of the individual spin packets: the magnetization is "bunched up" and tipped into the $x'y'$ plane (from now on we will omit the primes and call it the xy plane) as a single vector. As time passes the spin packets fall out of phase since each is characterized by its own Larmor frequency. In addition the individual spins within each packet fall out of phase mainly because of inhomogeneity in the B_0 field. We say that the spins "fan out": they spread like a chinese fan (Fig.3B).

The FID is made up of the sum of the signals of all the spin packets present. After some time the magnetization in the xy plane becomes zero because the magnetization vectors of the spin packets cancel each other as they are distributed in all four quadrants. The signal in the time domain contains all the information we need but in a form not easily recognisable. By applying a Fourier transformation we obtain the signal in the frequency domain which is the familiar NMR spectrum.

Table 1

NMR parameters pertinent to the study of biological macromolecules

Parameter	Physical Relevance	Application
δ Chemical shift	Nuclear shielding	Chemical structure
J Scalar coupling constant	Interaction of nuclear spins through bonds	Molecular geometry Recognition of spin systems
T_1, T_2 Time constants	Spin-lattice and spin-spin relaxation processes	Molecular motions, intramolecular flexibilities, conformational changes, distances of molecular approach, diffusion coefficients
NOE Nuclear Overhauser Effect	Dipolar coupling of nuclear spins	Intra - and intermolecular distances Sequential assignements
Lineshape	Physical or chemical dynamic events occuring at the NMR time scale	Kinetic processes causing the nuclei to sample more than one environment during the NMR pulse sequence

The basic NMR experiment described above follows the simple pulse sequence (relaxation delay-90_x-acquisition)n (Scheme I). It is called a 1D experiment because it involves a single time dimension. Often it is repeated n times to improve the signal to noise ratio by a factor $\sqrt{n}$. A relaxation delay may be introduced before each pulse to allow the return of magnetization to equilibrium. Other delays may be introduced according to need. The Fourier transformation is performed on the sum of the FID signals. Chemical shifts and coupling constants are obtained from the frequency spectrum in the conventional way.

More sophisticated experiments are necessary to extract information about the chemical environment from very complex spectra and/or to obtain additional NMR parameters (Table 1) used for the study of molecular structure and molecular motions. Two types of 1D experiments, based on spin-echoes and polarization transfer will be briefly described.

Spin-echo experiments

Spin-echoes represent magnetization appearing as a response to a series of pulses at a specific time after the last pulse. In the experiment described in Scheme II a 90_x pulse is applied followed after an interval τ sec by a 180_x or a 180_y pulse. At exactly τ sec after the second pulse an echo-signal appears. The origin of the echo is understood by following the evolution of the magnetization during the experiment (Fig.4) . The 90_x pulse turns the macroscopic magnetization in the y direction. After the pulse the spin packets start precessing with their characteristic Larmor frequencies. Some will be faster and some slower. Let us follow the course of only two packets one slow, s, and one fast, f. At the end of the first τ interval the spin packets will have a phase difference. At that point the 180_x pulse is applied. The pulse reverses the signs of the fast and slow components. After the pulse the spin packets continue to precess with the same frequencies respectively. The fast component will catch up with the slow after exactly the same time τ that was needed to create the difference in phase existing at the time of application of the 180 pulse. Thus, the total magnetization is refocused along the y axis but in the -y direction. If a 180_y pulse were used, the alignment would be along +y. If the receiver is turned on an echo signal will be detected originating from the precession of the refocused magnetization. If

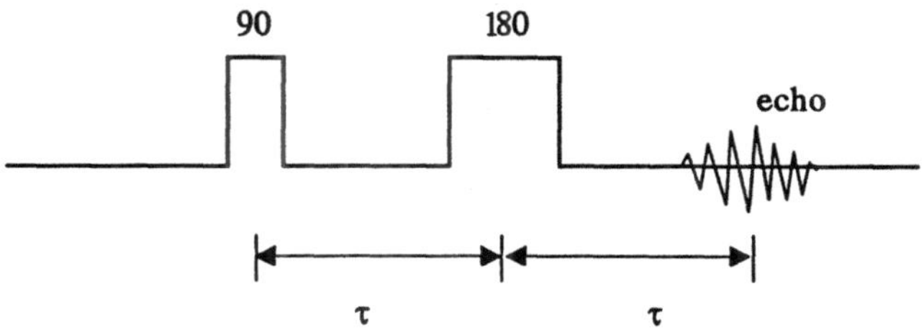

Scheme II

the pulse sequence is repeated many times the intensity of the echo signal will diminish because of spin-spin relaxation and, possibly, diffusion of the nuclei between regions of the sample characterized by different magnetic environments. Thus, spin-echo experiments are used among other applications for the measurement of T_2 and diffusion coefficients.

Differences in precession frequency may originate from chemical shift or scalar coupling. The chemical shift effect is always refocused. The effect of coupling is different for homonuclear and heteronuclear spin systems. A heteronuclear spin system IS, e.g. ^{1}H, ^{13}C, gives normally one doublet in each rotating frame. If the observed nucleus is proton, the pulses applied affect only protons. The doublet appearing in the proton spectrum has two components due to the scalar coupling to carbon, one fast and one slow which will refocus at time τ after the 180 pulse as described above. Thus the effect of the scalar coupling will be removed at the echo maximum. In the homonuclear case though, the pulses affect both I and S spins. The 180 pulse reverses the identity of the spin packets: the fast becomes slow and the slow fast. The nuclei which precess faster before the second pulse because they are coupled to certain spins α are now precessing slower because those spins have flipped and are now β. The packets continue to diverge after the second pulse. As a consequence they are not refocused at time τ. The scalar coupling is not removed and is manifest as a time modulation of the echo

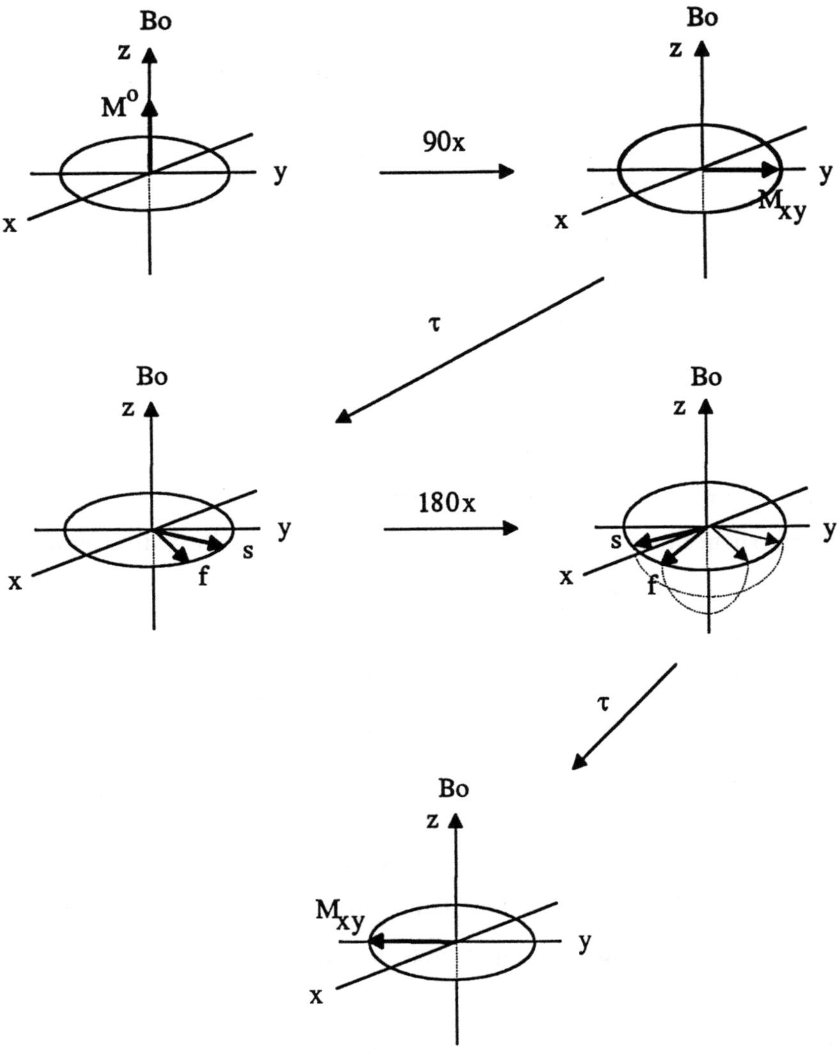

Figure 4. Evolution of the magnetization in a spin echo experiment.

maximum by the coupling constant J. Accurate proton-proton couplings can be evaluated in biomolecules by taking advantage of this effect. Other applications include spin-echo difference spectroscopy which has been used to simplify crowded proton spectra in biological tissues.

Spin echo modulation by heteronuclear scalar coupling can be introduced by applying a 180 pulse to the heteronuclei at the same time or immediately after the second pulse. A well known application is the Attached Proton Test (APT) experiment used for the assignment of carbon spectra: depending on the number of attached protons carbons give positive or negative signals.

Polarization transfer experiments

In polarization transfer experiments magnetic order or coherence is

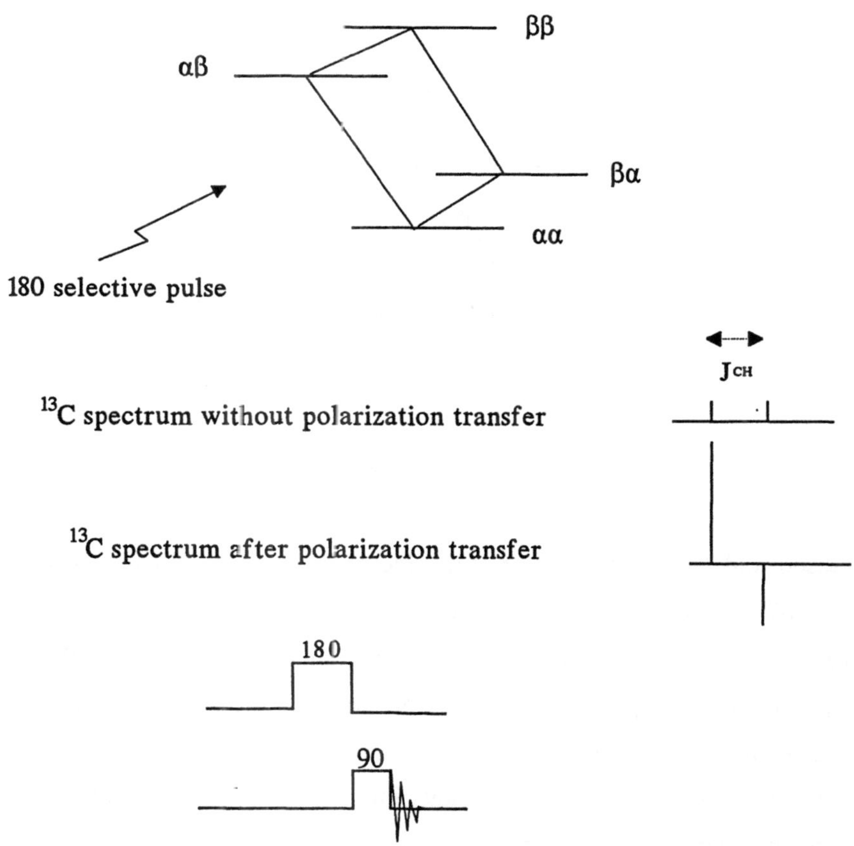

Figure 5. Selective population transfer in a CH system (SPT experiment).

transferred from a nuclear species of high γ to another of low γ. The transfer results in an increase of the signal intensity of the species with low magnetogyric ratio.

Consider a pair of heteronuclear coupled spins, ^{1}H ^{13}C. The energy levels of the system are given in Figure 5 where the proton is the polarizing nucleus and carbon is the observed nucleus. Polarization will be transferred via the scalar coupling. A 180 selective pulse, affecting only one transition, is applied to the proton transition $\alpha\alpha$ to $\alpha\beta$. The pulse reverses the direction of the magnetization and interchanges the spin states of the protons coupled to carbons in the $\alpha\beta$ spin state. Thus, the populations of the states connected by the irradiated transition are exchanged. The population of the upper state is

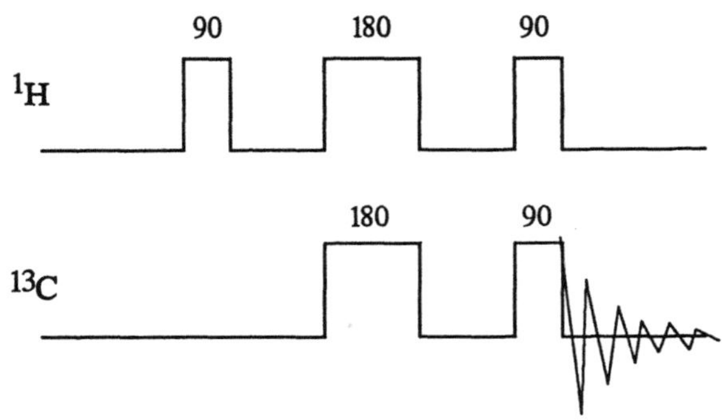

Scheme III

increased and that of the lower state decreased. Increase in the population of the $\alpha\beta$ state increases the probability of the $\alpha\beta$ to $\beta\beta$ ^{13}C transition: the corresponding line in the carbon spectrum will be enhanced. On the other hand, a decrease in population of the $\alpha\alpha$ state increases the probability of the $\beta\alpha$ to $\alpha\alpha$ ^{13}C transition : there will be an emission line at the carbon frequency corresponding to this transition. The relative intensities of the carbon lines can be derived from the differences in energy of the states in Fig.5. The experiment described above is called Selective Polarization Transfer (SPT). A widely used experiment for the observation of heteronuclei is the Insensitive Nuclei Enhancement by Polarization Transfer (INEPT). The sequence of INEPT (Scheme III) uses non-selective pulses. It is so designed that polarization transfer occurs in all multiplets with similar carbon-proton J values with corresponding increase in intensity of the carbon lines involved .

More complex pulse sequences and higher time-dimension experiments will be discussed in the following chapters.

General References

Brey WS (1988) Basic Methods and simple pulsed experiments. In : Brey WS(ed.) Pulse methods in 1D and 2D liquid-phase NMR. Academic Press, Inc.. Harcourt Brace Jovanovich, Publishers: 1-109.

Derome AE (1987) Modern NMR techniques for chemistry research. Pergamon Press.

Kessler H, Gehrke M and Griesinger C (1988) Two-dimensional NMR spectroscopy : Background and overview of the experiments. Angew. Chem. Int. Ed. Engl. *27*: 490 - 536.

Sanders JKM and Hunter BK (1988) Modern NMR spectroscopy, a guide for chemists. Oxford University Press.

Schwartz LJ (1988) A step-by-step picture of pulsed (time-domain) NMR. J.Chem.Educ. *65*: 752-756.

Wuthrich K (1986) NMR of proteins and nucleic acids. Wiley-Interscience, John Wiley & Sons.

PRINCIPLES OF MULTIDIMENSIONAL NMR

Gerhard Wagner
Department of Biological Chemistry and Molecular Pharmacology
Harvard Medical School
240 Longwood Avenue
Boston, MA 02115

1. Introduction

Multidimensional NMR has become a key tool in chemistry and structural biology. One of the reasons for this interest is that the method can be used to solve structures of small proteins or nucleic acids in solution. Thus, NMR has evolved from an esoteric technique to a major analytical tool. NMR experiments are built from a combination of various radio frequency pulses separated by delays, and the oscillation of the radiofrequency of the pulses may have different phases with respect to a spectrometer reference phase. This allows for an essentially unlimited number of different pulse sequences, many of which can provide interesting information about molecular structure and mobility. The advent of the operator product formalism (Sørensen et al., 1983) and a general familiarity with this formalism makes the design of new and interesting pulse sequences rather simple and has lead to an avalanche of new NMR techniques. In this article, the basic principles of 1D and nD NMR experiments are introduced on a basic and pictorial level. No explanation of the concept of coherence and net magnetization, or the operator product formalism is given since these are treated in a different chapter of this book.

2. Basics of pulsed NMR spectroscopy.

2.1 One-pulse sequence

The simplest pulsed NMR experiment is the so-called one-pulse sequence (Figure 1). It consists of a single 90° pulse after which the signal is detected in the receiver coil. The 90^o_x pulse turns the equilibrium z-magnetization of a nucleus (M^A) into the -y direction. If the frequency of the rotating frame (spectrometer reference frequency or carrier) is not exactly on resonance, the magnetization vector will precess in the x,y plane and relax back to the equilibrium z-orientation. The receiver coil can be adjusted so that either M_y or (and) M_x are detected. The signals are decaying cosine and sine functions, respectively. The decaying oscillation is called free induction decay (FID). The FID can be converted to a spectrum by a Fourier transformation.

NATO ASI Series, Vol. H 87
NMR of Biological Macromolecules
Edited by C. I. Stassinopoulou

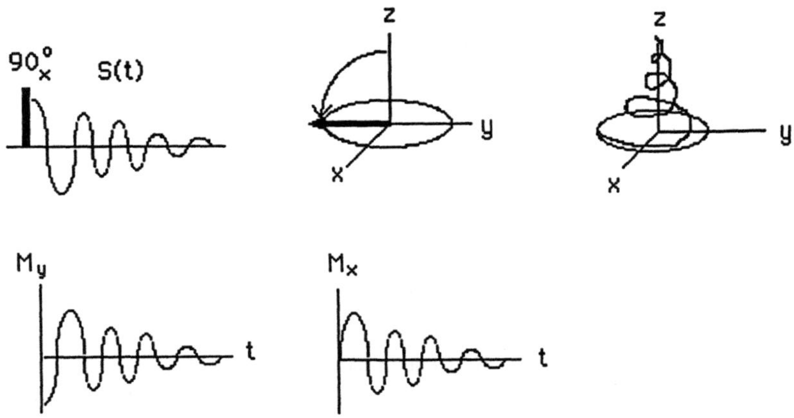

$$M_y^A(t) = -M_o^A \cos(\omega_A t)\, e^{-t/T_2}$$
$$M_x^A(t) = +M_o^A \sin(\omega_A t)\, e^{-t/T_2}$$

Fig. 1: Effect of a 90° pulse on nuclear magnetization in equilibrium.

2.2 Fourier Transformation - Quadrature Detection

The Fourier transformation is essentially a mathematical frequency filter. There are two ways to create an absorption line from an FID. A cosine transformation of the decaying cosine function, S_C^C, and a sine transformation of the decaying sine function, S_S^S, both yield absorptive Lorentzian lines.

$$S_c^c(\omega) = \int M_y(t) \cos \omega t \, dt = \int -M_o^A \cos\omega_A t \cos\omega t \, e^{-t/T_2^A} \, dt$$
$$S_s^s(\omega) = \int M_y(t) \sin \omega t \, dt = \int -M_o^A \sin\omega_A t \sin\omega t \, e^{-t/T_2^A} \, dt$$

The decaying cosine function is multiplied with a cosine function of variable frequency ω which is incremented from 0 to ω_{max}. Only if ω is equal or close to ω_A or $-\omega_A$ the integral of the function is not zero because the integrand is a squared cosine function. Otherwise, it is a function oscillating between positive and negative values, and the integral is zero. Similarly, the sine transformation of the sine modulated decaying signal yields absorption lines, however, the line at negative frequency has opposite sign.

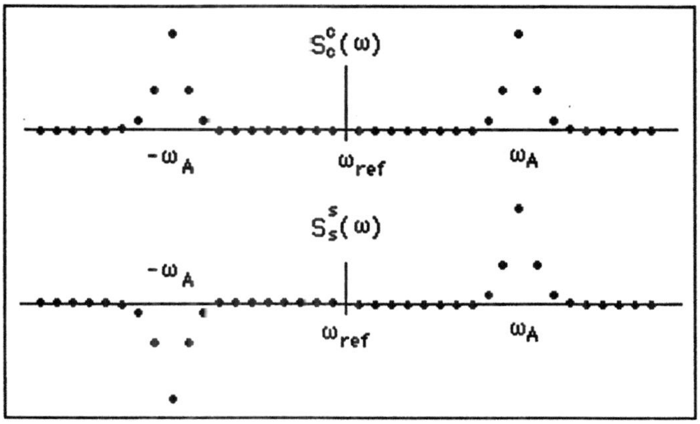

Fig. 2: The effect of a cosine Fourier transformation on a decaying cosine function and sine Fourier transformation on a decaying sine function. Addition of both transforms yields the determination of the sign of the frequency relative to the spectrometer reference frequency.

Therefore, both transforms enable us to determine the sign of the frequency ω_A relative to ω_{ref} by adding the two transforms. $S_c^c(\omega) + S_s^s(\omega)$ gives a single resonance at $\omega = \omega_A$. This is usually called quadrature detection and allows to place the carrier into the center of the spectrum and to consider positive and negative test frequencies in the Fourier transform. Alternatively, the correct spectrum could be obtained by placing the carrier outside of the spectrum and to consider only positive test frequencies in the Fourier transformation.

2.3 Time-proportional phase incrementation (TPPI)

In this method for determination of the sign of a resonance frequency, the phase of the nth sampling point measured at $n * \Delta t$ is incremented by $n * (\pi/2)$. This has the effect that the frequency of the signal is artificially shifted away from the carrier frequency (Fig. 3) by $\frac{1}{4\,\Delta t}$. In the Fourier transformation only positive frequencies are considered.

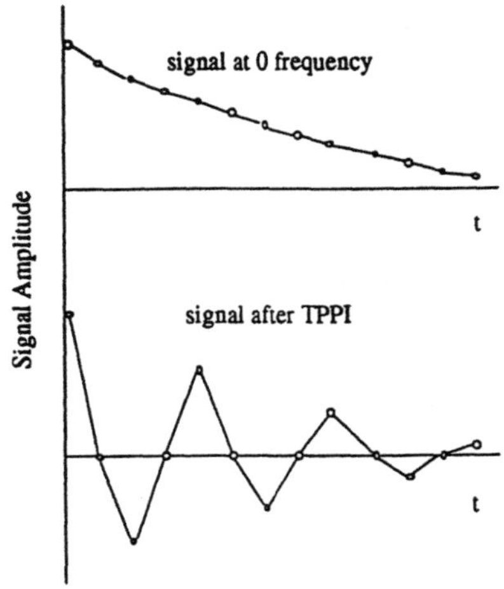

Fig. 3: Time-proportional phase incrementation imposes an oscillation on the decaying exponential

3. Principle of nD NMR

A 1D NMR experiment consists of a preparation sequence (in the simplest case a single 90° pulse) and a detection period (t) during which the time dependence of a signal is recorded, this means a signal in the receiver coil is measured at the times $t_n = t_o + n * \Delta t$ $(n = 0,1,...n_{max})$. A 2D experiment consists, in the simplest case, of a sequence of two 90° pulses separated by the evolution time t_1. The first pulse, lets say a $90°_x$ pulse, converts the z-magnetization into the transverse plane (-y direction). The transverse magnetization precesses at the Larmor frequency of that particular spin. The effect of the second pulse, for example a $90°_x$ pulse, depends on how far the magnetization has already precessed. A 2D experiment consists of a series of two-pulse sequences where the delay t_1 is systematically incremented. Thus the effect of the second pulse (mixing pulse) is modulated with the frequency of precession during t_1. The experiment contains two incremented delays, an evolution and a detection period, that contain both oscillations with characteristic frequencies which are then identified by a 2D Fourier transformation. A n dimensional NMR experiment contains n-1 evolution and mixing periods and one detection period. The general scheme of a n-dimensional NMR experiment is:

Preparation - (Evolution - Mixing $)_{n-1}$ - Detection

4. NOE

The nuclear Overhauser effect (NOE) is among the most useful features of NMR spectroscopy. Most importantly, it allows to estimate intramolecular distances that can be used for structure calculations. It is based on the effects of dipolar interactions between protons (Abragam, 1961; Solomon, 1955).

4.1. Hamiltonian for a Two-Spin System, I and S

To understand some principles of the NOE, it is sufficient to consider Zeeman and Dipole-Dipole Interactions only. The Hamiltonians for these interactions are listed below.

$$H_o = - \hbar \gamma_I B_o I_z - \hbar \gamma_S B_o S_z + H'$$

$$H' = - \frac{\hbar^2 \gamma_I \gamma_S}{b^3} [3 \cdot (\mathbf{I} \cdot \mathbf{r})(\mathbf{S} \cdot \mathbf{r}) - \mathbf{I} \cdot \mathbf{S}]$$

The Zeeman term, H_o, is a coherent Hamiltonian, it is the same for all molecules of the sample. It only leads to coherent precession. The dipolar term, H', is an incoherent Hamiltonian. It is different for all the molecules of the sample because it depends on the orientation of the molecules relative to the external field, B_o. This interaction leads to transitions between different energy levels and thus to relaxation. Other interactions may also cause relaxation or NOE. However, for proton NMR in macromolecules they are usually not important and shall be neglected in this context.

4.2. Energy Levels and Transition Probabilities

Time-dependent perturbation theory with expansions up to second order terms is generally used to derive the principles of relaxation theory (Abragam, 1961). The transition probabilities per unit time between levels i and j are:

$$w_{ij} = \frac{1}{t} \frac{1}{\hbar^2} \left| \int_0^t < m_j | H'(t) | m_i > e^{-i \omega_{ij} t'} dt' \right|^2$$

It is obvious that transitions between the levels i and j occur only if $H'(t)$ fluctuates with the resonance frequency ω_{ij}. Thus, only motions with time constants corresponding to transition frequencies will have an influence to relaxation and NOE's.

4.3. Pictorial Description of Transition Probabilities

The spin I in an external magnetic field, B_0, produces a dipole field at the site of the spin S (dipole-dipole interaction). Rotational motion of the molecule causes the direction of the dipole field vectors to change. This leads to an oscillation of the dipole-dipole interaction with the frequency of the rotation of the molecule. This is analogous to irradiation of a radio frequency field with the frequency of the rotation of the molecule. If there are rotational motions with frequencies of energy differences in the energy level diagram transitions may be induced and lead to relaxation and NOE.

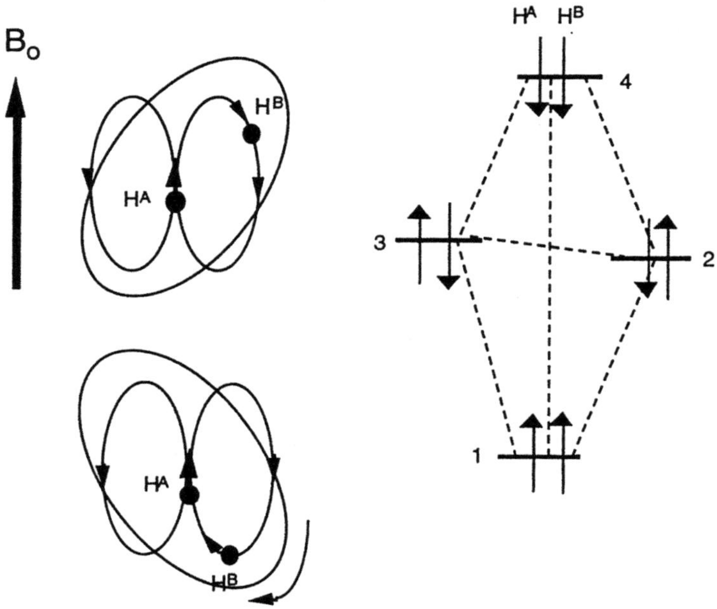

4.4. A Numerical Example - Positive and Negative NOE's

It is informative to construct a highly simplified numerical example of populations of spin states to understand positive and negative NOE's. This is based on the two-spin energy level diagram discussed above. For example, in equilibrium, the four states could have populations of 60, 52, 48 and 40 spins. (In reality, these populations at ambient temperature will be nearly identical). The intensity of a signal is proportional to the population difference. This is 12 for the I-spin and 8 for the S-spin (a). Saturation of a resonance creates a non-equilibrium situation (b). For example, saturation of the I-spin equalizes these populations to 54 and 46, respectively. Due to the coupling of the molecular motions with the nuclear states the system will tend to relax back to equilibrium(c).

a

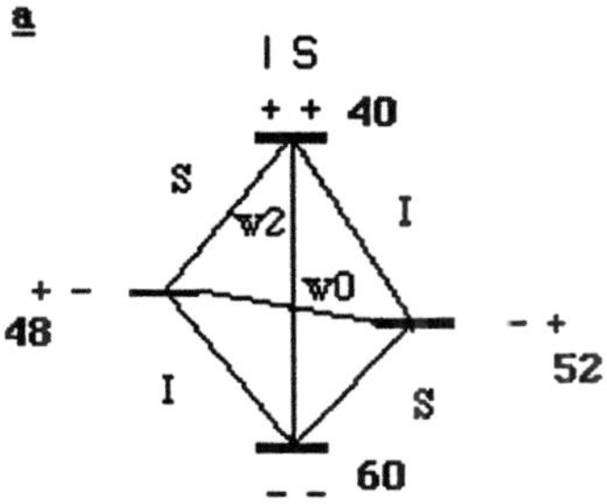

$\Delta_{1I} = 12$

$\Delta_{1S} = 8$

$\Delta_{2} = 20$

$\Delta_{0} = 4$

b

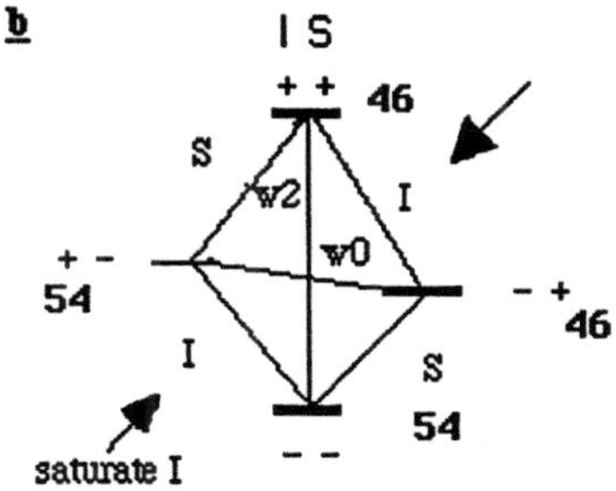

$\Delta_{1I} = 0$

$\Delta_{1S} = 8$ unchanged

$\Delta_{2} = 8$

$\Delta_{0} = -8$

c

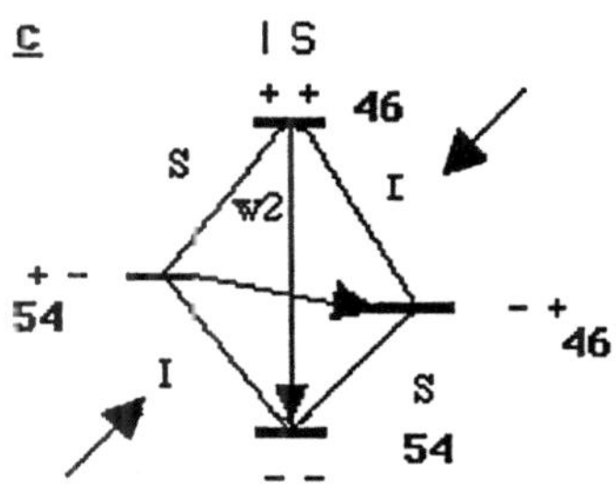

flow of spins towards equilibrium

d

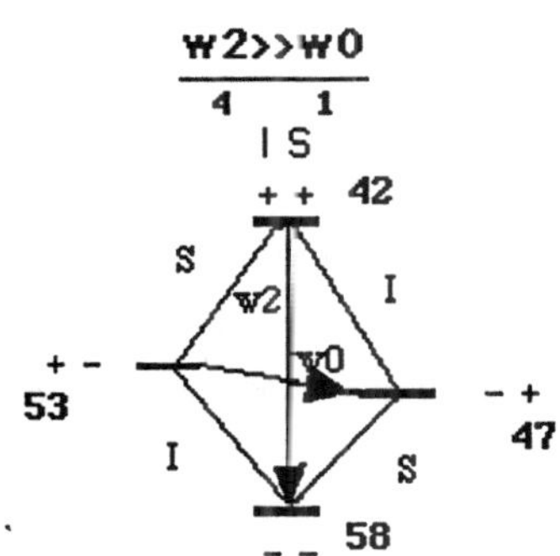

small molecules

$\Delta_{1I} = 11$ (vs.8) positive NOE

e

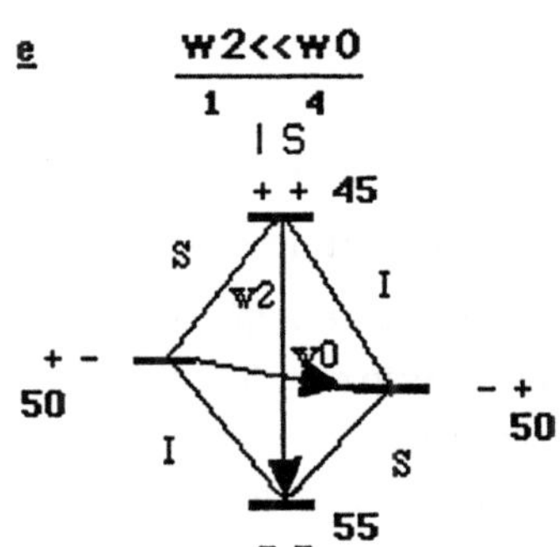

large molecules

$\Delta_{1I} = 5$ (vs.8) negative NOE

Only w_2 and w_o transitions play a role for cross relaxation rates (Solomon, 1955). If $w_2 >> w_o$ (small molecules) we have positive enhancements, if $w_2 << w_o$ (large molecules) we have negative enhancements.

5. NOESY

In a NOESY experiment we create a non-equilibrium situation with a sequence of two 90° pulses. Let us consider two spins, A and B. Spin A is close to the carrier (spectrometer reference frequency) and will thus precess slowly in the rotating frame (frame rotating with the frequency of the carrier). Spin B is farther away and will thus precess more rapidly.

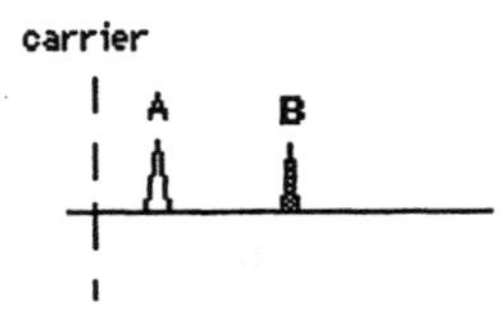

The NOESY experiment consists of three 90° pulses:

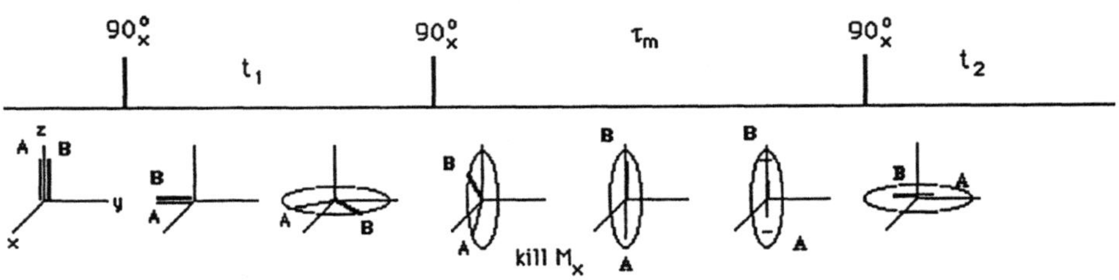

The first 90°_x pulse turns the magnetization of both spins in the -y direction (all rotations are defined as anti-clockwise rotations). They precess with different rates. Application of a second 90°_x pulse at time t_1 leads, in the example depicted, to a situation where the B-spin has a positive and the A-spin a negative z-component. This is a classical non-equilibrium situation, and the length of the vectors will be influenced by dipole-dipole interaction between the two spins. In macromolecules ($\omega\tau_c > 1$), the vectors become shorter, due to relaxation and cross relaxation. The transverse component of the magnetization is not interesting in this context and even undesired. It is eliminated by some tricks (phase cycling or homospoil pulses).

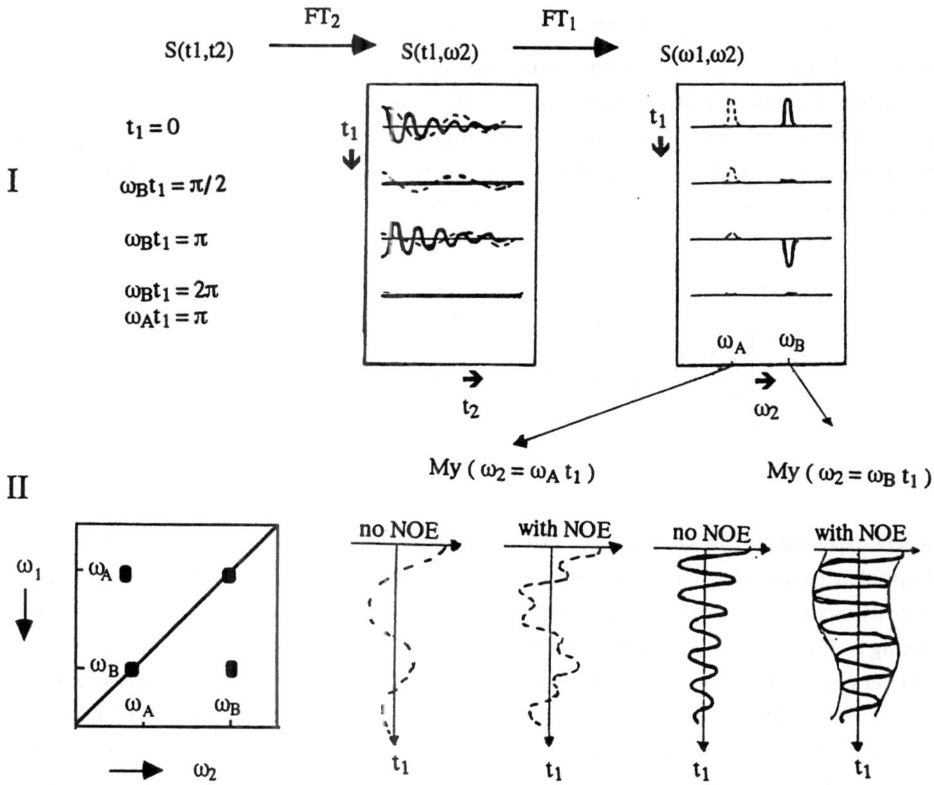

Fig. 4 : Panel I: Examples of free induction decays at selected t_1 values and corresponding 1D spectra. Panel II: NOE modulation of FID'S in the time/frequency domain (right) and 2D spectrum (left).

The third 90^o_x pulse turns the magnetization back into the transverse plane for detection. In this case, which describes the situation for one particular t_1 value, the A-spin has a positive, the B-spin a negative x-component at the beginning of t_2. The signals precess and decay during t_2. Fourier transformation with respect to t_2 yields a positive resonance for the A-spin and a negative signal for the B-spin, since the intensity of the frequency-domain signal is equal to the first point of the free induction decay. The vector picture depicts the situation at one particular t_1 value.

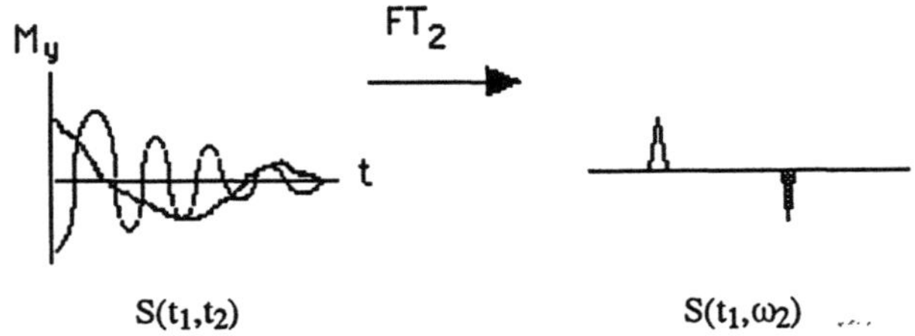

A 2D experiment consists of a series of experiments where t_1 is incremented from 0 to t_1^{max}. Depending on the value of t_1, the magnetization present for spin A or spin B has different values, it actually oscillates with their respective Larmor frequencies, due to the different phase accumulation during t_1. Fig. 4I shows some examples of free induction decays for some selected t_1 values. After Fourier transformation with respect to t_2, we have a stack of 1D spectra, the mixed time/frequency domain $S(t_1,\omega_2)$ where the intensities of the peaks oscillate with the precession frequencies during t_1 (Fig. 4II). Fourier transformation with respect to t_1 yields the frequency domain data set, $S(\omega_1,\omega_2)$. If there is no NOE transfer, the dependence on t_1 of the A-peak in the $S(t_1,\omega_2)$ data set is modulated with ω_A, that of the B-peak with ω_B. Thus this yields diagonal peaks. If there is NOE, the peak intensities are modulated with the frequencies of the interacting spin. This yields the cross peaks. Thus, the peaks above and below the diagonal mean NOE transfers from B to A and from A to B, respectively.

The elimination of the transverse component of the magnetization can be readily obtained with cycling the phases of the 90 pulses, as is shown in the Fig. 5. The phase cycling leads to opposite signs of the transverse components which cancel if the two scans are added together. This is the simplest case of a phase cycling. Usually, additional steps are necessary to eliminate unwanted signals.

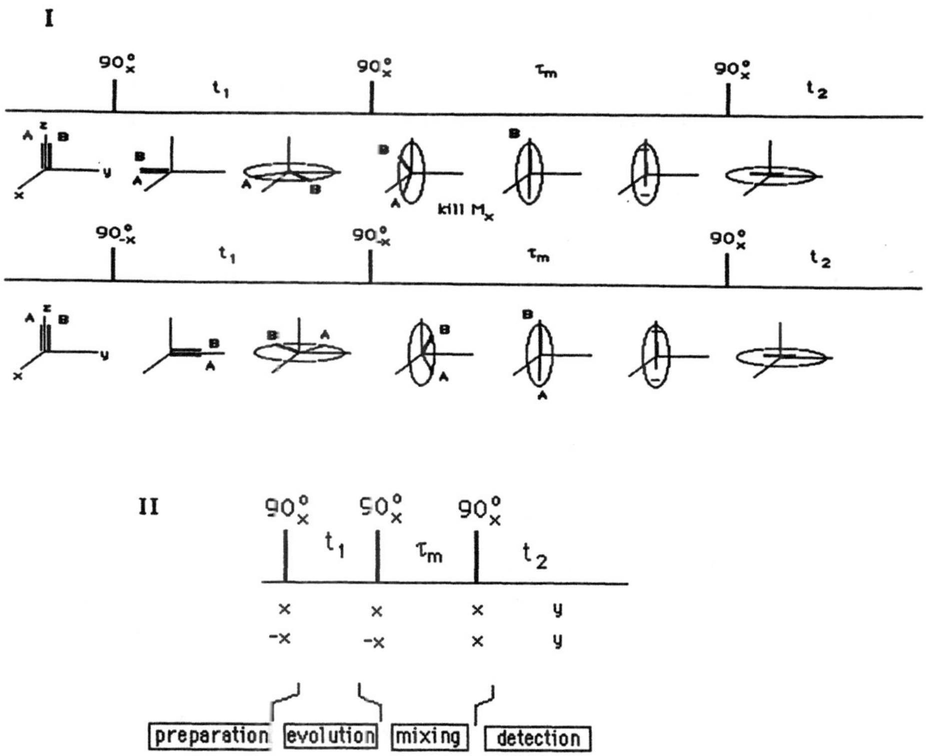

Fig. 5 : Panel I: Elimination of the transverse component of the magnetization (single quantum transition: 1QC). Panel II: Short notation for the experiment described in A. This phase cycle eliminates 1QC present during time τ_m . In addition it eliminates axial peaks arising due to relaxation during t_1.

6. The effect of a two pulses sequence on a group of two coupled spins.

The effect of a sequence of two 90° pulses on a system of coupled spins, leading to coherence transfer and two-quantum coherence can be understood with vector pictures. Two spins coupled

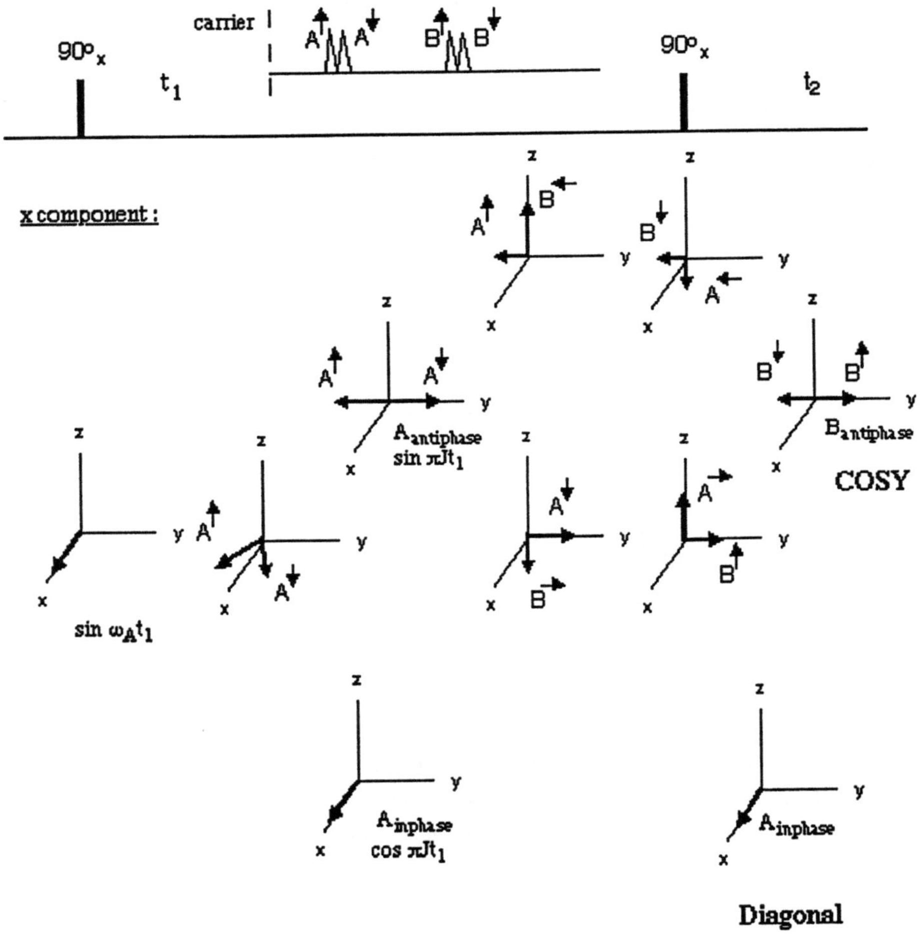

Fig. 6: Vectorial representation of the effect of a two-pulse sequence on two coupled spins.

by scalar interaction show two doublets (Fig. 6, x component). For simplicity, we consider initially only the doublet of the A-spin. The two components of the signal correspond to either 50% of the molecules in which the B-spin is either up or down, respectively. The same is true for the B-spin signal. However, the two doublet components represent different halves of the molecule population, there is no coherence between the A and B spin initially, but it can be created with the two-pulse sequence.

The first 90^o_x pulse turns A into the -y direction. During t_1 it starts to precess. Since a macroscopic magnetization vector (a coherence) is constituted from a large ensemble of microscopic nuclear magnetic moments, it can always be decomposed into its x and y components, each of which is constituted from a subgroup of nuclear magnetic moments. We first consider the x component (Fig. 6). The ensemble of spins that constitutes this magnetization can be further divided into two groups of A-spins, in the one group the spins of the coupled nuclei B are up, in the other group the B-spins are down. Immediately after the first 90^o_x pulse, the two groups of spins have the same transverse magnetization, pointing both in the +x direction, they are in phase. However, the two groups of A spins see slightly different effective fields due to the effect of scalar coupling and precess apart from each other, with a precession (secession) rate of πJ. The inphase coherence decreases with cos $\pi J t_1$, the antiphase coherence increases with sin $\pi J t_1$. At $t_1 = (2J)^{-1}$ they are completely anti parallel (antiphase). The two vectors of the antiphase coherence represent two ensembles of spins (corresponding to the two doublet components) . The first group has a net coherence pointing into the -y direction. All B-spins in this ensemble of spins are up, this group of spins has a net z-magnetization of the B-spins. The same applies for the other vector of the antiphase coherence, with opposite signs. If we apply a 90^o_x pulse in this situation, we convert the z-magnetization of the B-spin into y-coherence and the y-coherence of the A-spin into z-magnetization. We have created antiphase coherence of the B-spin. This refocuses into in-phase coherence of the B-spin. We have achieved a coherence transfer from the A-spin to the B-spin. During the evolution period, this coherence was precessing with ω_A, during t_2 it is precessing with ω_B. This pathway leads to COSY cross peaks. The component of the coherence that is in-phase at the time of the second 90^o_x pulse leads to diagonal peaks in the COSY experiment.

Let us consider the pathway originating from the y-component of the in-phase coherence which evolves with $\cos\omega_A t_1$. This component also evolves into two subgroups of spins, an inphase and an antiphase component. The antiphase coherence of the A spin is, however, along the x-axis. The second 90^o_x pulse converts the net z-magnetization of th B-spins into y-coherence without affecting

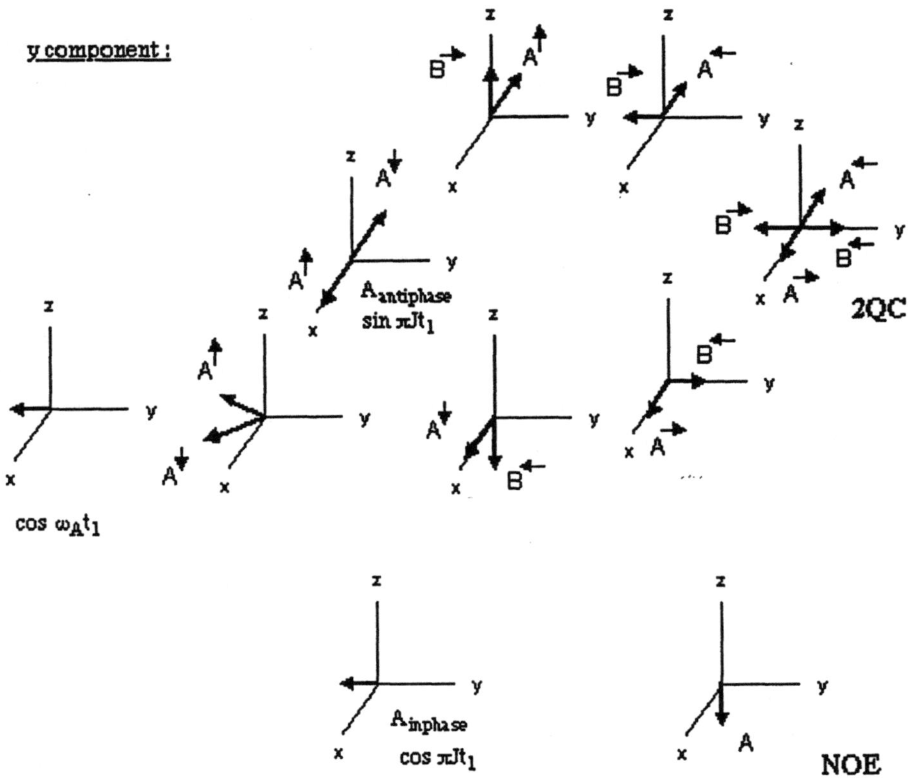

if we detect the x-component

COSY cross peaks	$\sin \omega_A t_1$.	$\sin \pi J t_1$.	$\cos \omega_B t_2$.	$\sin \pi J t_2$
diagonal peak	$\sin \omega_A t_1$.	$\sin \pi J t_1$.	$\cos \omega_A t_2$.	$\cos \pi J t_2$
double quantum coherence	$\sin \omega_A t_1$.	$\sin \pi J t_1$		
NOE	$\sin \omega_A t_1$.	$\cos \pi J t_1$		

Fig. 6 (continued) : Vectorial representation of the effect of a two-pulse sequence on two coupled spins.

the x-coherence of the A-spins, and two-spin coherence (2QC) between the A and B-spins is created. The net magnetization of the two-quantum coherence is zero. Thus, this will never induce a current in the receiver coil and cannot be detected. Only if another 90^o_x pulse is applied, this coherence can be converted into antiphase coherence which then refocuses into detectable inphase coherence. The conversion of the 2QC into antiphase coherence works only if the two double headed arrows are perpendicular. Of course, they precess with different Larmor frequencies. The evolution of the 2QC follows $\cos\omega_A t \cdot \cos\omega_B t$. According to the multiplication rules of trigonometric functions we have:

$$\cos\omega_A t \cdot \cos\omega_B t = \frac{1}{2}\{\cos(\omega_A+\omega_B)t + \cos(\omega_B-\omega_B)t\}$$

This means that the possibility to convert 2QC back into antiphase single quantum coherence occurs with a cosine dependence of the sum and difference frequencies of the two spins. These two aspects are called double-quantum and zero-quantum coherences.

7. Mixing elements

The second 90^o_x pulse of the two-pulse sequence described above, represents the simplest mixing element, leading to coherence transfer. The second and third 90^o_x pulse of the NOESY sequence represent the simplest incoherent magnetization transfer. Both mixing elements are the most fundamental ones for the construction of more complicated pulse sequences. In the case of heteronuclear scalar coupling, the coherence transfer can be optimized since heteronuclear one-bond couplings are uniform in size. This is utilized in numerous heteronuclear coherence transfer schemes. Other mixing schemes use complex sequences of pulses, such as TOCSY or ROESY sequences. Covering of these sequences is beyond the scope of this article.

8. Heteronuclear pulse sequences - 3D NOESY

The possibility to uniformly enrich proteins and nucleic acids with NMR-active isotopes ^{15}N and ^{13}C has led to the development of numerous heteronuclear pulse sequences. Fig. 7 shows two basic sequences which occur in many heteronucleral experiments. Fig. 7I is an HSQC sequence with an antiphase evolution period. The preparation period consists of a period of two delays of length a, separated by a 180° pulse. The delay a has the length $(4J)^{-1}$; thus, at the end of the second delay, the proton coherence is antiphase with respect to the heteronuclear coupling. The 180°(H) pulse has the purpose to refocus the chemical shift evolution of the proton. The 180°(X) pulse has the purpose to maintain heteronuclear coupling. This sequence is called INEPT (Incredible Nuclear Enhancement by Polarization transfer) (Morris and Freeman, 1979). The pair of 90° pulses at the end of the preparation period achieve the heteronuclear coherence transfer. During t_1, the antiphase x coherence evolves with the X-nucleus chemical shift. The 180°(H) pulse in the center of the evolution period eliminates the heteronuclear coupling. After the evolution period, a reverse INEPT sequence achieves a coherence transfer back to the proton and a refocusing to inphase proton coherence.

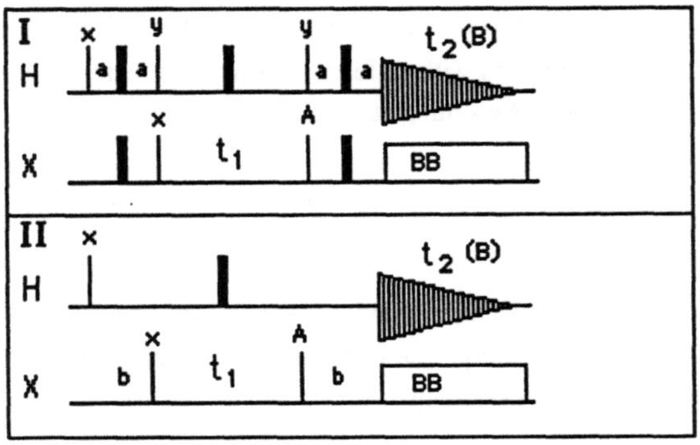

Fig. 7: HSQC and HMQC pulse sequences. The simplest phase cycles are: A = x, -x; B = y, -y.

The second sequence (Fig. 7II) is the Heteronuclear Multiple Quantum Coherence (HMQC) spectroscopy. The delay b is tuned to $(4J)^{-1}$ so that we have heteronuclear antiphase coherence at the end of the delay. The 90°(X) pulse creates now heteronuclear multiple quantum coherence which evolves with the single quantum chemical shift of the X-nucleus, since the proton chemical shift is refocused by the 180°(H) pulse in the center of the evolution period. The 90°(X) pulse at the end of the evolution period converts the multiple quantum coherence back into antiphase proton coherence which then refocuses to inphase coherence at the beginning of the detection period.

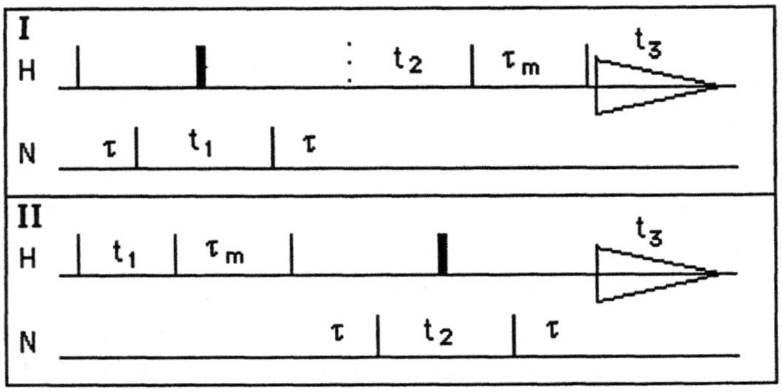

Fig. 8: HMQC-NOESY and NOESY-HMQC pulse sequences.

These two pulse sequences are commonly combined with NOESY or TOCSY sequences to yield 3D NOESY and 3D TOCSY experiments (Fesik & Zuiderweg, 1988; Marion et al., 1989). Fig. 8 shows two ways to combine the NOESY with the HMQC sequence. Sequence II is the preferred

NOESY-HMQC sequence. It detects HN resonances and yields an ^{15}N dispersed NOESY. All cross peaks are downfield of the water signal (in the ω_3 dimension). Overlap with the t_1-noise of the water is thus minimized. This principle of heteronuclear dispersed NOESY experiments is used to disperse NOESY spectra with respect to ^{15}N and ^{13}C. If an HMQC sequence is inserted on either side of the NOESY, we have 4D dispersed NOESY experiments (Kay et al., 1990) which are important for assigning NOESY cross peaks in crowded spectra of large macromolecules.

9. Heteronuclear multiple resonance experiments

With the availability of uniformly ^{13}C and ^{15}N labeled proteins, it became feasible to achieve resonance assignments in proteins just by one- and two-bond coherence transfer experiments. This required construction of new probes that can pulse ^1H, ^{13}C and ^{15}N. The first triple resonance experiments were conducted for measurements of homonuclear coupling constants (Montelione & Wagner, 1989) but yielded also sequential H_i-N_i-C^{α}_{i-1} connectivities (Montelione & Wagner, 1990). These experiments were further developed (Kay et al., 1990) and numerous triple resonance experiments are used today for resonance assignments.

10. Strategies for resonance assignments in proteins.

The initial strategy for sequential resonance assignments in proteins (Dubs et al., 1978; Wagner & Wüthrich, 1982) was based on COSY experiments for establishing intraresidue connectivities and sequential NOE connectivities between the amide protons and the H^N, H^{α} or H^{β} resonances of the preceding residues (Fig. 9). Long-range NOE's are used to obtain constraints for structure calculations. Later on, this set of experiments was expanded with the TOCSY technique which provides easier spin system assignments.

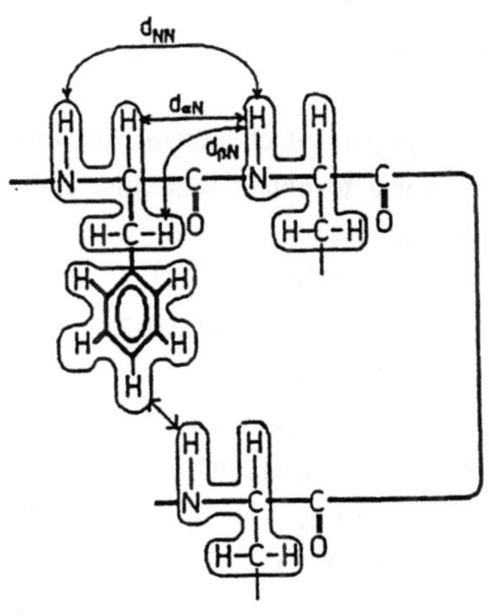

Fig. 9 : COSY-NOESY strategy for assignments of protein resonances.

Triple resonance experiments yield sequence specific assignments independent of the conformation of the protein. The experiments are usually abreviated according to the nuclei that are connected by the experiment (Fig. 10). A 3D HNCA experiment, for example, connects the HN(ω_3) with the N(ω_1) and the C^{α}(ω_2) of the same residue, as well as the C^{α}(ω_2) of the preceding residue. This is complemented with the HN(CO)CA experiment which connects the HN(ω_3) with the N(ω_1) and the C^{α}(ω_2) of the preceding residue.

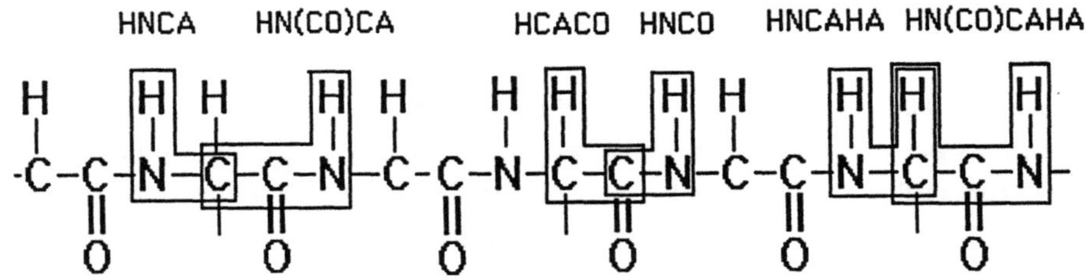

Fig. 10: Selection of connectivities that can be obtained with triple resonance experiments for sequential assignments in proteins.

11. A representative (minimal) set of experiments used for assignments and structure analysis of proteins

For assignment of proteins, the following set of experiments (or a subset thereof) is usually sufficient to obtain sequential assignments and a large set of NOE constraints for structure calculations.

2D NOESY
2D TOCSY (DIPSI-2)
^{1}H-^{15}N HSQC
^{15}N dispersed 3D NOESY-HSQC

^{1}H-^{13}C HSQC	ct-H-^{13}C HSQC
HNCA	HNCOCA
HCACO	HNCO
HCCH-COSY	HCCH-TOCSY

^{13}C dispersed 3D NOESY-HSQC
^{13}C/^{15}N dispersed 4D NOESY

These experiments are complemented with techniques to measure vicinal coupling constants to obtain dihedral angle constraints. Numerous experimental techniques have been developed in the last couple of years to achieve this task. They are related to techniques for stereospecific assignments of methylene and methyl groups. These techniques are important for refining structures of proteins.

12. Conclusion

An almost infinite number of pulse sequence has been developed over the last 1 1/2 decades to analyze structures of proteins by NMR, and a large number of protein structures have been solved with this method. It is important to make the right choice of the best and the minimum number of experiments to achieve the task of solving a protein structure. There are still many unexplored possibilities to push the limits of protein structure analysis, in particulat in the area of isotope labeling, data acquisition and data processing. Also higher field instruments and progress in hardware development will further push the limits of the technique. Now, since the technique has matured, the focus is slowly changing towards solving structures of biologically relevant proteins and towards applications for rational drug design. With the mapping of the human genome, many new proteins within the size range that can be handled by NMR appear in the literature, and NMR scientists will have a wide field of structural biology to explore.

ACKNOWLEDGEMENTS

Preparation of this article was supported by funds from NIH and NSF.

REFERENCES

Abragam, A.: The principles of nuclear magnetism. Oxford, Oxford University Press (1961).

Braunschweiler, L., Ernst, R.R. Coherence transfer by isotropic mixing. J. Magn. Reson. **53**, 521-528 (1983).

Dubs, A., Wagner, G. and Wüthrich, K. : Individual assignments of amide proton resonances in the proton NMR spectrum of the basic pancreatic trypsin inhibitor. Biochim. Biophys. Acta **577**, 177-194 (1979).

Ernst, R.R., Bodenhausen, G. and Wokaun, A. Principles of nuclear magnetic resonance in one and two dimensions. Clarendon Press, Oxford, 1987.

Fesik, S. W. and Zuiderweg, E.R.P.: Heteronuclear 3D NMR spectroscopy; A strategy for the simplification of homonuclear 2D NMR spectra. J. Magn. Reson. **78**, 588-593 (1988).

Kay, L.E., Clore, G.M., Bax, A. and Gronenborn, A.M. 4D heteronuclear triple resonance NMR spectroscopy of Interleukin 1β in solution. Science **249**, 411-414 (1990).

Kay, L.E, Ikura, M., Tschudin, R., Bax, A. Three-dimensional triple resonance spectroscopy of isotopically enriched proteins. J, Magn.. Reson. **89**, 496-514 (1990).

Marion, D. Driscoll, P.C., Kay, L.E., Wingfield, P.T., Bax, A. Biochemistry **28**, 6150-6156 (1989).

Montelione, G.T. and Wagner, G.: Accurate Measurements of Homonuclear H^N-H^α Coupling Constants in Polypeptides Using Heteronuclear 2D NMR Experiments. J. Am. Chem. Soc., **111**, 5474-5475 (1989).

Montelione, G.T. and Wagner, G.: Triple Resonance Experiments For Establishing Conformation-Independent Sequential NMR Assignments in Isotope-Enriched Polypeptides., J. Magn. Reson.; **87**, 183-188 (1990).

Morris, G.A. & Feeman, R.: Enhancement of Nuclear Magnetic Resonance Signals by Polarization Transfer. J. Am. Chem. Soc. **101**, 760-762 (1979).

Solomon, I.: Relaxation processes in a system of two spins. Phys. Rev. **99**, 559 -565 (1955).

Sørensen, O.W., Eich, G.W., Levitt, M.H., Bodenhausen, G., Ernst, R.R.: Operator Product description of NMR Experiments, Progr. NMR Spectroscopy, **16**, 163-192 (1983).

Wagner, G. and Wüthrich, K.: Sequential Resonance Assignments in Protein ^{1}H NMR Spectra: Basic Pancreatic Trypsin Inhibitor J. Mol. Biol. **155**, 347-366 (1982).

UNDERSTANDING PULSED FTNMR EXPERIMENTS

Ernest D. Laue
Department of Biochemistry
University of Cambridge
Tennis Court Road
Cambridge, UK

1. PRODUCT OPERATORS

(a) Introduction

It is not possible to describe many modern NMR experiments with the vector model. A full quantum treatment is needed. One such treatment is the 'product operator' approach which combines quantum mechanical rigour with a formalism that is easy to visualize in terms of rotations and vectors. Product operators can be used to analyse NMR experiments using a 'recipe' approach which does not require a deep understanding of the quantum mechanics but rather relies on using a simple set of rules.

Most elementary quantum mechanics is concerned with finding the wavefunctions ψ, which are the solutions to the Schrödinger equation $H\psi=E\psi$. H is the Hamiltonian for the system and varies from problem to problem. H is also described as an operator because it 'operates on' ψ. In quantum mechanics operators represent physically observable quantities such as energy, angular momentum, position etc. Since what we are trying to predict is the outcome of an experiment in the sense of what is observable, it turns out to be convenient to concentrate on the operators themselves, rather than the solutions to the Schrödinger equation. This is the approach used in the 'product operator' treatment.

Of particular interest are the angular momentum operators I_X, I_Y and I_Z which represent the X, Y and Z components of the spin angular momentum. They also represent the X, Y and Z magnetizations of the system. For a single spin the position of a magnetization vector can be specified by giving the amounts of X, Y and Z magnetization. In the same way, the quantum mechanical state of the spin can be described by specifying the amount of the operators I_X, I_Y and I_Z that are present at any time.

NATO ASI Series, Vol. H 87
NMR of Biological Macromolecules
Edited by C. I. Stassinopoulou

Formally, the state of the system is specified by the density operator (or sometimes density matrix), denoted σ, which is expressed as a linear combination of operators. For a single spin these operators are I_X, I_Y and I_Z.

This approach becomes more useful for the case of two coupled spins, traditionally denoted I and S. It turns out that the following operators, or products of operators are needed to express the state of the two spin system:

$$I_X \; I_Y \; I_Z \quad S_X \; S_Y \; S_Z$$
$$2I_XS_Z \; 2I_YS_Z \; 2I_ZS_Z \quad 2I_ZS_X \; 2I_ZS_Y$$
$$2I_XS_X \; 2I_YS_Y \; 2I_XS_Y \; 2I_YS_X$$

Some of these products have distinct physical meanings, that can be represented either by vectors (see (e) (iv), below) or by sketches of the spectra that would result if such a term were to be observed. For example:

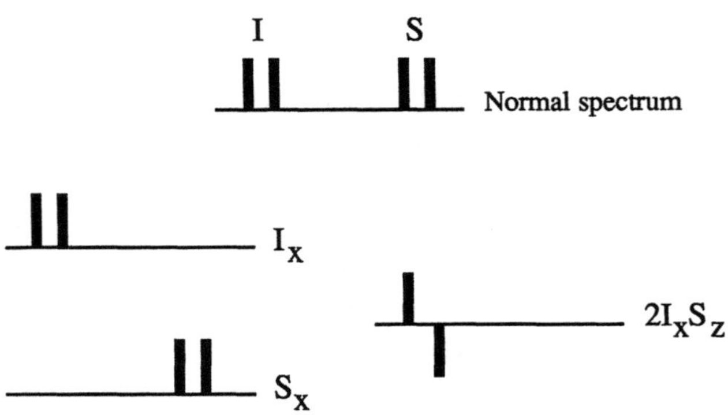

I_X corresponds to an 'in-phase' I spin multiplet aligned along the X axis. $2I_XS_Z$ is called an 'anti-phase' state and corresponds to an up-down magnetization on the X axis. Note how this term is anti-phase with respect to the I-S coupling and that the S_Z operator thus appears in the product. The factors of 2 are present for normalization purposes. The terms I_Y and $2I_YS_Z$ have similar interpretations to I_X and $2I_XS_Z$ except that they correspond to magnetization aligned along the Y axis.

Formally speaking, the anti-phase terms are not observable, in the sense that at a particular instant $2I_XS_Z$ does not contribute to the observed X magnetization. However, such an anti-phase term will evolve under the influence of a coupling to an in-phase state which is observable, and if a free induction decay is recorded during this process, subsequent Fourier transformation would give an anti-phase doublet in the spectrum. We shall often use the device of 'pretending' that anti-phase terms are observable.

The operators with only Z terms are not directly observable. I_Z corresponds to Z magnetization on the I spin and S_Z likewise on S. The expression $2I_ZS_Z$ is a special non-equilibrium distribution of populations often described as J-order.

$2I_XS_X$, $2I_YS_Y$, $2I_XS_Y$ and $2I_YS_X$ correspond to multiple quantum coherence and do not give rise to any observable signals. This will be discussed later.

(b) Rotations of Operators

A pulse sequence consists of periods of free precession and short radio-frequency pulses. Each of these can cause the product operators present to change into different operators, for example we shall see later that a pulse can cause the following transformation:

$2I_XS_Z$ (anti-phase I magnetization) $\Rightarrow$ $2I_XS_Y$ (multiple quantum coherence)

Likewise, a period of free precession can cause the following transformation:

I_X (in-phase I magnetization along X) $\Rightarrow$ $2I_YS_Z$ (anti-phase I magnetization along Y)

Each of these transformations is formally described as a rotation of an initial density operator, σ-, to a new operator, σ+ under the effect of a particular Hamiltonian, H:

$$\sigma_+ = e^{-iHt}\sigma_- e^{iHt} \qquad (1)$$

The Hamiltonian will vary according to whether a pulse or a delay of duration t is causing the rotation. The nice thing is that if H and σ are both expressed in terms of product operators the solutions to all likely versions of equation (1) have already been worked out. These known solutions can then be applied using simple rules. For simplicity the arrow notation will be adopted:

$$\sigma_- \xrightarrow{Ht} \sigma_+$$

which is a shorthand for (1).

For various historical reasons, the convention adopted for product operator calculations is that an X pulse takes Z magnetization on to -Y, and a positive precession takes X towards Y. Apart from the labelling of axes and trivial sign changes, the choice of convention is essentially arbitrary.

(i) Chemical Shift Evolution

For spin I, H has the form $\Omega_I I_Z$, where Ω_I is the offset of spin I. For example I_X rotates as:

$$I_X \xrightarrow{\Omega_I I_Z t} \cos(\Omega_I t) I_X + \sin(\Omega_I t) I_Y$$

which is just like a vector rotating in the X-Y plane. Likewise, I_Y rotates as:

$$I_Y \xrightarrow{\Omega_I I_Z t} \cos(\Omega_I t) I_Y - \sin(\Omega_I t) I_X$$

which is again easily visualized as a rotation in the X-Y plane. Similar forms occur for the S spin, but here the offset is Ω_S. These rotations can be summarized in diagrammatic form:

$$\begin{array}{ccc} I_X & \Rightarrow & I_Y \\ \Uparrow & & \Downarrow \\ -I_Y & \Leftarrow & -I_X \end{array} \qquad \text{Angle} = \Omega_I t$$

Note how each rotation gives a cosine times the old term, and a sine times the new term.

(ii) Evolution under Couplings

For two spins the form of H for evolution under a scalar coupling J_{IS} is $H = 2\pi J_{IS} I_Z S_Z$. The magnetization I_X evolves as:

$$I_X \xrightarrow{2\pi J_{IS} I_Z S_Z t} \cos(\pi J_{IS} t)\, I_X + \sin(\pi J_{IS} t)\, 2 I_Y S_Z$$

This corresponds to in-phase magnetization evolving into anti-phase magnetization under the influence of a scalar coupling. ***Note once again that the result is in the form: cosine times the old term + sine times the new term.*** In the same way, anti-phase magnetization evolves into in-phase magnetization.

$$2 I_Y S_Z \xrightarrow{2\pi J_{IS} I_Z S_Z t} \cos(\pi J_{IS} t) 2 I_Y S_Z - \sin(\pi J_{IS} t)\, I_X$$

Note that in-phase magnetization along X becomes anti-phase along Y, and *vice versa*. The effect of scalar couplings can be summarized as before.

$$\begin{array}{ccc} I_X & \Rightarrow & 2I_YS_Z \\ \Uparrow & & \Downarrow \\ -2I_YS_Z & \Leftarrow & -I_X \end{array} \qquad \text{Angle} = \pi J_{IS}t$$

and for the case of Y magnetization:

$$\begin{array}{ccc} I_Y & \Rightarrow & -2I_XS_Z \\ \Uparrow & & \Downarrow \\ 2I_XS_Z & \Leftarrow & -I_Y \end{array} \qquad \text{Angle} = \pi J_{IS}t$$

(iii) Evolution Under Pulses

Pulses about the X axis have $H = \omega_1 I_X$ and about the Y axis have $H = \omega_1 I_Y$. The radio frequency field strength is ω_1 rad s^{-1}, and is defined such that $\omega_1/2\pi$ is the rate at which the magnetization is rotated by the radio-frequency pulse. Thus, a 360° pulse takes $2\pi/\omega_1$ seconds, a 90° pulse 1/4 $(2\pi/\omega_1)$ and so on.

For example, an X pulse on Z magnetization gives:

$$I_Z \xrightarrow{\omega_I I_X t} \cos(\omega_I t) I_Z - \sin(\omega_I t) I_Y$$

and a Y pulse on X magnetization gives:

$$I_X \xrightarrow{\omega_I I_Y t} \cos(\omega_I t) I_X - \sin(\omega_I t) I_Z$$

These have the familiar form ***(cosine times the old term + sine times the new term)*** and can be summarized in diagrams. For an X pulse:

$$\begin{array}{ccc} I_Z & \Rightarrow & -I_Y \\ \Uparrow & & \Downarrow \\ I_Y & \Leftarrow & -I_Z \end{array} \qquad \text{Angle} = \omega_1 t$$

And for a Y pulse:

$$\begin{array}{ccc} I_Z & \Rightarrow & I_X \\ \Uparrow & & \Downarrow \\ -I_X & \Leftarrow & -I_Z \end{array} \qquad \text{Angle} = \omega_1 t$$

(c) Practical Points

Before using the above rotation rules to describe some experiments, we will discuss some special features found in the manipulations of these operators.

(i) Cascades

Suppose a period of free precession is present, the effect of both the chemical shifts of the I and S spins as well as the IS scalar coupling must be taken into account. In the arrow notation:

$$\sigma_- \xrightarrow{\Omega_I I_Z t + \Omega_S S_Z t + 2\pi J_{IS} I_Z S_Z t} \sigma_+$$

This can be broken down into a series of rotations, or a *cascade*:

$$\sigma_- \xrightarrow{\Omega_I I_Z t} \sigma_1 \xrightarrow{\Omega_S S_Z t} \sigma_2 \xrightarrow{2\pi J_{IS} I_Z S_Z t} \sigma_+$$

The order in which the rotations due to chemical shifts and couplings appear is unimportant. Likewise, for a homonuclear system the effect of a pulse written as,

$$\sigma_- \xrightarrow{\omega_I I_X t + \omega_I S_X t} \sigma_+$$

can be broken down into a pulse on I and a pulse on S:

$$\sigma_- \xrightarrow{\omega_1 I_X t} \sigma_1 \xrightarrow{\omega_1 S_X t} \sigma_+$$

(ii) Rotations of Products

Suppose we have a term such as $2I_XS_Z$ and an X pulse is applied to the S spin, how are the above rules applied to find out the effect of the pulse? The rule is that rotations only affect those terms which are operators of the same spin. In other words the I_X part remains unaffected by the X pulse to the S spin, and the S_Z part is rotated on to -Y:

$$\{2I_X\}S_Z \xrightarrow{\omega_1 S_X t} \{2I_X\}(\cos(\omega_1 t)S_Z - \sin(\omega_1 t)S_Y)$$

or:
$$2I_XS_Z \xrightarrow{\omega_1 S_X t} \cos(\omega_1 t)2I_XS_Z - \sin(\omega_1 t)2I_XS_Y$$

(iii) Rotations Involving the Same Operators

Operators are unaffected by rotations about themselves. For example:

$$I_X \xrightarrow{\omega_1 I_X t} I_X$$
$$2I_YS_Z \xrightarrow{\Omega_S S_Z t} 2I_YS_Z$$

(d) Equilibrium

The situation consists of simple Z magnetization, and is described in terms of operators by:

$$\sigma_e = I_Z + S_Z$$

Sometimes a calculation involving two spins just starts with I_Z so that we can follow the fate of one spin's magnetization and obtain a simpler result. The above expression for the equilibrium density operator is something of an over simplification and some constants of proportionality must sometimes be included:

$$\sigma_e = k_I I_Z + k_S S_Z$$

where k_I and k_S are constants that depend on, among other things, the gyromagnetic ratios, γ_i of the two nuclei. For example, if I is a proton and S is a carbon-13, then k_I=$4k_S$. These constants can become important in analysing experiments which transfer magnetization from one spin to another.

(e) Examples

The best way to see the formalism working is to try some examples.

(i) Spin Echo for One Spin

$$\begin{array}{ccccccccc} 90(X) & - & t & - & 180(X) & - & t & - & \\ | & & | & & | & & | & & | \\ \sigma_e & & \sigma_1 & & \sigma_2 & & \sigma_3 & & \sigma_4 \end{array}$$

The first stage is, starting from equilibrium:

$$\sigma_e \xrightarrow{\omega_1 I_X t_p} \sigma_1$$

$$I_Z \xrightarrow{\omega_1 I_X t_p} \cos(\omega_1 t_p) I_Z - \sin(\omega_1 t_p) I_Y$$

$$I_Z \xrightarrow{\alpha I_X} \cos\alpha\, I_Z - \sin\alpha\, I_Y \qquad (2)$$

$$I_Z \xrightarrow{(\pi/2) I_X} \cos(\pi/2)\, I_Z - \sin(\pi/2) I_Y \qquad (3)$$

$$I_Z \xrightarrow{(\pi/2) I_X} -I_Y \qquad (4)$$

where line (2) has used the substitution $\alpha=\omega_1 t_p$, where α is the flip angle of a pulse of duration t_p and field strength ω_1. Line (3) uses the fact that the pulse has flip angle $\pi/2$ in this case, leading to the final result in (4). From now on, we will proceed directly to (4) without going through the intervening stages. The overall result of the first pulse is, of course, the generation of transverse magnetization.

Next follows a period of free evolution:

$$\sigma_1 \xrightarrow{\Omega_1 I_Z t} \sigma_2$$

$$-I_Y \xrightarrow{\Omega_I I_Z t} -\cos(\Omega_I t) I_Y + \sin(\Omega_I t) I_X$$

This is evolution in the X-Y plane. Next follows a 180° pulse:

$$\sigma_2 \xrightarrow{\omega_1 I_X t_p} \sigma_3$$

$$-\cos(\Omega_I t) I_Y + \sin(\Omega_I t) I_X \xrightarrow{\alpha I_X} \cos\alpha[-\cos(\Omega_I t)] I_Y + \sin\alpha[-\cos(\Omega_I t) I_Z + \sin(\Omega_I t) I_X$$

$$-\cos(\Omega_I t) I_Y + \sin(\Omega_I t) I_X \xrightarrow{\pi I_X} \cos\pi[-\cos(\Omega_I t)] I_Y + \sin\pi[-\cos(\Omega_I t)] I_Z + \sin(\Omega_I t) I_X$$

$$-\cos(\Omega_I t) I_Y + \sin(\Omega_I t) I_X \xrightarrow{\pi I_X} -1[-\cos(\Omega_I t)] I_Y + \sin(\Omega_I t) I_X$$

$$-\cos(\Omega_I t) I_Y + \sin(\Omega_I t) I_X \xrightarrow{\pi I_X} \cos(\Omega_I t) I_Y + \sin(\Omega_I t) I_X$$

The 180° pulse about the X axis leaves the term in I_X unaffected and reverses the sign of the term in I_Y, which is precisely the effect expected. Again, all the steps have been gone through above, but in practice for a 180° pulse it is easy to simply recognize those terms whose signs will be altered. Finally, a further period of free precession takes place. For simplicity the condensed nomenclature $c_I=\cos \Omega_I t$, $s_I=\sin \Omega_I t$ is used:

$$\sigma_3 \xrightarrow{\Omega_I I_Z t} \sigma_4$$

$$c_I I_Y + s_I I_X \xrightarrow{\Omega_I I_Z t} c_I c_I I_Y - s_I c_I I_X + c_I s_I I_X + s_I s_I I_Y$$

$$c_I I_Y + s_I I_X \xrightarrow{\Omega_I I_Z t} I_Y$$

where the last line has used the identity $\cos^2\phi + \sin^2\phi = 1$. Thus, the -t-180-t- sequence has simply taken an initial state $-I_Y$ to a final state I_Y. Apart from a trivial sign change the initial state has been restored, or in other words the chemical shift has been refocussed, which is precisely the result expected for a spin echo.

(ii) Spin Echo for Two Coupled Spins

$$90(X) - t - 180(X) - t -$$

σ_e	σ_1	σ_2	σ_3	σ_4

pulses applied to I and S

For simplicity we shall start with an equilibrium density operator I_Z. In the previous example it was shown that this sequence refocusses the chemical shift term, and the same is equally true for the two spin case. Thus, to simplify the calculation, the chemical shift term will not be included as we know it will have no effect. The evolution during the delays t is simply that due to the scalar coupling. As before $\sigma_1 = -I_Y$ so the first delay gives:

$$\sigma_1 \xrightarrow{2\pi J_{IS} I_Z S_Z t} \sigma_2$$

$$-I_Y \xrightarrow{2\pi J_{IS} I_Z S_Z t} -\cos(\pi J_{IS} t) I_Y + \sin(\pi J_{IS} t) 2 I_X S_Z$$

The 180° pulse can be broken down into a 180° pulse to I followed by a 180° pulse to S, both about the X axis:

$$\sigma_2 \xrightarrow{\pi I_X} \sigma' \xrightarrow{\pi S_X} \sigma_3$$

$$-\cos(\pi J_{IS} t) I_Y + \sin(\pi J_{IS} t) 2 I_X S_Z \xrightarrow{\pi I_X} -\cos(\pi J_{IS} t)\{-I_Y\} + \sin(\pi J_{IS} t) 2 I_X S_Z$$

$$\xrightarrow{\pi S_X} -\cos(\pi J_{IS} t)\{-I_Y\} + \sin(\pi J_{IS} t) 2 I_X \{-S_Z\}$$

$$-\cos(\pi J_{IS} t) I_Y + \sin(\pi J_{IS} t) 2 I_X S_Z \xrightarrow{\pi I_X} \xrightarrow{\pi S_X} \cos(\pi J_{IS} t) I_Y - \sin(\pi J_{IS} t) 2 I_X S_Z$$

Note how the 180°(X) pulse to I does not affect the S spin operators, nor the operator I_X, as was discussed above. Likewise, the 180°(X) pulse to S does not affect the I spin operators. Braces { } have been added to emphasize the terms that are inverted. Again the shorthand nomenclature $c_{IS} = \cos(\pi J_{IS} t)$ and $s_{IS} = \sin(\pi J_{IS} t)$ is introduced, and the second delay of t gives:

$$\sigma_3 \xrightarrow{2\pi J_{IS} I_Z S_Z t} \sigma_4$$

$$c_{IS} I_Y - s_{IS} 2I_X S_Z \xrightarrow{2\pi J_{IS} I_Z S_Z t} c_{IS} c_{IS} I_Y - s_{IS} c_{IS} 2I_X S_Z - c_{IS} s_{IS} 2I_X S_Z - s_{IS} s_{IS} I_Y$$

$$c_{IS} I_Y - s_{IS} 2I_X S_Z \xrightarrow{2\pi J_{IS} I_Z S_Z t} \cos(\pi J_{IS} 2t) I_Y - \sin(\pi J_{IS} 2t) 2I_X S_Z$$

where the last line uses the identities $\cos 2\phi = \cos^2\phi - \sin^2\phi$ and $\sin 2\phi = 2 \sin\phi \cos\phi$. Hence the overall effect of the -t-180-t- sequence on the initial magnetization $-I_Y$ is:

$$-I_Y \xrightarrow{-t-180(X)-t-} \cos(2\pi J_{IS} t) I_Y - \sin(2\pi J_{IS} t) 2I_X S_Z$$

Setting the delay t to $1/(4J_{IS})$ generates the pure anti-phase state $2I_X S_Z$ and setting t to $1/(2J_{IS})$ simply results in an inversion of the initial magnetization. Note, however, that these predictions are based upon a quantum mechanical calculation, and not a vector model with some arbitrarily added extra rules.

This generation of anti-phase states by a spin echo is a basic part of many more complex pulse sequences.

(iii) Heteronuclear Spin Echo

$$90(X) - t - 180(X) - t -$$

	|	|	|	|	|
I spin only	σ_e	σ_1	σ_2	σ_3	σ_4

The sequence is as before, except that the 180° pulse is only applied to the I spins. This sequence can be analysed as follows, again ignoring the I spin chemical shift which will be refocussed. Thus, as before:

$$\sigma_2 \xrightarrow{\pi I x} \sigma_3$$

$$-\cos(\pi J_{IS} t) I_Y + \sin(\pi J_{IS} t) 2I_X S_Z \xrightarrow{\pi I_X} -\cos(\pi J_{IS} t)\{-I_Y\} + \sin(\pi J_{IS} t) 2I_X S_Z$$

$$-\cos(\pi J_{IS} t) I_Y + \sin(\pi J_{IS} t) 2I_X S_Z \xrightarrow{\pi I_X} \cos(\pi J_{IS} t) I_Y + \sin(\pi J_{IS} t) 2I_X S_Z$$

Note how this time only the term in I_Y has been inverted, and the other term in S_Z is unaffected because the 180° pulse is only applied to I. The second delay then gives:

$$\sigma_3 \xrightarrow{2\pi J_{IS} I_Z S_Z t} \sigma_4$$

$$c_{IS} I_Y + s_{IS} 2I_X S_Z \xrightarrow{2\pi J_{IS} I_Z S_Z t} c_{IS} c_{IS} I_Y - s_{IS} c_{IS} 2I_X S_Z + c_{IS} s_{IS} 2I_X S_Z + s_{IS} s_{IS} I_Y$$

$$c_{IS} I_Y + s_{IS} 2I_X S_Z \xrightarrow{2\pi J_{IS} I_Z S_Z t} I_Y$$

In contrast to the case (ii), if the 180° pulse only affects one of the spins, both the coupling and chemical shifts are refocussed.

(iv) Coherence Transfer

Suppose that the state $2I_X S_Z$ has been created in some way. The effect of applying a pulse $\alpha°(Y)$ to both spins is:

$$\sigma_- \xrightarrow{\alpha I_Y} \sigma' \xrightarrow{\alpha S_Y} \sigma_+$$

$$2I_X S_Z \xrightarrow{\alpha I_Y} \sigma' \xrightarrow{\alpha S_Y} \sigma_+$$

$$2I_X S_Z \xrightarrow{\alpha I_Y} \cos\alpha\ 2I_X S_Z + \sin\alpha\ 2\{-I_Z\} S_Z$$

$$\xrightarrow{\alpha S_Y} \cos\alpha\ \cos\alpha\ 2I_X S_Z + \sin\alpha\ \cos\alpha\ 2I_X \{S_X\}$$

$$-\cos\alpha\ \sin\alpha\ 2I_Z S_Z - \sin\alpha\ \sin\alpha\ 2I_Z \{S_X\}$$

For the case $\alpha = 90°$ the overall result is simply:

$$2I_X S_Z \xrightarrow{(\pi/2) I_Y} -2I_Z S_Z \xrightarrow{(\pi/2) S_Y} -2I_Z S_X$$

In other words, anti-phase magnetization on the I spin is transferred to anti-phase magnetization on the S spin by the 90° pulse. This process, usually called coherence transfer, is a central feature of modern NMR spectroscopy. Anti-phase magnetization is created by free evolution for a time comparable with 1/2J, as described above for the spin echo sequence.

The generation of the anti-phase state $2I_X S_Z$ from an initial state $-I_Y$, and the subsequent coherence transfer step, in a homonuclear spin system, can also be analysed in terms of the vector model. In a rotating frame, of frequency equal to the chemical shift Ω_I, these states can be understood as a sum of two equal populations depending on whether the S spin (black) is either aligned along the +Z (S_α) or -Z (S_β) axis. The two transverse I spin components (grey), initially aligned along the -Y axis in the $-I_Y$ state, move in opposite directions because their chemical shifts differ from the frequency of the rotating frame by ±J/2 cycles s^{-1} (±πJ rad s^{-1}). Thus after a time equal to 1/2J s they are aligned along the ±X axis in the $2I_X S_Z$ state:

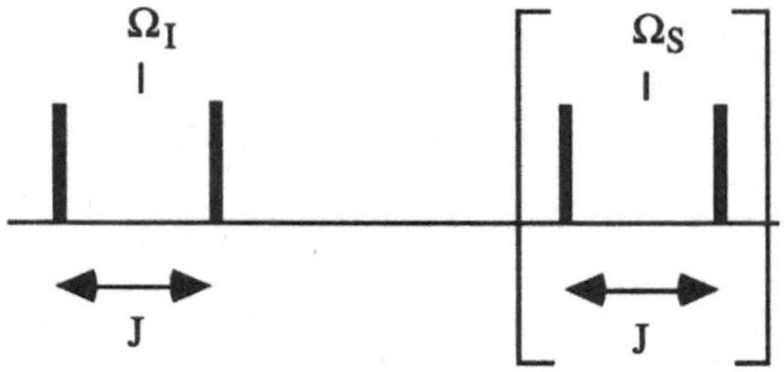

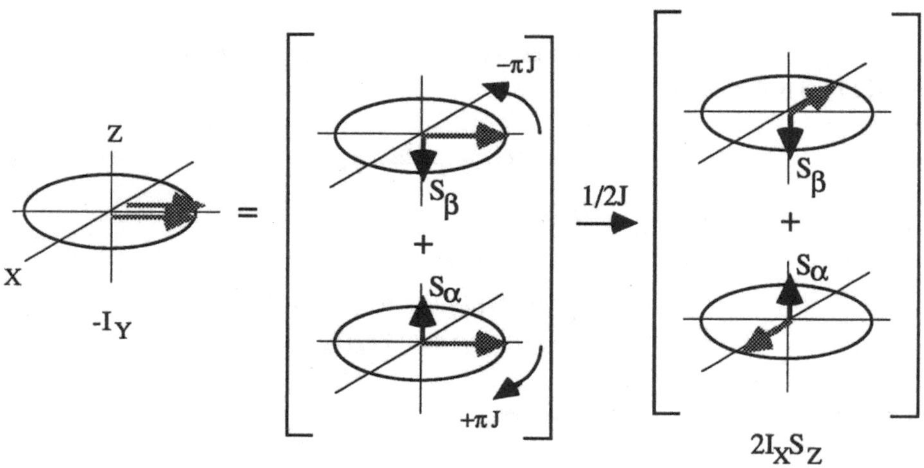

The effects of the 90° degree pulses to both the I and S spins can be predicted using the same rules as in the product operator calculations, ie. an X pulse takes Z magnetization on to -Y etc. The effects of the 90° pulse can, as before, be broken down into separate pulses on the I and S spins. The I_Y pulse affects only the I spin components and generates the J order state $-2I_ZS_Z$. The 90° S_Y pulse affects the S spin components to generate the desired $-2I_ZS_X$.

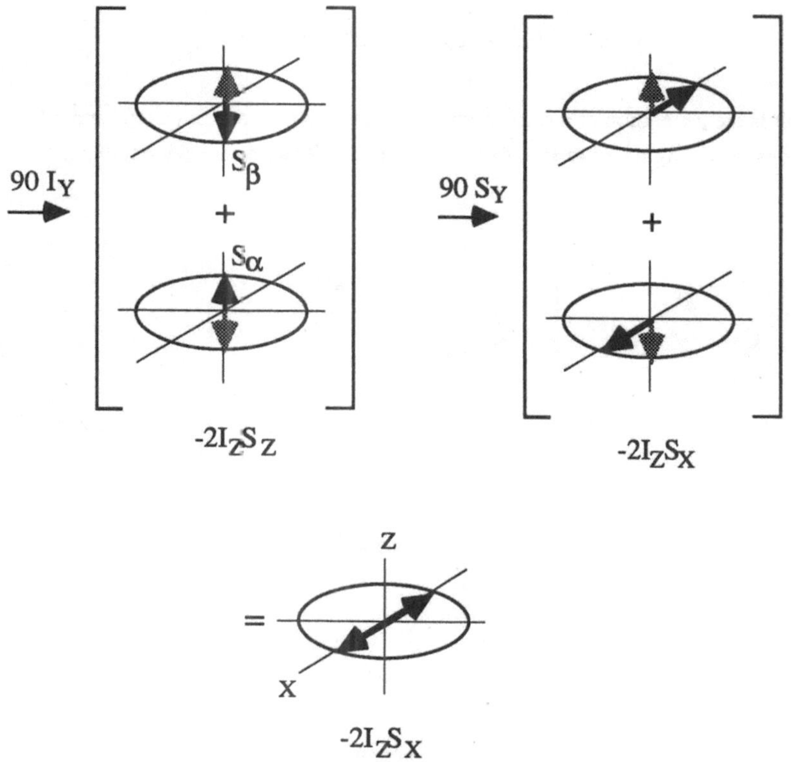

Although the generation of the anti-phase state $2I_XS_Z$, and the subsequent coherence transfer step, can be understood using the vector model it is more difficult to gain a complete understanding. In particular the effects of pulses on relative populations (eg. in heteronuclear experiments) and, in general, phase coherences between energy levels is difficult to understand.

(v) Generation of Multiple Quantum Coherence

Suppose that the state $2I_XS_Z$ has been prepared in some way. A 90° (X) pulse to both I and S gives:

$$\sigma_- \xrightarrow{\alpha I_X} \sigma' \xrightarrow{\alpha S_X} \sigma_+$$

$$2I_XS_Z \xrightarrow{\alpha I_X} \sigma' \xrightarrow{\alpha S_X} \sigma_+$$

$$2I_XS_Z \xrightarrow{\alpha I_X} 2I_XS_Z \xrightarrow{\alpha S_X} \cos\alpha \, 2I_XS_Z + \sin\alpha \, 2I_X\{-S_Y\}$$

For a 90° pulse the result is simply:

$$2I_XS_Z \xrightarrow{(\pi/2)I_X} 2I_XS_Z \xrightarrow{(\pi/2)S_X} -2I_XS_Y$$

The result is the generation of a state of multiple quantum coherence, which cannot be directly observed. Again, in terms of the vector model, it can be considered to be:

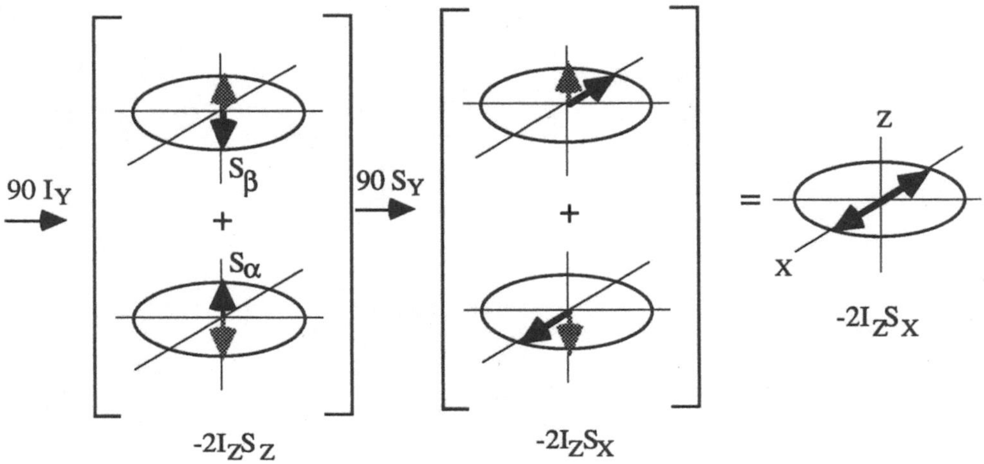

This time the 90° pulse only affects the S spin components.

(f) INEPT (Insensitive Nuclei Enhanced by Polarization Transfer)

The basic elements described above can be used to construct a large number of pulse sequences. One such is the INEPT experiment, the aim of which is to transfer proton magnetization to a coupled nucleus, such as carbon-13. The gyromagnetic ratio of protons is four times that of carbon-13 and so the magnetization is correspondingly large. By transferring this magnetization to the carbon-13 nuclei, the detected signal from carbon will be stronger. The signal enhancement is even greater for lower γ nuclei such as nitrogen-15. The simplest INEPT pulse sequence is:

I spin	$90(X) - t - 180(X)$		$- t - 90(Y)$
S spin		$180(X)$	$90(X)$ − *acquire*

$$| \quad | \quad |$$
$$\sigma_1 \quad \sigma_2 \quad \sigma_3$$

Up to the point σ_1 the sequence is simply the spin echo analysed in section (e) (ii). Any chemical shift evolution during t is refocussed. Starting with an equilibrium density matrix of $k_I I_Z$, σ_1 is:

$$k_I I_Z \xrightarrow{90-t-180,180-t-} k_I(\cos(2\pi J_{IS}t)I_Y - \sin(2\pi J_{IS}t)2I_X S_Z)$$

A 90(Y) pulse is applied to the I spin, which rotates the term $-2I_X S_Z$ to $2I_Z S_Z$.

$$k_I(\cos(2\pi J_{IS}t)I_Y - \sin(2\pi J_{IS}t)2I_X S_Z) \xrightarrow{(\pi/2)I_Y} k_I(\cos(2\pi J_{IS}t)I_Y + \sin(2\pi J_{IS}t)2I_Z S_Z)$$

The 90(X) pulse to the S spin rotates $2I_Z S_Z$ into $-2I_Z S_Y$.

$$k_I(\cos(2\pi J_{IS}t)I_Y + \sin(2\pi J_{IS}t)2I_Z S_Z) \xrightarrow{(\pi/2)S_X} k_I(\cos(2\pi J_{IS}t)I_Y - \sin(2\pi J_{IS}t)2I_Z S_Y)$$

If we choose the optimum condition $t = 1/(4J_{IS})$ the final signal observed on the S spin is:

$$k_I(-2I_Z S_Y)$$

This is a doublet, anti-phase with respect to the IS coupling, through which the transfer has taken place (the 'active' coupling). In contrast, if a simple 90(X) pulse had been applied to equilibrium S spin magnetiation $k_S S_Z$, the result would be:

$$k_S S_Z \xrightarrow{(\pi/2)S_X} k_S(-S_Y)$$

The result is an in-phase doublet, but of intensity proportional to k_S. Thus the intensity ratio between the INEPT and conventional experiment is:

$$\frac{\textit{INEPT observation}}{\textit{conventional observation}} = \frac{k_I}{k_S} = \frac{\gamma_I}{\gamma_S}$$

The advantage of INEPT observation becomes greater and greater the lower the γ of the S spin. A further advantage is that the repetition rate of the experiment is set by the relaxation times of the high γ spins, which are typically much shorter than those of the low γ spins.

Often, it is desirable to decouple the S spins from the I spins by irradiating the latter during data acquisition. However, this is not possible for a simple INEPT experiment as decoupling the anti-phase S spin doublet collapses the two lines on top of one another so that mutual cancellation occurs and no signal is observed. If decoupling is required, the anti-phase doublet has to be allowed to re-phase, and this is achieved with a further spin echo sequence. The result is the refocussed INEPT experiment:

I spin	$90(X)-t-180(X)$	$-t-90(Y)$	**180(X)**	***decouple***
S spin	$180(X)$	**90(X) – t – 180(X)**		**– t – *acquire***

The new part of the sequence is indicated in boldface. Suppose that the delay t has been set to the optimum value $1/(4J_{IS})$, so that the density matrix just before the first τ delay is $k_I(-2I_ZS_Y)$, it is straightforward to show that the overall effect of the -τ-180,180-τ- sequence is:

$$k_I(-2I_ZS_Y)\xrightarrow{-\tau-180,180-\tau-}k_I(-\cos(2\pi J_{IS}\tau)2I_ZS_Y-\sin(2\pi J_{IS}\tau)S_X)$$

In fact, this is simply the reverse of the J modulated spin echo sequence analysed in section (ii). Again, we chose the optimum condition $\tau = 1/(4J_{IS})$ so that just before data acquisition starts the density matrix is:

$$k_I(-S_X)$$

This is an in-phase doublet, and we may now safely switch on the decoupler and observe the spectrum. Note how the spin echo sequence has been used twice over, once to create anti-phase magnetization and once to turn anti-phase into in-phase magnetization. In each case, the effects of chemical shifts have been refocussed.

Let us return for a moment to the simple INEPT experiment. The observed signal would be a sum of that from the coherence transfer step and that from the original equilibrium magnetization of the S spin:

$$k_I(-2I_ZS_Y)\ +\ k_S(-S_Y)$$

For the case of protons and carbon-13, where $k_I = 4k_S$, the resulting addition of the in-phase and anti-phase doublets gives a rather strange looking multiplet with relative intensities +5 and -3.

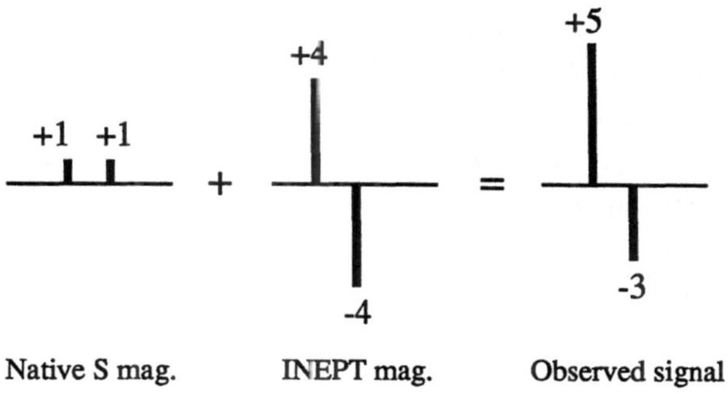

Often, it is desired to suppress the signal from the S spin magnetization. This can be achieved by repeating the experiment with the first I spin 90° pulse phase-shifted to the -X axis. A similar analysis to that above shows that this has the effect of reversing the sign of the transferred anti-phase signal but, of course, leaves the signal from the S spin equlibrium magnetization unaffected. Subtraction of these two experiments suppresses this latter signal. This is an example of a phase cycle (see later).

2. TWO-DIMENSIONAL NMR

(a) Historical

It was a Belgian physicist, Jean Jeener, who introduced the revolutionary idea for two-dimensional spectroscopy, but his work was never published, only reported at a Summer School. The idea was generalized and extended by Ernst. The essential new ingredient is a second time variable t_1 called the *evolution time*. Two-dimensional spectroscopy always involves the repetition of a basic pulse sequence n times, t_1 being incremented each time. A free induction decay $S(t_2)$ is acquired for each new value of t_1, so at the end of the series of n experiments we have acquired a two-dimensional data array $S(t_1,t_2)$. After two stages of Fourier transformation this yields a two-dimensional spectrum $S(f_1,f_2)$.

What transpires during t_1 can be arranged to be quite different from what happens during t_2, and consequently the two frequency dimensions, f_1 and f_2, may contain different kinds of NMR information. For example, we can easily arrange to have coupling constants displayed in the f_1 dimension (using modulated spin echoes) and chemical shifts in f_2 (using a decoupling technique). This is an example of the use of two-dimensional spectroscopy to separate parameters.

A more important idea is that of *correlation*. Suppose that we set up a coherence transfer experiment where some I magnetization is transferred to the S spins in a coupled IS spin system. This coherence may precess at the I-spin chemical shift δ_I during t_1, but at the S-spin chemical shift δ_S during t_2. There will therefore be a peak in the two-dimensional spectrum with coordinates (δ_I,δ_S) reflecting the fact that I is coupled to S. We say that we have *correlated* the I and S chemical shifts. Structurally this means that I and S are close neighbours in the molecule.

(b) Outline

The information presented in a two-dimensional spectrum depends on the pulse sequence and phase cycling used to record it. Expressed in the most general way a two-dimensional experiment has the form:

preparation - evolution during t_1 - mixing - detection during t_2

Note that signals are only detected during the evolution time t_2, which is the same time as in conventional pulsed 1D NMR experiments. Thus, it is not necessary for the coherence present during t_1 to be directly observable, it could be, for example, a multiple quantum coherence or a signal not at the spectrometer frequency.

During the preparation time, some particular coherence of interest is generated. At its simplest, the preparation sequence could be a single pulse, but in more complex experiments a series of carefully chosen pulses and delays might be used. The prepared coherence evolves during time t_1 and during the mixing period is then transferred to observable single quantum coherence. Finally the signal is observed during t_2. The nature of the mixing process is crucial in determining the information content of the spectrum.

Recording a two-dimensional data set involves setting t_1=0 and then acquiring a signal as a function of t_2 in the usual way with whatever time averaging is necessary. This signal is then stored away. The evolution time t_1 is then incremented by one sampling period Δt_1 and the experiment repeated with the signal again recorded as a function of t_2. This signal is then also stored away. The evolution time t_1 is again incremented by Δt_1 and the process repeated. In this way a series of free induction decays is recorded for different values of t_1. As in t_2, the increment Δt_1 sets the spectral width in the new dimension (f_1), according to $SW_1=1/\Delta t_1$.

In a conventional NMR experiment we might record several thousand data points during a free induction decay. In two-dimensional NMR a similar number of data points is recorded during t_2, but the number recorded during t_1 is usually much smaller, typically no more than 512, and often very many less. The reasons for recording a small number of t_1 increments are to restrict the amount of data generated by the experiment and to keep the experiment time within reasonable bounds.

Data processing then proceeds by first Fourier transforming the individual free induction decays recorded for different values of t_1 to give what are usually called the 'f_2 spectra'. As in conventional data processing, weighting functions might be applied before Fourier transformation. These f_2 spectra are then re-arranged in the following way. The first data point from each f_2 spectrum is extracted and assembled into an 'interferogram'. The data points in this interferogram correspond to increasing values of the time t_1, and when Fourier transformed yield a spectrum which is a function of f_1. Next, the second data point from each spectrum is extracted, assembled into an interferogram, and then Fourier transformed. The whole process is repeated until all the data points in the f_2 spectra have been treated. Ultimately, this gives a series of f_1 spectra, each of which corresponds to a different f_2 frequency. 'Viewed from above' in the form of a contour plot, this is the familiar two-dimensional spectrum of intensity *vs.* the two frequency axes.

This rather formal description of two-dimensional NMR does not address the question of how the evolution during t_1 is 'transferred' to the observed signal in t_2. This very important process will now be examined in some detail for a simple case.

(c) The Two Pulse Experiment

$$90(X) - t_1 - 90(X) - t_2$$

Consider the above pulse sequence applied to a single spin; the situation can be analysed exactly using the vector model:

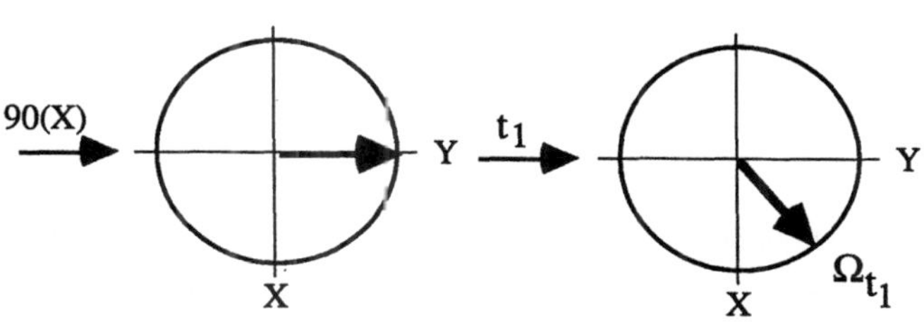

At the end of t_1 the vector has moved such that its X component is sin Ωt_1 and the Y component is cos Ωt_1. The second 90(X) pulse leaves the X component unaffected and rotates the Y component on to the Z axis where it is unobservable:

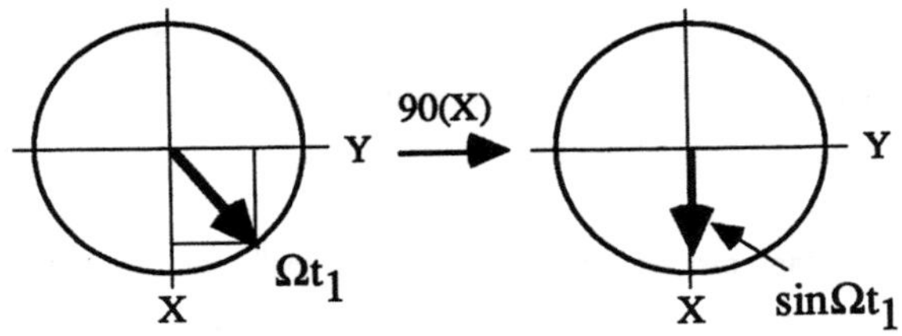

The component remaining along the X axis then evolves during t_2:

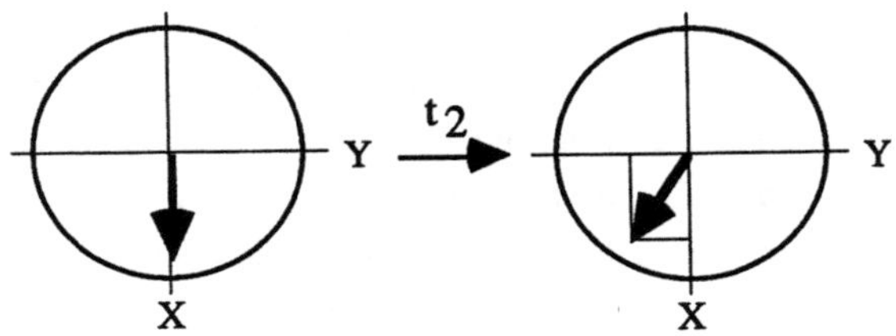

Giving the following observed X and Y components:

$$M_X = \sin\Omega t_1 \cos\,\Omega t_2$$
$$M_Y = -\sin\Omega t_1 \sin\,\Omega t_2$$
$$\textit{or, if } M \textit{ is defined as } M = M_X - iM_Y$$
$$M = \sin\Omega t_1 e^{i\Omega t_2}$$

The observed signal during t_2 is, as usual, an oscillating function with frequency Ω. *Note that the amplitude of the observed signal depends on the evolution during t_1 via the sin Ωt_1 term.* It is through this amplitude dependence that the information about the evolution during t_1 is transferred to the signal observed during t_2. The detected signal is said to be *amplitude modulated* by the evolution during t_1.

(d) COSY

The prototype two-dimensional NMR experiment is COSY (COrrelation SpectroscopY). It will be shown in this section that if two spins I and S are coupled a 'cross-peak' will appear in the two-dimensional spectrum at $(f_1,f_2)=(\Omega_I,\Omega_S)$. The pulse sequence for COSY is:

$$90(X) - t_1 - 90(X) - t_2 \; acquire$$

$$\begin{array}{cccc} | & | & | & | \\ \sigma_e & \sigma_1 & \sigma_2 & \sigma_3 \end{array}$$

Starting with an equilibrium density matrix of I_Z the first step has been calculated many times before:

$$\sigma_e \xrightarrow{(\pi/2)I_X} \sigma_1$$
$$I_Z \xrightarrow{(\pi/2)I_X} -I_Y$$

Chemical shift evolution for time t_1 gives:

$$-I_Y \xrightarrow{\Omega_I I_Z t_1} -\cos(\Omega_I t_1) I_Y + \sin(\Omega_I t_1) I_X$$

Evolution under scalar coupling gives, with the usual abbreviations for the sin and cos terms,

$$-c_I I_Y + s_I I_X \xrightarrow{2\pi J_{IS} I_Z S_Z t_1} \sigma_2$$
$$-c_I I_Y + s_I I_X \xrightarrow{2\pi J_{IS} I_Z S_Z t_1} -c_J c_I I_Y + s_J c_I 2I_X S_Z + c_J s_I I_X + s_J s_I 2I_Y S_Z$$

The final pulse acts on both I and S and affects each of the four terms in the previous expression in different ways:

$$\sigma_2 \xrightarrow{(\pi/2)I_X} \sigma' \xrightarrow{(\pi/2)S_X} \sigma_3$$
$$-c_J c_I I_Y \xrightarrow{(\pi/2)I_X} -c_J c_I I_Z \xrightarrow{(\pi/2)S_X} -c_J c_I I_Z$$
$$s_J c_I 2I_X S_Z \xrightarrow{(\pi/2)I_X} s_J c_I 2I_X S_Z \xrightarrow{(\pi/2)S_X} -s_J c_I 2I_X S_Y \quad C$$
$$c_J s_I I_X \xrightarrow{(\pi/2)I_X} c_J s_I I_X \xrightarrow{(\pi/2)S_X} c_J s_I I_X \quad A$$
$$s_J s_I 2I_Y S_Z \xrightarrow{(\pi/2)I_X} s_J s_I 2I_Z S_Z \xrightarrow{(\pi/2)S_X} -s_J s_I 2I_Z S_Y \quad B$$

Only the terms A and B are observable. Consider first the term A. It has the operator I_X and so corresponds to an in-phase I spin doublet centred at $f_2 = \Omega_I$. Its amplitude depends on $\Omega_I t_1$, so that the term is modulated by Ω_I in the way described above. The modulation coefficient can be expanded as:

$$c_J s_I = \cos(\pi J_{IS} t_1)\sin(\Omega_I t_1) = 1/2\,\{\sin(\Omega_I + \pi J)t_1 + \sin(\Omega_I - \pi J)t_1\}$$

The modulation is therefore expected to give rise to two peaks at frequencies $\Omega_I \pm \pi J$, which are just the frequencies of the two components of the I spin doublet. The appearance of the two-dimensional spectrum can be found by 'multiplying together' the multiplets expected in the f_1 and f_2 dimensions. This is best illustrated by a sketch:

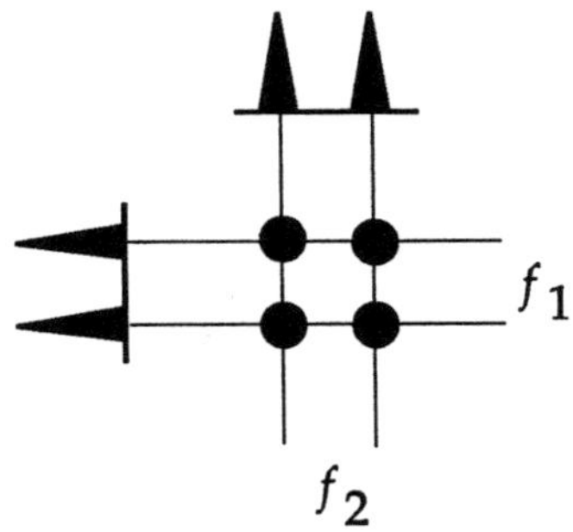

Note that there are four components to this two dimensional multiplet, and they all have the same sign. This multiplet is called a diagonal peak as it is centred at the same frequencies in f_1 and f_2. It contains very little useful information.

The term B is altogether more interesting. It has the operator $2I_ZS_Y$ which corresponds to an anti-phase doublet on the S spin and has arisen by magnetization transfer from the I spin. The modulation in t_1 can be expressed as:

$$s_J s_I = \sin(\pi J_{IS} t_1)\sin(\Omega_I t_1) = 1/2\,\{-\cos(\Omega_I + \pi J)t_1 + \cos(\Omega_I - \pi J)t_1\}$$

Fourier transformation with respect to t_1 will yield two peaks at $\Omega_I \pm \pi J$ but note that this time they have opposite signs, they are anti-phase in f_1. The doublet is also anti-phase in the f_2 dimension, giving the two-dimensional pattern:

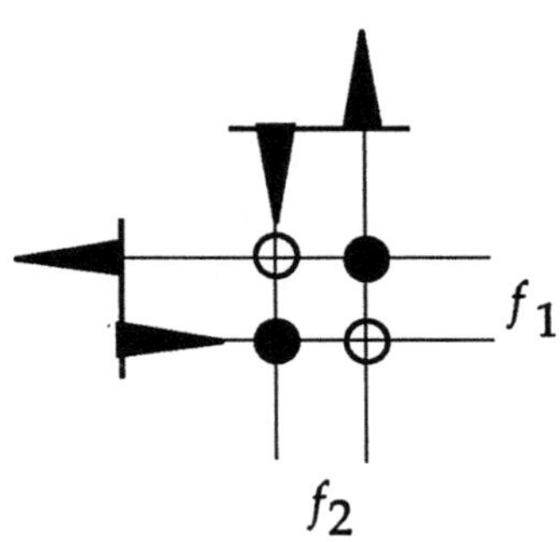

Note that this time the four components are in what is called an 'anti-phase square array'. This array is centred at (Ω_I, Ω_S) and is called a cross peak. The cross peak indicates the presence of a coupling between I and S and constitutes the useful part of the information from a COSY spectrum. The calculation can be repeated with an initial condition of S_Z which yields one more diagonal and one more cross peak. The overall result is the beautifully symmetrical COSY spectrum.

(e) Lineshapes in COSY

Careful inspection of the above calculation will reveal some significant differences between the cross and diagonal peaks.

Type of Peak	Operator in t_2	Modulation in t_1
Diagonal	I_X	$\sin(\Omega_I \pm \pi J)t_1$
Cross	$2I_ZS_Y$	$\pm\cos(\Omega_I \pm \pi J)t_1$

The diagonal peak arises from X magnetization whereas the cross peak arises from Y magnetization. If the former is phased so as to give a dispersion-mode lineshape, the cross peak will have an absorption-mode lineshape. Likewise in t_1, the modulations are 90° out of phase with one another. Often this is inconvenient, as the dispersive peaks are very broad and spread into the spectrum in such a way as to obscure nearby resonances.

(f) Double Quantum Filtered (DQF)-COSY

This is a simple modification of the COSY experiment which has the very considerable advantage of giving the diagonal and cross peaks with the same lineshape (e.g. both absorption), and also giving the diagonal peaks the same anti-phase structure as the cross peaks. This latter point is important, since the anti-phase structure of crosspeaks can lead to their cancellation when the linewidth is comparable to the coupling constant. In conventional COSY the diagonal peaks are in-phase, so that no such cancellation occurs and this can result in a very large intensity difference between the cross and diagonal peaks. Double quantum filtered COSY spectra have both diagonal and crosspeaks in anti-phase, so cancellation affects each in the same way and does not result in a large intensity difference between them. The pulse sequence for double quantum filtered COSY is:

$$\begin{array}{ccccc} 90(X) - & t_1 - & 90(X) & 90(\Phi) - & t_2\ acquire(\psi) \\ | & | & | & | & | \\ \sigma_e & \sigma_1 & \sigma_2 & \sigma_3 & \sigma_4 \end{array}$$

The phase of the last pulse is *phase cycled* in such a way that only double quantum coherences at point σ_3 contribute to observed signals. From the previous calculation the relevant term is C:

$$-s_J c_I \; 2I_X S_Y \quad C$$

A similar term arises from magnetization starting on the S spin,

$$-s_J c_S \; 2I_Y S_X$$

which we will not consider further. To understand experiments involving multiple quantum coherences, it is often useful to re-express in terms of the so called raising and lowering operators I^+ and I^-, where:

$$I_X = 1/2(I^+ + I^-) \text{ and } I_Y = 1/2i(I^+ - I^-)$$

Re-expressing term C gives:

$$\begin{aligned}\{-s_J c_I\}\; 2I_X S_Y &= \{-s_J c_I\} 1/2i(I^+ + I^-)(S^+ - S^-) \\ &= \{-s_J c_I\} 1/2i\big[(I^+S^+ - I^-S^-)(-I^+S^- + I^-S^+)\big]\end{aligned}$$

DQC ZQC

from which it can be seen that term C is a mixture of double and zero quantum coherence (DQC and ZQC respectively). If all other terms, except the DQC, are rejected by the phase cycling, and we re-express in terms of the cartesian operators:

$$\begin{aligned}\{-s_J c_I\} 1/2i(I^+S^+ - I^-S^-) &= \{-s_J c_I\} 1/2i[(I_X + iI_Y)(S_X + iS_Y) - (I_X - iI_Y)(S_X - iS_Y)] \\ &= \{-s_J c_I\} 1/2(2I_X S_Y - 2I_Y S_X)\end{aligned}$$

Then the final 90(X) pulse applied to σ_3 gives:

$$\sigma_3 \xrightarrow{(\pi/2)I_X} \sigma' \xrightarrow{(\pi/2)S_X} \sigma_4$$

$$\begin{aligned}\{-s_J c_I\} 1/2(2I_X S_Y - 2I_Y S_X) &\xrightarrow{(\pi/2)I_X} \{-s_J c_I\} 1/2(2I_X S_Y - 2I_Z S_X) \\ &\xrightarrow{(\pi/2)S_X} \{-s_J c_I\} 1/2(2I_X S_Z - 2I_Z S_X)\end{aligned}$$

Now, the final observable terms on I and S are both X magnetization, and the modulations in t_1 are both of the same form. The two terms correspond to the diagonal and cross peaks respectively and both multiplets are in anti-phase. Although the sensitivity is reduced, for most purposes there is little point in not taking advantage of these very useful properties, and the double quantum filtered COSY experiment is the method of choice.

3. PHASE CYCLING AND THE USE OF PULSED FIELD GRADIENTS

(a) Background

We saw earlier that we needed, in both the INEPT and DQF-COSY experiments, to select magnetisation arising from the required coherence transfer pathway. The most widely used method is so called *phase cycling*. Here the pulse sequence is repeated with a systematic variation in the phase of the radio-frequency pulses and the signals recorded with each repetition are combined in such a way that the signals arising from the wanted pathway add, whilst all others cancel. This, however, is a difference method, and as such it places great demands on the stability of the spectrometer. This is particularly true if the signals from the wanted pathway are much smaller than those from an unwanted pathway. A further difficulty is that the total experiment time is determined in part by the need to complete the phase cycle, rather than by the required resolution and sensitivity. This is particularly problematic in 3D/4D NMR experiments.

An alternative method involves the use of *pulsed field gradients* (PFG). During a PFG, the main magnetic field (B_0) is made spatially inhomogenous for a short time. As a result, transverse magnetisation and other coherences dephase across the sample and are lost. However, this loss can be reversed by a subsequent PFG which 'undoes' the dephasing process thus restoring the magnetisation or coherence. The crucial property is that different coherences dephase at different rates eg. DQC dephases twice as fast as single quantum coherence (SQC). Thus by applying PFGs of different lengths or strengths it is possible to refocus coherences which have been changed from say SQC to DQC by a radio-frequency pulse. Gradients are introduced so as to retain only the wanted signals in any experiment. Thus phase cycling is not required and there is no reliance on complete subtraction in NMR experiments. As a consequence 't_1 noise' and artifacts are much reduced. Again, in contrast to phase cycling, no repetitions are required and importantly for macromolecular NMR, pre-saturation of the water signal can be avoided.

(b) Coherence Transfer Pathways

Coherence is a generalisation of transverse magnetisation. The latter has a coherence order of $\rho=\pm1$ and is called SQC. Similarly DQC has $\rho=\pm2$ and coherence order zero has ZQC or Z magnetisation. The density operator can be separated in to contributions from different coherence orders $\sigma^{(\rho)}$,

$$\sigma = \sum_{\rho=-\rho_m}^{\rho=+\rho_m} \sigma^{(\rho)}$$

where ρ_m is the maximum coherence order (equal to the number of coupled spins, for spin =1/2). A pulse sequence can be defined in terms of the sequence and the desired coherence transfer pathway. Consider again the DQF-COSY experiment:

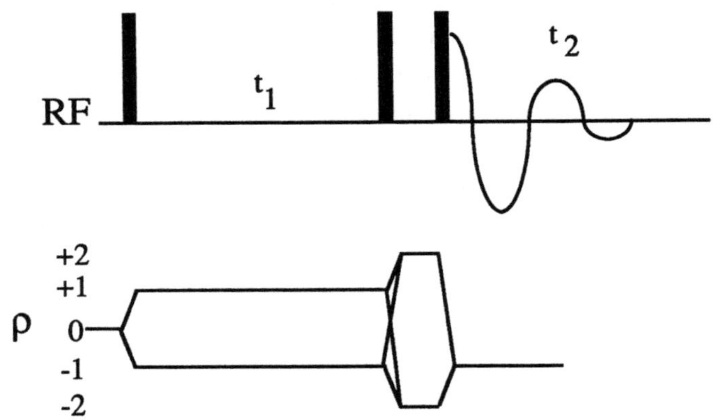

Before the first pulse equilibrium Z magnetisation is present and the coherence order is zero. The first pulse generates SQC with, $\rho=\pm1$. This evolves during t_1 and is converted into DQC by the second pulse. The final observable magnetisation is created by the last pulse. Only SQC is observable, and if we use quadrature detection we detect coherence of either order +1 or -1 (conventionally we consider observable SQC to be -1). However, it must be emphasised that this is the desired pathway and others will also contribute in the absence of some form of selection. It also pre-supposes that the spin system can support the required coherence eg. that we have two coupled spins to generate DQC.

(c) Coherence Pathway Selection

Coherence of different orders responds to Z rotations at different rates. The effect of a Z rotation on a coherence of order ρ, $\sigma^{(\rho)}$ is,

$$\sigma^{(p)} \xrightarrow{\phi F_Z} \sigma^{(p)} \exp^{(-p\phi)} \tag{5}$$

where F_Z represents the total Z component of the spin angular momentum. A Z rotation can be brought about by either Larmor precession induced by a magnetic field (PFG) or by a phase shifted pulse (phase cycling); eg. X magnetisation of a single spin evolves as:

$$I_X \xrightarrow{\phi F_Z} \cos\phi I_X + \sin\phi I_Y$$

The connection with coherence order is clearer if the cartesian operators I_X and I_Y are again expressed in terms of the raising and lowering operators (see above). Thus the evolution is now:

$$1/2(I^+ - I^-) \xrightarrow{\phi F_Z} 1/2(\exp^{(-i\phi)} I^+ + \exp^{(i\phi)} I^-) \tag{6}$$

Comparison of (6) with (5) (above) shows that I^+ corresponds to coherence order +1 (and I^- to -1). Likewise X magnetisation is a mixture of coherence order ±1 and I^+S^+ is a coherence of order +2 etc.

(d) Selection Using Pulsed Field Gradients

If the B_0 field is made spatially inhomogenous, after a period of time, spins in different parts of the sample will precess to different degrees (they have different Larmor frequencies) such that the net magnetisation, obtained by summing up all the vectors, will be zero. The Larmor frequency at a particular point in the sample depends on the applied field:

$$\omega_{L(Z)} = \gamma(B_0 + B_G)$$

After a time (t), the phase at any position in the sample is,

$$\phi_Z = \gamma(B_0 + B_G)t$$

where $\gamma B_g t$ represents the *spatially dependent phase* which is also dependent on the coherence order and any different gyromagnetic ratios for each spin such that:

$$\phi_Z = B_G t \sum_i p_i \gamma_i$$

The method of selection can be illustrated by again considering the DQF-COSY experiment.

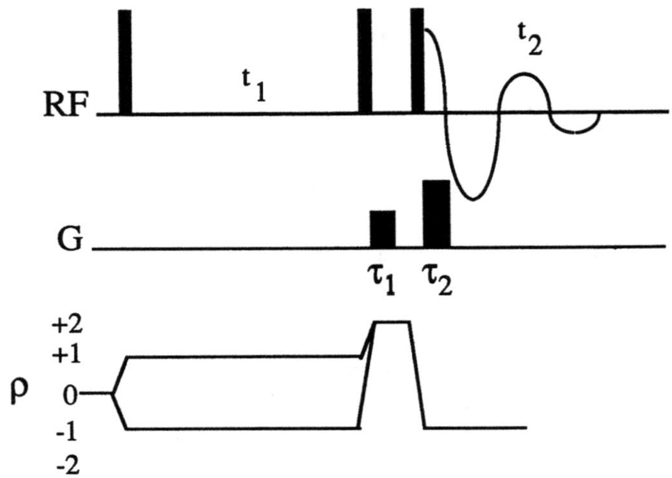

The first and second gradients encode a phase,

$$\phi_1 = \rho_1 B_{G,1}\tau_1 \quad \text{and} \quad \phi_2 = \rho_2 B_{G,2}\tau_2$$

respectively. After the two gradients, the net phase is $\phi_1 + \phi_2$ and to select a pathway involving transfer of coherence order ρ_1 to ρ_2 this net phase must be zero. Thus:

$$\frac{B_{G,1}\tau_1}{B_{G,2}\tau_2} = \frac{-\rho_2}{\rho_1}$$

So if ρ_1=+2 and ρ_2=-1, refocussing is achieved if the second gradient is either twice as strong, or twice as long, as the first. Other pathways remain dephased, eg. the water magnetisation which cannot support the DQ state is removed without the need for pre-saturation. To refocus a pathway in which there is no change in sign of coherence order, the second gradient must be applied in the opposite sense to the first (eg. if ρ_1=-2 and ρ_2=-1). Notice that, in contrast to the phase cycled experiment, only one pathway can be refocussed at a time and the gradient experiment is less sensitive. This is unfortunately a very common feature of gradient experiments.

AKNOWLEDGEMENTS

Parts of this lecture were adapted from lecture material given in a post-graduate course by Prof. R. Freeman and Dr. J. Keeler in the Department of Chemistry, University of Cambridge and that fact is gratefully acknowledged.

FURTHER READING

A few selected references for further reading are listed below. Most of the material given in this lecture came from these books or papers.

(a) General texts

A.E. Derome, 'Modern NMR Techniques for Chemistry and Research', Pergamon, 1987.
R. Freeman, 'A Handbook of Nuclear Magnetic Resonance', Longman, 1987.

(b) Advanced texts

R.R. Ernst, G. Bodenhausen and A. Wokaun, 'Principles of Nuclear Magnetic Resonance in One and Two Dimensions, OUP, 1987.
M. Goldman, 'Quantum Description of High-Resolution NMR in Liquids', OUP, 1988.

(c) Original papers or review articles

(i) Product operators
O.W. Sørensen, G.W. Eich, M.H. Levitt, G. Bodenhausen and R.R. Ernst, *Prog. NMR Spectrosc.* **16**, 163 (1983).

(ii) Coherence level diagrams and pathways
G. Bodenhausen, H. Kogler and R.R. Ernst, *J. Magn. Reson.*, **58**, 370 (1984).
A.D. Bain, *J. Magn. Reson.*, **56**, 418, (1984).

(iii) Experiments involving pulsed field gradients
J. Keeler, R.T. Clowes, A.L. Davis and E.D. Laue, *Methods in Enzymology*, in the press.

COMPUTER–ASSISTED STRUCTURE DETERMINATION OF BIOMACROMOLECULES BY NMR

Martin Billeter
Institut für Molekularbiologie und Biophysik
Eidgenössische Technische Hochschule - Hönggerberg
CH-8093 Zürich
Switzerland

OVERVIEW

In the past years the use of multidimensional heteronuclear techniques has pushed the limit in size of the molecular systems studied by NMR well beyond 100 amino acids (Wagner 1993). The resulting growth in complexity of the NMR spectra places an increasingly important role on computer programs that support the interpretation of the spectra and the management of the assignment data. When the earliest NMR structure determinations were accomplished about ten years ago (Wüthrich 1986), this was not evident for the sequence-specific assignment of the spectral cross peaks (Billeter et al. 1982). In contrast, the distance geometry calculations were always performed in an almost exclusively automated manner (Havel and Wüthrich 1984). Yet, still today two different uses of computers are observed in NMR structure determinations. One consists of the interactive process of obtaining cross peak assignments and extracting structural information from the NMR spectra with the help of a computer graphics package running on a workstation. The other typically consists of the submission of lists with structural constraints to a "black box", *e.g.* a program running on a remote supercomputer, in order to obtain a three-dimensional structure. Both approaches are subject to possible errors. The wealth of spectral information is difficult to interpret in a consistent way, and distance geometry approaches are known to be biased in various ways.

The flowchart of Fig. 1 summarizes the individual steps of a protein structure

NATO ASI Series, Vol. H 87
NMR of Biological Macromolecules
Edited by C. I. Stassinopoulou

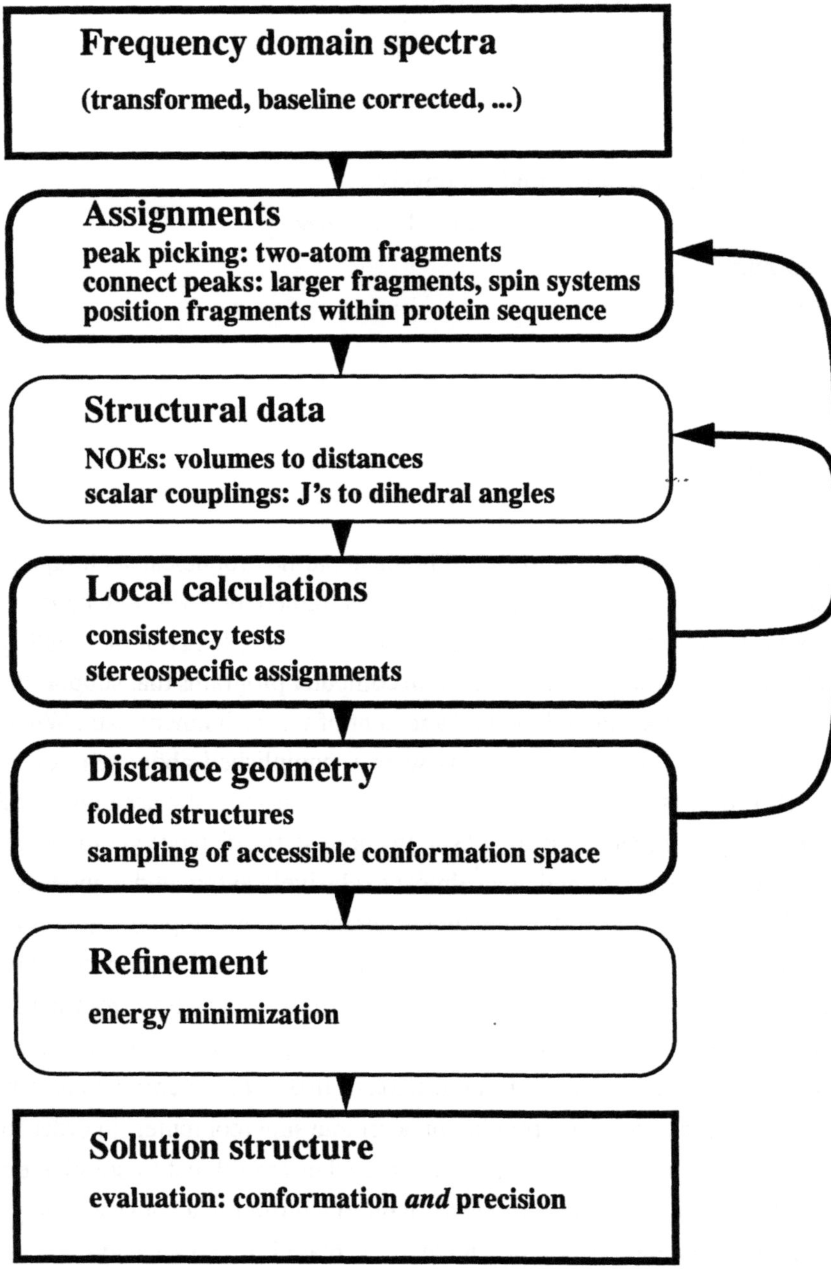

Fig. 1: Flowchart illustrating the individual steps of the structure determination of a protein using NMR and distance geometry.

determination by NMR and distance geometry. Computers already play an essential role in the preparation of the spectra prior to the collection of individual resonance assignments: Recording of time-domain data, transformation to frequency-domain spectra and baseline corrections are computer-intensive procedures. The following two sections will focus on the main steps in a structure determination, the resonance assignments and the structure calculations. The other steps of the flowchart, the conversion of spectral information to structural constraints and the final refinements, are less critical and are therefore only shortly addressed. The last box of the flowchart of Fig. 1 emphasizes the importance of a critical evaluation of the new structure; it deserves a separate discussion in a final section. Another feature of any NMR structure determination is illustrated in Fig. 1 by the arrows pointing from boxes in the lower part of the flowchart to boxes in the upper part. They indicate the presence of feed-back loops, *i.e.* the necessity of using preliminary results, *e.g.* from distance geometry calculations, for the revision and improvement of the data obtained from earlier steps, *e.g.* the assignment of NOE cross peaks.

COMPUTER–ASSISTED ASSIGNMENTS

The goal of the interpretation of a series of NMR spectra recorded for a given protein is to assign all cross peaks in these two-, three- or four-dimensional spectra to individual pairs, triples or quadruples of protein atoms (mostly hydrogens), respectively, and to extract data that allow the determination of the protein structure. In general, the assignment starts with the identification of cross peaks. These define a certain relation among the atoms involved: separation by only few covalent bonds in the case of scalar coupling, or spatial proximity in the case of dipolar coupling. In a next step, cross peaks are grouped based on common chemical shifts; this yields larger sets of atoms that correspond to molecular fragments, *e.g.* spin systems in the case of through-bond couplings (Wüthrich 1986). Finally, these fragments have to be located within the protein sequence. Fig. 2 illustrates these steps in a short polypeptide fragment. Dashed lines indicate through-bond proton-proton couplings, thick arrows dipolar couplings between protons that were found to be in close spatial proximity for many of the occurring backbone conformations

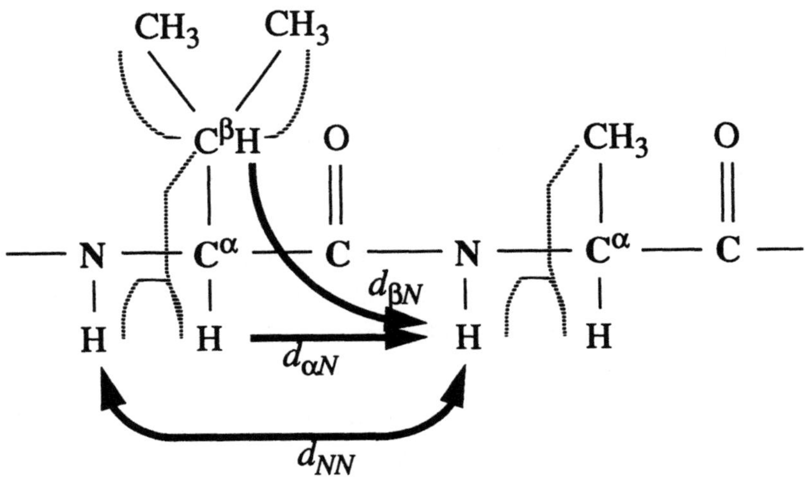

Fig. 2: Polypeptide fragment (Val-Ala) with indication of through-bond scalar couplings (dashed lines) and sequential NOEs (thick arrows labelled d_{NN}, $d_{\alpha N}$ and $d_{\beta N}$). The scalar couplings allow the identification of spin systems, *i.e.* the grouping of protons from the same residue, while the NOEs are used for the sequential assignment of these spin systems.

(Billeter et al. 1982). Both the presence of several NOEs connecting sequentially neighbouring residues and the fact that already short fragments, like the Val-Ala dipeptide of Fig. 2, are often unique in a protein sequence allow the resolution of ambiguities caused by NOEs between protons of sequentially distant residues.

In addition to the ability of making and modifying cross peak assignments, a software tool for the interpretation of NMR spectra should also allow for consistency checks between the calculated structures and the experimental data. The support of these procedures by a computer program requires the representation of spectra, cross peaks, assignments and structure-relevant information in a computer-suitable form that allows easy manipulation of these data: Zooming, display of cross-sections, extraction of all available information on a cross peak and establishment or modification of assignments are examples of operations that should be supported by bookkeeping routines. These interactive tools may be complemented by automated routines which are best suited for repetitive operations such as peak picking, peak integration or systematic comparisons of different sets of cross peaks. Today, many programs for workstations are available, and their possibilities are constantly being improved. The following is a presentation of selected aspects of the program XEASY, an interactive program for the efficient analysis of multidimensional NMR spectra (C. Bartels, T.-H. Xia, P. Güntert, M. Billeter and K. Wüthrich, in preparation; the predecessor EASY was described in Eccles et al. 1991).

XEASY offers a wide variety of commands that cover the handling of one or several NMR spectra, the selective display of parts of these spectra, the analysis of cross peaks and their sequence-specific assignment. Two possible applications are addressed in the following sections: the identification or the "picking" of cross peaks and the characterization of spin systems.

In XEASY cross peaks are identified by two to four chemical shifts, depending on the dimension of the spectrum. They may be further characterized by a list of possible assignments, a display colour for their classification into groups, a user comment, pointers for each dimension to an atom list (providing a unique assignment of the peak) and the peak volume. Several options are implemented for the automatic localization of cross peaks; their reliability is however reduced by overlap problems, especially in the case of in-phase peaks. Thus many interactive tools such as zooming or display of cross sections support the manual inspection and modification of the data describing cross peaks.

A typical use of XEASY is the collection of cross peaks that belong to the same spin system. Fig. 3 illustrates, with a 3D HCCH-TOCSY (Bax et al. 1990), a procedure for this assignment step which involves several XEASY options. After picking of the peaks, every peak on the diagonals of the ^{1}H-^{1}H planes is used to define a strip. (These peaks identify individual C–H fragments.) The strips, which are distributed over several ^{1}H-^{1}H planes, are then compared to each other. The similarity between two strips, *i.e.* the presence of cross peaks at identical ^{1}H shifts on the strips, indicates that the C–H fragments defining the two strips are likely to belong to the same spin system. After selection of a strip, XEASY automatically extracts all other strips from all ^{1}H-^{1}H planes and scores them according to their similarity with the selected strip. (The scoring function is discussed in C. Bartels and K. Wüthrich, in preparation.) By displaying the strips with highest scores next to the selected strip, all relevant information of the 3D-spectrum for the assignment of the spin system to which the selected C–H fragment belongs is reproduced in a two-dimensional presentation.

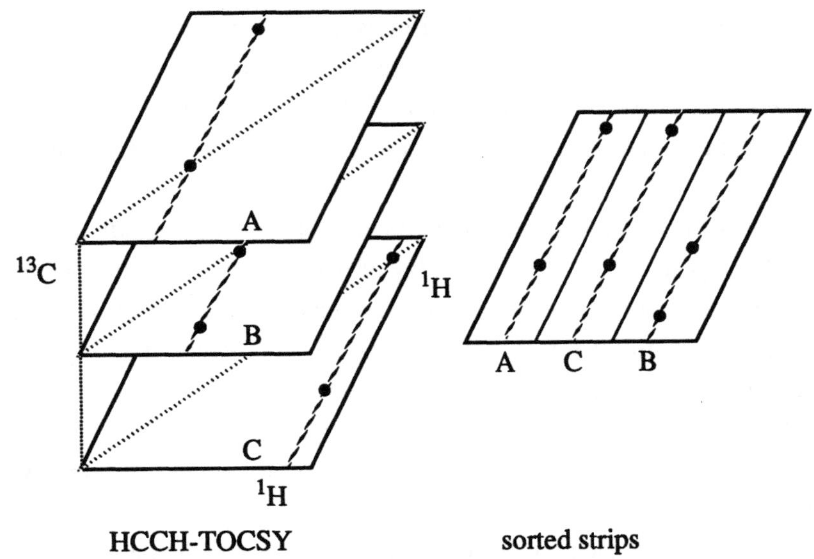

Fig. 3: The left part shows schematically three ^{1}H-^{1}H planes of a HCCH-TOCSY (Bax et al. 1990). Peaks are marked by black dots. In each plane a cross section with a peak on the diagonal (dotted lines) is indicated by a dashed line; these diagonal peaks identify C–H fragments. After selection of the strip with the cross section shown in plane A, XEASY extracts all strips with cross sections containing a diagonal peak and sorts them according to their similarity with the selected strip (right). The strip from plane C identifies a C–H fragment belonging to the same spin system as the selected fragment from plane A, while the C–H fragment from plane B belongs to another spin system.

As the ultimate goal of the software development for spectral assignments one might envision a fully automated algorithm. Even if such an algorithm would provide only half of the assignments, it would greatly facilitate the assignment step, since large amounts of the data would already be interpreted, thus leaving for the user only gaps in the sequential assignments to fill (*e.g.* Billeter 1991). These gaps would be well defined and starting points for the missing assignments would be provided by the assigned protons immediately adjacent to the gap. The difficulty with automatic routines lies in the abstraction inherent to the peak lists that are normally used. Many peaks are weak or in overlapping regions. Thus the initial peak lists obtained prior to any assignments will miss some peaks and/or include artifactual peaks arising from strong spectral noise. Because most automatic routines have no direct access to the spectral data, they cannot test the reliability of the peaks in their input and thus they cannot properly handle erroneous input.

STRUCTURE CALCULATIONS

Once a sufficient number of NOEs have been assigned to unique pairs of hydrogen atoms (or methyl groups), one may start with initial structure calculations. The volumes, V, of the NOE cross peaks are translated into distances, d, using the following equation:

$$V = k \langle d^{-6} \rangle f(\tau_c).$$

Here, k is a calibration constant which has to be estimated using known (fixed) distances or, better, from a fit of all NOE volumes to the corresponding distances in preliminary structures. The correlation time τ_c describes the motion of the atoms that give rise to the NOE. This quantity, which may vary within the protein, is difficult to assess. Therefore, and for other reasons that might influence the size of the NOEs (Wüthrich 1986), the information from a NOE is usually translated into an upper limit of the distance between the two atoms involved. Unless spin diffusion effects are explicitly taken care of, it is advisable to use only distance constraints for distances shorter than about 5 Å. With longer distances, other hydrogen atoms are likely to be found along the line connecting the two hydrogens for which the NOE is measured. For example, in an ideal α-helix the distance from the HN of residue i to the H^{α} of residue i+1 is 5.3 Å; the distances from both of these atoms to the HN of residue i+1 are 2.8 Å. Thus, this latter HN is located almost in the middle of the other two atoms and presents an efficient spin diffusion pathway. Additional structure relevant information comes from vicinal coupling constants. These can be translated with the use of Karplus-type curves into constraints on dihedral angles, mostly for ϕ-angles and χ^1-angles (Karplus 1959, DeMarco et al. 1978, Pardi et al. 1984).

At this stage, erroneous assignments are usually still around, and a good calibration of the distance constraints from NOEs is not yet available. Because the calculation of a structure involves many small steps (typically about 10,000 iterations of the distance geometry algorithms are required for one structure calculation), errors of the input are spread over a sizeable part of the output structure and are thus difficult to identify. It is therefore very useful to apply, prior to the actual structure calculations, some tests to the input constraints. One approach consists of selecting small fragments of the polypeptide, and of systematically checking all possible conformations of these for their consistency with the structural constraints. The program HABAS (Güntert et al. 1989) thoroughly examines the following dipeptide segment

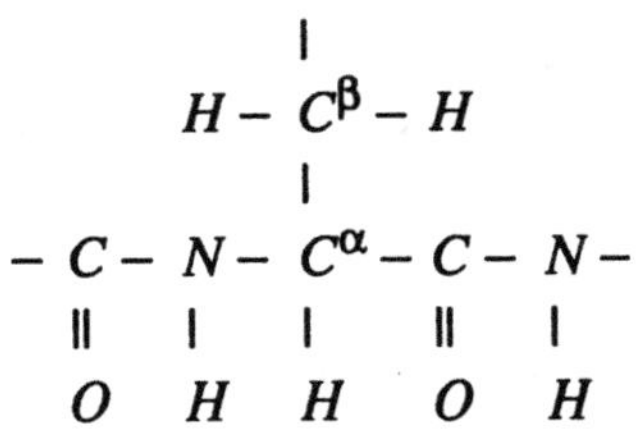

by checking the structural constraints that depend solely on the three dihedral angles ϕ, ψ and χ^1 for simultaneous consistency with all possible combinations of these three dihedral angles. This procedure not only detects errors among the short-range constraints (intraresidual and sequential NOEs, vicinal coupling constants) and improves the calibration of the distance constraints, it also provides for many residues stereospecific assignments of the individual β-methylene protons. These latter assignments can in turn improve the local structure of the backbone or the conformation of the side chains.

The general approach taken by the distance geometry programs consists of finding a conformation of the protein that minimizes a target function. This function consists of the sum of violations of all structural constraints derived from the NMR data, *i.e.* of constraints on selected distances and dihedral angles, and of further constraints ensuring a proper covalent geometry and minimal non-bonding distances (Table 1). The NMR-independent terms may be replaced by a complete empirical energy function (*e.g.* Weiner et al. 1986), although usually a simplified function is used. Several distance geometry procedures for the calculation of three-dimensional structures from NMR data have been presented (for a review see Brünger and Nilges 1993). The major problem in these calculations is the occurrence of a large number of minima in the target function; only the lowest minima correspond to structures with negligible residual violations of the experimental

Table 1: Terms of a typical target function used for the calculation of protein structures from NMR data.

term[†]	symbols[‡]	explanation[§]
$\Sigma K_i (b_i - b_i^o)^2$	b_i: bond length b_i^o: ideal bond length	enforce bond lengths
$\Sigma K_i (\theta_i - \theta_i^o)^2$	θ_i: bond angle θ_i^o: ideal bond angle	enforce bond angles
$\Sigma K_i (1 + \cos(\omega_i - \omega_i^o))$	ω_i: torsion angle ω_i^o: low-energy torsion angle	torsion energy
$\Sigma A_{ij}/r_{ij}^{12} - B_{ij}/r_{ij}^6$	r_{ij}: interatomic distance	Lennard-Jones energy
$\Sigma (q_i q_j) / (\varepsilon r_{ij})$	q_i, q_j: atomic charges ε: dielectric constant r_{ij}: interatomic distance	Coulomb energy
$\Sigma K_i (d_i - d_i^o)^2$	d_i: proton-proton distance d_i^o: upper distance limit	distance constraints from NOEs (applied only if $d_i > d_i^o$)
$\Sigma K_i (1 + \cos(c_i\omega_i - \omega_i^o))$	c_i: scaling to allowed range ω_i: torsion angle ω_i^o: centre of allowed range	torsion angle constraints from coupling constants (applied only if ω_i outside allowed range)

[†] The sums with running indices *i*, or double sums with running indices *i* and *j*, include all bond lengths, bond angles, torsion angles, atom pairs, NMR distance constraints and NMR torsion angle constraints, respectively.

[‡] The symbols K_i, A_{ij} and B_{ij} represent force constants.

[§] For a description of an empirical force field as described by the first five terms see Weiner et al. (1986). The two terms enforcing the NMR constraints are further explained in Widmer et al. (1989).

constraints and low conformational energy. During the search of these low minima, most algorithms risk to get trapped in a higher minimum. Fig. 4 illustrates one approach to overcome this problem, the variable-target-function algorithm (Braun and Go 1985; Güntert et al. 1991). The calculation consists of minimizations of actual target functions at various "levels", where the levels define the maximal range

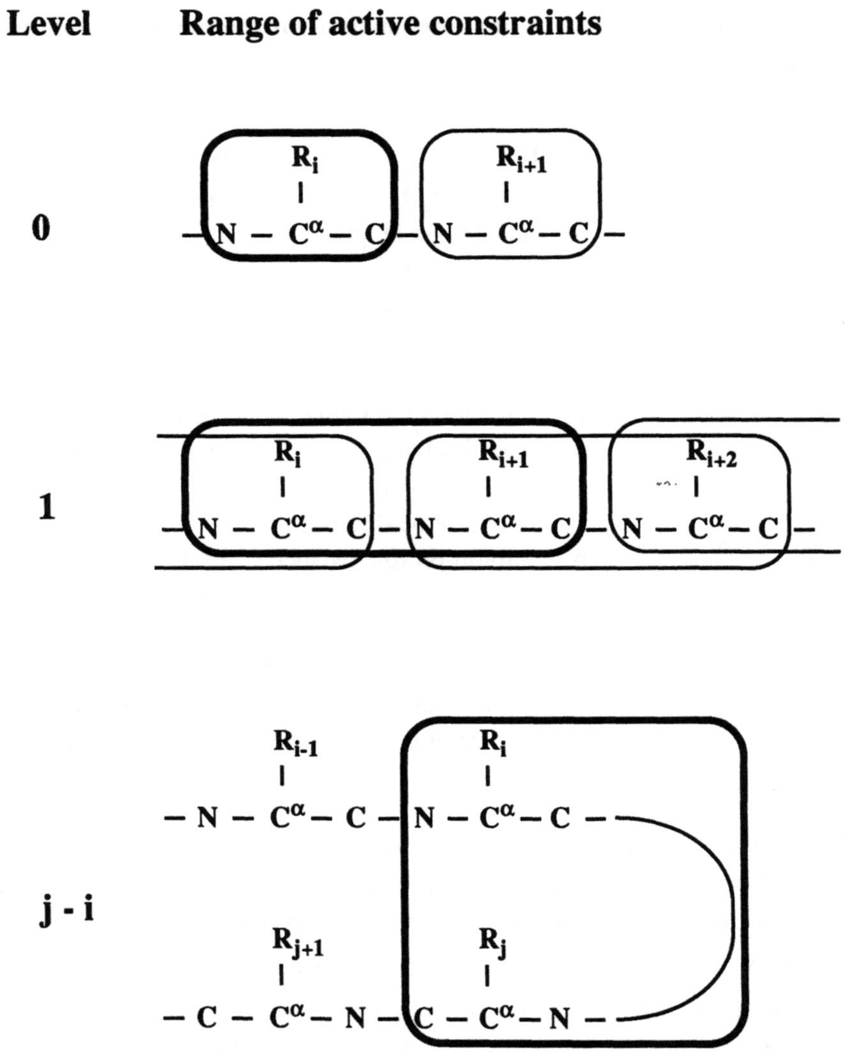

Fig. 4: Illustration of the variable-target-function algorithm as implemented in the program DIANA (Güntert et al. 1991). The boxes indicate examples of allowed ranges of active constraints at various levels (see text).

in terms of residue numbers over which the selected constraints that form the actual target function may extend. Thus, at level 0 only intraresidual constraints are included in the target function, at level 1 constraints between sequentially neighbouring residues are added etc. Eventually, all constraints are included at level *n*-1, where *n* is the number of residues in the protein. The gradual addition of constraints with increasing sequential range has the effect that helices are folded first, at levels three and four, while β-sheets are formed only late in the calculation. A special strategy has been described to partially overcome this disadvantage for non-helical structures (Güntert and Wüthrich 1991). A particular feature of the most widely used implementations of this algorithm is that the conformation space in which the protein is folded is spanned by the dihedral angles; thus the initial (correct) covalent geometry is always preserved (Braun and Go 1985; Güntert et al. 1991).

Table 2: Analysis of the NMR solution structure of the mixed disulfide between *E. coli* glutaredoxin(C14S) and glutathione.

quantity†	**average value**‡
largest residual distance constraint violation	0.10 ± 0.01 Å
largest residual torsion angle constraint violation	2.1 ± 0.3°
AMBER energy	-3270 ± 91 kcal/mol
RMSD: N, C^α, C' of all residues (1-85)	0.86 ± 0.11 Å
N, C^α, C' of residues 2-6, 12-25, 31-36, 43-81	0.53 ± 0.10 Å
same as above plus 46 best defined side chains§	0.77 ± 0.08 Å

† These quantities were calculated for the 20 conformers that describe the NMR solution structure of the mixed disulfide between *E. coli* glutaredoxin(C14S) and glutathione (Bushweller et al. 1993). The largest residual violations describe the consistency with the experimental constraints from NMR, the AMBER energy lists the intramolecular energy according to the AMBER force field (Weiner et al. 1986) using about 800 explicit water molecules and a constant dielectric permittivity, and the RMSD values the structural deviations between the 20 conformers and their average coordinates.

‡ The average value and the standard deviation is given for the quantities of the individual 20 conformers.

§ Side chain heavy atoms of the residues 2-6, 12, 14-15, 17-22, 24-25, 31, 33-36, 43-44, 47-49, 51-52, 55-56, 58-60, 62-64, 67-69, 72-79.

As an example, some aspects of the recent structure determination by NMR and distance geometry of the mixed disulfide between *E. coli* glutaredoxin(C14S) and glutathione (Bushweller et al. 1993) are discussed in this section. *E. coli* glutaredoxin is a small enzyme of 85 residues involved in electron transfer reactions (for a review see Holmgren 1989). Fig. 5 illustrates the density of the distance constraints derived from NOEs for this protein. The structure of the glutaredoxin-glutathione complex was determined based on 1010 distance constraints and 116 dihedral angle constraints. As with most NMR structures, several individual conformers were calculated independently (20 conformers in Fig. 6). The quality of the resulting structure is normally described by the extent of residual violations of the experimental constraints and by atomic root mean square deviations (RMSD) among the resulting conformers. Table 2 list these quantities for the NMR solution structure shown in Fig. 6. As indicated by the presence of several RMSD values in this table, which were calculated for various selections of atoms, different parts of the protein structure are often determined with different precision. This is also illustrated in a qualitative way by the spread among the set of 20 conformers in Fig. 6.

The description of residual violations of the NMR constraints in the resulting structures should indicate the consistency with the measured NMR data; however, the input constraints represent already an interpretation of the NMR spectra. This, and the fact that the NMR measurements are made on dynamic molecules whereas most distance geometry approaches implicitly assume a rigid structure, makes it difficult to define a single quantity describing the consistency between the structure and the experimental data, such as the crystallographic R-factor (see for example Glusker and Trueblood 1985). The RMSD values provide a measure of the precision achieved. This precision depends not only on the experimental data, but also on the type of the distance geometry algorithm. For example, a thorough energy minimization may fix individual methyls in a staggered rotamer state, although it is known that these groups are rapidly rotating. Similarly, degeneracy of the chemical shifts of the ortho and the meta protons in the rings of Phe and Tyr is usually observed indicating rapid ring flips (Wüthrich and Wagner 1978), whereas many NMR structures provide a well-defined ring orientation. Thus a NMR structure, like a crystal structure, should be understood as an average structure.

A

B

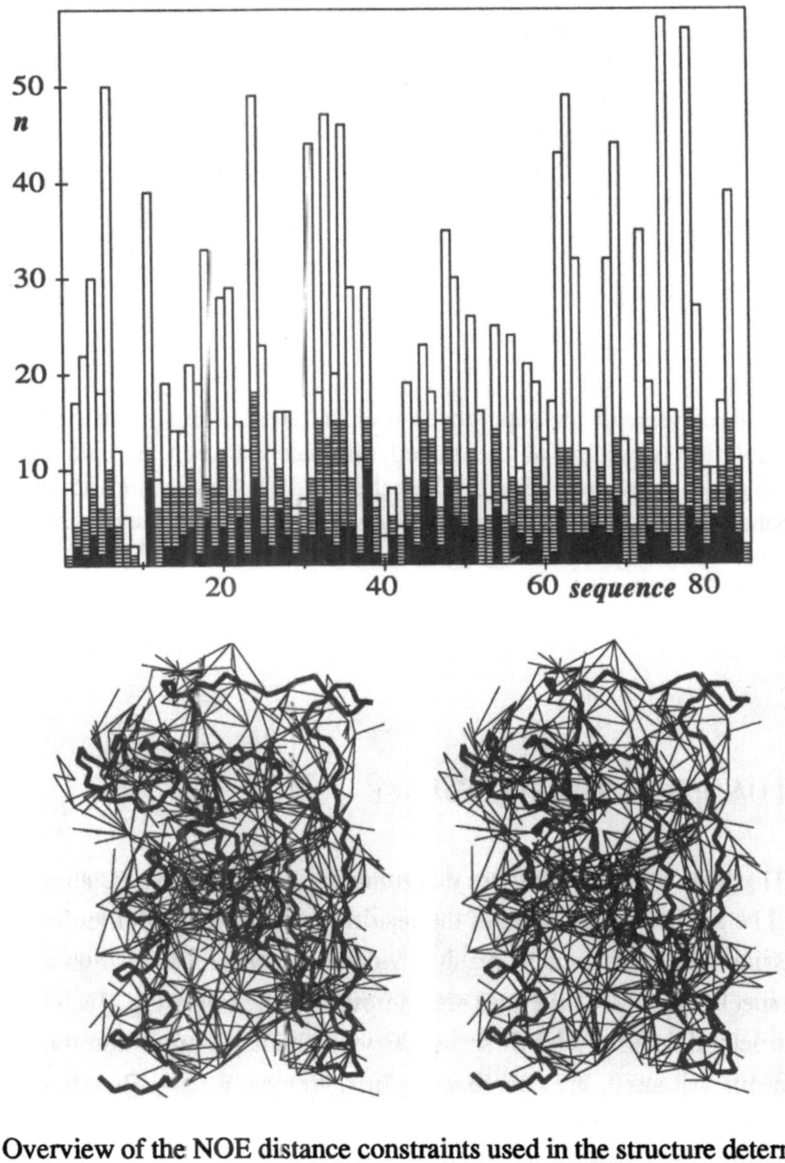

Fig. 5: Overview of the NOE distance constraints used in the structure determination of the mixed disulfide between *E. coli* glutaredoxin(C14S) and glutathione (Bushweller et al. 1993). A: Number *n* of distance constraints *vs.* the amino acid sequence of glutaredoxin. Intraresidual constraints are indicated by black bars, sequential constraints by hatched bars and the remaining constraints by empty bars. B: Density of the distance constraints in the final structure of the protein. The backbone of the structure is displayed with heavy lines. Each thin line represents one constraint; these connect hydrogen atoms which are not displayed.

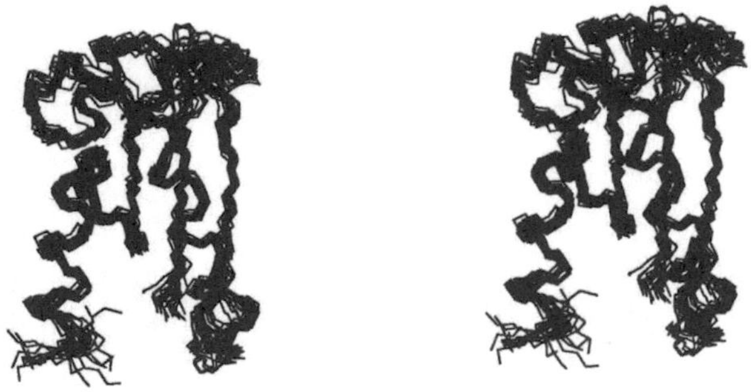

Fig. 6: Stereo view of the superposition of the 20 conformers that represent the NMR solution structure of the mixed disulfide between *E. coli* glutaredoxin(C14S) and glutathione (Bushweller et al. 1993). Shown are the polypeptide backbones of the glutaredoxin conformers as obtained with the program DIANA (Güntert et al. 1991) and energy-refined with the program OPAL (P. Luginbühl, P. Güntert, M. Billeter and K. Wüthrich, in preparation) using the AMBER force field (Weiner et al. 1986).

EVALUATION OF THE STRUCTURES

The final step in a structure determination by NMR and distance geometry should be a critical evaluation of the resulting structure. In addition to the above discussion on the meaning of residual violations and of RMSD values, some aspects specific to NMR structures are addressed in the following. The most important structural data from NMR spectra are constraints on proton-proton distances; this means that atoms are positioned *relative* to other atoms. Therefore, errors in the position of one proton will spread out to other proton positions via distance constraints to the first proton. This is in contrast to crystal structures where the electron density determines the position of each atom independently. However, short range NOEs and coupling constants may determine the conformation of individual side chains very accurately. As an example, intraresidual NOEs and the coupling constants $J_{\alpha\beta}$ of Leu 14 in Tendamistat clearly defined the position of each of the two methyls groups of this residue. In contrast, the X-ray data did not

A

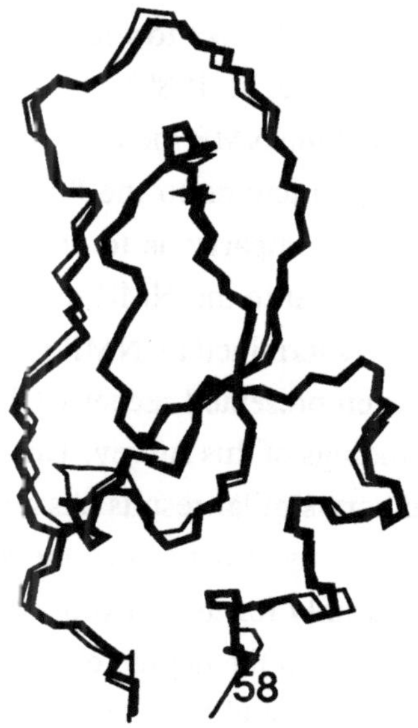

B

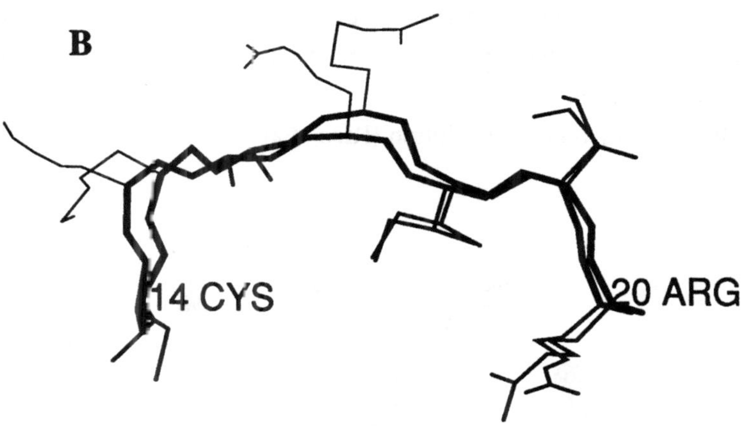

Fig. 7: Superpositions of the NMR (Berndt et al. 1992) and X-ray (Wlodawer et al. 1984) structures of BPTI. A: The complete backbones of the NMR (thick line) and the X-ray (thin line) structures are superimposed. B: Using the same superposition as in A, all heavy atoms of the fragments 14-20 (sequence CKARIIR) are displayed. The backbones of both structures are drawn with heavy lines. In both structures, side chains with a well defined conformation (small atomic displacements among the NMR conformers, low B-factors in the X-ray structure) are represented by medium lines, and side chains with poorly defined conformation by thin lines.

properly distinguish between this conformation and the one with the two methyl groups interchanged (Billeter et al. 1989).

The best evaluation of the NMR method of structure determination would be to compare the resulting structures to the "true" structures. Since these "true" structures are not known, comparisons to crystal structures may provide some clues about problems inherent to the NMR method (and vice versa). A thorough comparison of structures determined by NMR in solution and by X-ray diffraction in single crystals has been presented recently (Billeter 1992). This paragraph illustrates some of the findings of this survey. The first and major conclusion is that the two methods yield very similar results. As an example, Fig. 7 shows the similarity of the NMR and X-ray structures of the small proteinase inhibitor BPTI. This similarity holds not only for the backbone fold (Fig. 7A), but also for many side chains; it extends even to the degree of local precision to which the conformation of these side chains could be determined. As shown in Fig. 7B, the side chains of Ile 18, Ile 19 and Arg 20 have a well defined conformation in both structures, *i.e.* small atomic displacements are observed among the NMR conformers and low B-factors in the X-ray structure. As expected, the conformations of these side chains coincide.

The comparison of the solution (Rico et al. 1991) and crystal (Wlodawer et

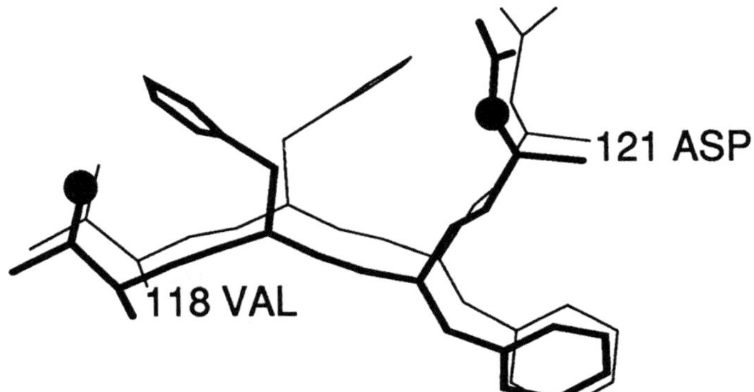

Fig. 8: Superposition of the fragments 118-121 (sequence VHFD) from the NMR (Rico et al. 1991, thick line) and the X-ray (Wlodawer et al. 1988, thin line) structures of ribonuclease A. The two filled circles indicate positions to which NOEs were observed from the ring of His 119 at different pH values (see text).

al. 1988) structures of bovine ribonuclease A yielded an interesting structural difference for the conformation of His 119 (Fig. 8). The two structures were determined at different pH values: pH 5.3 for the X-ray structure and pH 4.0 for the NMR structure. Additional NMR experiments with increased pH, resulting in conditions more similar to the ones used for the crystal structure, showed an NOE from the ring of His 119 to Asp 121; this NOE indicates a conformation similar to the X-ray structure. Moreover, in an independent crystal structure of the same protein by Borkakoti et al. (1982), electron density was reported for two alternative conformations of His 119. Thus, experimental conditions such as the pH value may well exceed the structural uncertainty of the NMR (and X-ray) method.

ACKNOWLEDGMENTS

Financial support by the Schweizerischer Nationalfonds (project 31.32022.91) is gratefully acknowledged. The author expresses his thanks for stimulating discussions to Drs. C. Bartels, J.H. Bushweller, P. Güntert, P. Luginbühl, K. Wüthrich and T.-H. Xia.

REFERENCES

Bax A, Clore GM and Gronenborn AM (1990) ^{1}H-^{1}H correlation via isotropic mixing of ^{13}C magnetization, a new three-dimensional approach for assigning ^{1}H and ^{13}C spectra of ^{13}C-enriched proteins. J Magn Res *88*:425-431

Berndt KD, Güntert P, Orbons LPM and Wüthrich K (1992) Determination of a high-quality nuclear magnetic resonance solution structure of the bovine pancreatic trypsin inhibitor and comparison with three crystal structures. J Mol Biol *227*: 757-775

Billeter M (1991) New interactive and automatic algorithms for the assignment of NMR spectra. In: Hoch JC, Poulsen FM and Redfield C (eds). Computational aspects of the study of biological macromolecules by nuclear magnetic resonance spectroscopy. NATO ASI Series, Plenum Press, New York: 279-290

Billeter M (1992) Comparison of protein structures determined by NMR in solution and by X-ray diffraction in single crystals. Quart Rev Biophys *25*: 325-377

Billeter M, Braun W and Wüthrich K (1982) Sequential resonance assignments in protein ^{1}H nuclear magnetic resonance spectra. Computation of sterically allowed proton-proton distances and statistical analysis of proton-proton distances in single crystal protein conformations. J Mol Biol *155*: 321-346

Billeter M, Kline AD, Braun W, Huber R and Wüthrich K (1989) Comparison of the high-resolution structures of the α-amylase inhibitor tendamistat determined by nuclear magnetic resonance in solution and by X-ray diffraction in single crystals. J Mol Biol *206*: 677-687

Borkakoti N, Moss DS and Palmer RA (1982) Ribonuclease-A: least-squares refinement of the structure at 1.45 Å resolution. Acta Crystallogr. sect. B *38*: 2210-2217

Braun W and Go N (1985) Calculation of protein conformations by proton-proton distance constraints. A new efficient algorithm. J Mol Biol *186*: 611-626

Brünger AT and Nilges M (1993) Computational challenges for macromolecular structure determination by X-ray crystallography and solution NMR-spectroscopy. Quart Rev Biophys *26*: 49-125

Bushweller JH, Billeter M, Holmgren A and Wüthrich K (1993) The NMR solution structure of the mixed disulfide between *E. coli* glutaredoxin(C14S) and glutathione. in press

DeMarco A, Llinás M and Wüthrich K (1978) Analysis of the ^{1}H NMR spectra of ferrichrome peptides: I. The non amide protons. Biopolymers *17*: 617-636

Eccles C, Güntert P, Billeter M and Wüthrich K (1991) Efficient analysis of protein 2D NMR spectra using the software package *EASY*. J Biomol NMR *1*: 111-130.

Glusker JP and Trueblood KN (1985) Crystal Structure Analysis. Oxford University Press, New York

Güntert P and Wüthrich K (1991) Improved efficiency of protein structure calculations from NMR data using the program DIANA with redundant dihedral angle constraints. J Biomol NMR *1*: 447-456

Güntert P, Braun W, Billeter M and Wüthrich K (1989) Automated stereospecific

^{1}H NMR assignments and their impact on the precision of protein structure determinations in solution. J Amer Chem Soc *111*: 3997-4004

Güntert P, Braun W and Wüthrich K (1991) Efficient computation of three-dimensional protein structures in solution from nuclear magnetic resonance data using the program DIANA and the supporting programs CALIBA, HABAS and GLOMSA. J Mol Biol *217*: 517-530

Havel TF and Wüthrich K (1984) A distance geometry program for determining the structures of small proteins and other macromolecules from nuclear magnetic resonance measurements of intramolecular ^{1}H-^{1}H proximities in solution. Bull Math Biol *46*: 673-698

Holmgren A (1989) Thioredoxin and glutaredoxin systems. J Biol Chem *264*: 13963-13966

Karplus M (1959) Contact electron-spin coupling of nuclear magnetic moments. J Chem Phys *30*: 11-15

Pardi A, Billeter M and Wüthrich K (1984) Calibration of the angular dependence of the amide proton - C^{α} proton coupling constants, $^3J_{HN\alpha}$, in a globular protein. Use of $^3J_{HN\alpha}$ for identification of helical secondary structure. J Mol Biol *180*: 741-751

Rico M, Santoro J, González C, Bruix M, Neira JL, Nieto JL and Herranz J (1991) 3D structure of bovine pancreatic ribonuclease A in aqueous solution: An approach to tertiary structure determination from a small basis of ^{1}H NMR NOE correlations. J Biomol NMR *1*: 283-298

Wagner G (1993) Prospects for NMR of large proteins. J Biomol NMR *3*: 375-385

Weiner SJ, Kollman PA, Nguyen DT and Case DA (1986) An all atom force field for simulations of proteins and nucleic acids. J Comp Chem *7*: 230-252

Widmer H, Billeter M and Wüthrich K (1989) Three-dimensional structure of the neurotoxin ATX Ia from anemonia sulcata in aqueous solution determined by nuclear magnetic resonance. Proteins *6*: 357-371

Wlodawer A, Walter J, Huber R and Sjölin L (1984) Structure of bovine pancreatic trypsin inhibitor. Results of joint neutron and X-ray refinement of crystal form II. J Mol Biol *180*: 301–329

Wlodawer A, Svensson LA, Sjölin L and Gilliland GL (1988) Structure of phosphate-free ribonuclease A refined at 1.26 Å. Biochemistry *27*: 2705–2717

Wüthrich K (1986) NMR of proteins and nucleic acids. Wiley, New York

Wüthrich K and Wagner G (1978) Internal motion in globular proteins. Trends Biochem Sci *3*: 227-230

NUCLEAR SHIELDING AND BIOMOLECULAR STRUCTURE

Ioannis P. Gerothanassis
University of Ioannina, Department of Chemistry, Ioannina GR-45110, Greece

THEORETICAL BACKGROUND

The origin of the chemical shift is easily understood in qualitative terms (Jardetzky and Roberts 1981; Harris 1983). Since electrons are moving charges, they are subject to the laws of electromagnetic induction. The net effect of the magnetic field is to superimpose a motion of electrons around the field axis on the orbital motion of the electron. In addition, the applied field will polarize the electron cloud. The motion of the electron generates a local magnetic field at the nucleus opposing the external induction field. The magnitude of the induced field will be proportional to the imposed field B_o so that the effective field B_{eff} at the nucleus will be:

$$B_{eff} = B_o(1 - \sigma) \tag{1}$$

The nucleus is said to be *shielded*, the extend of shielding being given by the shielding constant σ. This scalar shielding constant represents the isotropic part of the chemical shift tensor σ:

$$\sigma = \frac{1}{3} \text{Trace}(\sigma) \tag{2}$$

The anisotropic components of σ are averaged by fast reorientational motions. The non zero scalar shielding constant is often defined as the sum of contributions from a diamagnetic, σ^d, and a paramagnetic*, σ^p, term:

$$\sigma = \sigma^d + \sigma^p \tag{3}$$

* This term is somewhat misleading since it represents a temperature independent paramagnetism which is entirely different in origin from the temperature dependent paramagnetism of the electron spin.

NATO ASI Series, Vol. H 87
NMR of Biological Macromolecules
Edited by C. I. Stassinopoulou

The diamagnetic part relates to the concept of the free electron rotation about the nucleus and involves only the ground state electronic wave function of the molecule. In molecules the presence of the other nuclei hinders rotation of the electron cloud about the nucleus whose shielding is being considered. This indicates that in molecular systems there exists another contribution to shielding which normally opposes the diamagnetic term. This paramagnetic contribution to shielding, σ^p, may be estimated by finding the way in which the electronic wave functions are modified by the magnetic field. When a field is applied new energy terms appear and the electronic wave functions calculated in the absence of the field are no longer eigenfunctions of the electronic Hamiltonian. The new electronic wave functions may be obtained as linear combinations of the eigenfunctions for the field-absent case. Thus the field can be considered to mix 'excited' electronic states into the ground state wave function. The principal extra energy term which appears when the field is applied is the Zeeman energy of interaction of the electron orbital magnetic moment with the field. This energy is very much smaller than the total electronic energy and so the way in which it modifies the wave functions may be derived using perturbation theory.

The quantum mechanical formalism of the nuclear shielding tensor is due to Ramsey (Ando and Webb 1983; Webb 1986). A more suitable model for dealing with problems of molecular interest is one in which a number of localised shielding contributions are considered. Such an approach permits nuclear shielding to be considered as a sum of expressions, each of which can be related to particular atoms of a molecule. The total screening constant may be written as a sum of local effects and long-range or molecular effects. These are further broken down into separate terms as follows:

$$\sigma = \sigma^d\,(\text{local}) + \sigma^p\,(\text{local}) + \sigma^{ani} + \sigma^r + \sigma^e + \sigma^s + \sigma^h \qquad (4)$$

where σ^{ani} = neighbour magnetic anisotropy effect

σ^r = effects of ring currents

σ^e = electric field effects

σ^s = solvent (or medium) effects

σ^h = hydrogen bond effects

This scheme is arbitrary but useful particularly for large molecules.

Full ab initio computations on structures with few hundred atoms are not currently feasible. However, nuclear shielding is fundamentally a local phenomenon. As a result, one needs to take into account only the local geometry so that only a small number of atoms require basis functions. Consider sites of helical or sheet segments of proteins. They normally have characteristic ϕ and ψ torsion angles, and the changes in shielding of a particular nucleus due to these geometrical parameters are caused by the changes in the electronic wave functions near the site of interest. These effects propagate through the bonding framework and are therefore very short range. Strong hydrogen bonding and field effects would affect σ^h and σ^e respectively. The contributions σ^{ani} and σ^r are in principle independent of the nucleus being shielded (though not, of course, of its position in the molecule). Therefore, although they may be proportionately important for light atoms (especially hydrogen), where the total shielding is small, they are generally insignificant for heavier nuclei (see below).

THE LOCAL DIAMAGNETIC AND PARAMAGNETIC TERM

Pople (1962) developed a semi-empirical MO theory for nuclear shielding. The rotationally-averaged values of the local diamagnetic and paramagnetic terms are given by (Ebraheem and Webb 1977):

$$\sigma_A^d = \frac{\mu_o}{4\pi}\frac{e^2}{3m}\sum_{\mu}^{A} P_{\mu\mu}^{(0)} \langle \phi_\mu | r^{-1} | \phi_\mu \rangle_A \tag{5}$$

and

$$\sigma_A^p = -\frac{\mu_o}{4\pi}\frac{2e^2\hbar^2}{3m^2}\langle r^{-3}\rangle_{np_A} \sum_{j}^{\text{occ}} \sum_{k}^{\text{unocc}} ({}^1E_j^{k(0)} - {}^1E_o^{(0)})^{-1}$$

$$\times \Big[(C_{y_{Aj}}^{(0)} C_{z_{Ak}}^{(0)} - C_{z_{Aj}}^{(0)} C_{y_{Ak}}^{(0)}) \sum_B (C_{y_{Bj}}^{(0)} C_{z_{Bk}}^{(0)} - C_{z_{Bj}}^{(0)} C_{y_{Bk}}^{(0)}) + (C_{z_{Aj}}^{(0)} C_{x_{Ak}}^{(0)} - C_{x_{Aj}}^{(0)} C_{z_{Ak}}^{(0)}) \sum_B$$

$$\times (C_{z_{Bj}}^{(0)} C_{x_{Bk}}^{(0)} - C_{x_{Bj}}^{(0)} C_{z_{Bk}}^{(0)}) + (C_{x_{Aj}}^{(0)} C_{y_{Ak}}^{(0)} - C_{y_{Aj}}^{(0)} C_{x_{Ak}}^{(0)}) \sum_B (C_{x_{Bj}}^{(0)} C_{y_{Bk}}^{(0)} - C_{y_{Bj}}^{(0)} C_{x_{Bk}}^{(0)}) \Big] \tag{6}$$

where $C_{\chi}{}^{(0)}{}_{Aj}$ is the unperturbed LCAO coefficient of the np_{χ} orbital on atom A in MO j etc. It is apparent, from the ordering of the coefficients, that $\pi \rightarrow \pi^*$ transitions do not contribute to the summation. The term $<r^{-3}>_{npA}$ is the mean inverse cube radius for the p orbital on atom A with primary quantum number n. This is usually evaluated by means of the relationship

$$\langle r^{-3} \rangle_{np} = \frac{1}{3}\left(\frac{Z_{np}}{na_o}\right)^3 \tag{7}$$

where Z_{np} is the effective nuclear charge for the atomic orbital ϕ_{np} on A and α_o is the Bohr radius.

Eqn. (6) is often simplified by means of the average excitation energy (AEE) approximation. This involves replacing $({}^1E_j{}^{k(o)} - {}^1E_o{}^{(o)})^{-1}$ by a mean value $(\Delta E)^{-1}$. Because ΔE corresponds to some unknown weighted average of singlet transition energies it is usually not calculated a priori but treated as an empirical parameter. Since overlap is neglected in the semi-empirical methods used, it follows that (Ebraheem and Webb 1977):

$$\sum_j^{occ} C_{\mu j} C_{\lambda j} + \sum_k^{unocc} C_{\mu k} C_{\lambda k} = \delta_{\mu\lambda}. \tag{8}$$

By incorporating eqn (8) and the AEE approximation, eqn (6) becomes:

$$\sigma_A^P = -\frac{\mu_o}{4\pi}\frac{e^2\hbar^2}{2m^2}\frac{1}{\Delta E}\langle r^{-3} \rangle_{npA} \sum_B Q_{AB} \tag{9}$$

where the summation over B includes atom A and all other atoms in the molecule. The bond-order charge-density terms Q_{AB} are given by:

$$\begin{aligned} Q_{AB} = \; & {}^4/_3\delta_{AB}(P_{xAxB} + P_{yAyB} + P_{zAzB}) - {}^2/_3(P_{xAxB}P_{yAyB} + P_{xAxB}P_{zAzB} \\ & + P_{yAyB}P_{zAzB}) + {}^2/_3(P_{xAyB}P_{xByA} + P_{xAzB}P_{xBzA} + P_{yAzB}P_{yBzA}) \end{aligned} \tag{10}$$

where the P's are the elements of the charge density bond order matrix.

Most attempts at evaluating the paramagnetic contribution to nuclear shielding have employed the AEE approximation or an essentially equivalent valence-bond approach. Various procedures have been used to evaluate the bond-order matrix elements including π-electron calculations together with σ-bond polarizations and all-valence electron methods such as the extended Hückel, CNDO and INDO approaches. In general these methods are reasonably successful in accounting for gross chemical shift trends in series of related molecules. However, their usefulness is limited by the necessity of choosing a value for ΔE.

Equations (5) and (6) have been evaluated by means of various semi-empirical MO parameter sets. These calculations reveal that $\sigma_A{}^d$(local) usually varies by less than 5% for a given nucleus in a range of molecular environments (Webb 1986). Hence, for the majority of nuclei, shielding variations are primarily due to a change in the value of $\sigma_A{}^p$(local). For the ^{1}H nuclei the restricted shielding range of the hydrogen nuclei is not accounted for by a change of less than 5% in $\sigma_A{}^d$(local). Equation (6) reveals three potential sources of variation for $\sigma_A{}^p$(local), namely $\langle r^{-3}\rangle_{np}$, the excitation energy E_k-E_j, and the MO coefficients of the p orbitals. In general, the larger values of $\langle r^{-3}\rangle_{np}$ are those which are more susceptible to change giving rise to larger nuclear shielding ranges. The value of $\langle r^{-3}\rangle_{np}$ is related to the atomic number and to the position of the element in the Periodic Table. It is a maximum for the noble gases and drops to a minimum for each alkali metal (Fig.1). Its variation with atomic number accounts for the fact that shielding ranges for corresponding elements in the first two rows of the Periodic Table are comparable but there are sharp increases for the third and fourth rows (Jameson and Gutowsky 1964). An expression analogous to equation (7) has been derived for d orbital shielding contributions. Since the d orbital contribution to $\sigma_A{}^p$(local) also has a $\langle r^{-3}\rangle$ dependence it shows a periodic variation analogous to that of the p orbitals.

From the ordering of the LCAO coefficients in eqn.(6) it is apparent that $\sigma \rightarrow \sigma^*$, $\sigma \rightarrow \pi^*$, $n \rightarrow \sigma^*$ and $n \rightarrow \pi^*$ singlet transitions may be expected to contribute to $\sigma_A{}^p$(local). It is also clear that both atoms A and B must possess valence p electrons for $\sigma_A{}^p$(local) to be non zero. The very small shielding ranges found for the elements of groups I and II of the Periodic Table are principally due to the fact that, in addition to the unfavourable $\langle r^{-3}\rangle_{np}$ term, these atoms do not normally have valence

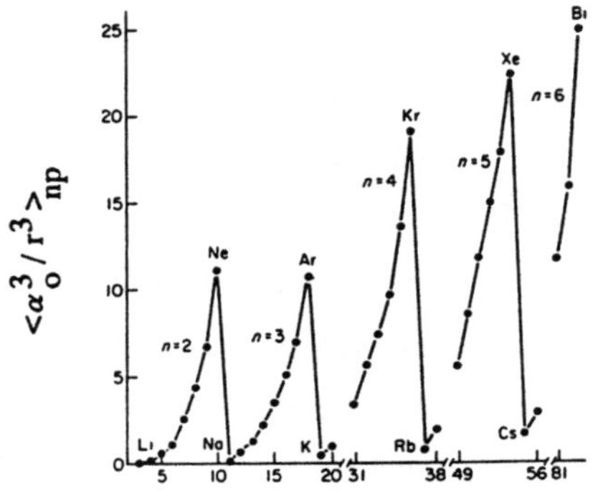

Fig.1 Variation with atomic number of the inverse cube radius of valence shell electrons (Jameson and Gutowsky 1964).

p electrons and form mainly ionic compounds. Consequently, the σ^p(local) term does not contribute to the shielding of these nuclei. ^{1}H is very probably the most important nucleus in biological NMR, however, ^{1}H shifts have a much smaller range (ca. 20 ppm with a few exceptions) than other nuclei (eg. ≥ 3000 ppm for ^{17}O). This is due to the fact that the total number of electrons in the vicinity of ^{1}H is smaller than for other nuclei so the magnitude of σ is smaller and consequently changes in σ from one molecule to another are also small. Further, the lowest excitation energy involved in σ_p for ^{1}H is of the type 1s → 2p (which implies a large value of ΔE and a small lower-state LCAO p-orbital coefficient), whereas for ^{17}O there are lower-lying orbitals and the ground-state bonding orbital involves p-electrons.

It is noteworthy that ^{14,15}N, ^{17}O and ^{19}F have larger ranges than ^{11}B and ^{13}C and that a parallel situation occurs amongst the third row elements. These ranges reflect the presence of non-bonding electrons on N, O and F. In general, these electrons give rise to low energy n → π^* transitions and thus to large values of σ^p(local) which cause nuclear deshielding. If the lone pairs are involved in bonding then this contribution to σ^p(local) is removed and the nuclear shielding increases. Thus the lone pair of electrons provide an extended range of nuclear shielding values. This effect is well demonstrated in the ^{17}O NMR spectra of the oxygenated synthetic haemoprotein model compounds of Fig.2. Two well separated resonances for the bridged and the terminal oxygens of the Fe-O_2 moiety are observed with chemical shifts of 1736 and 2488 ppm respectively for 1a (Gerothanassis et al 1992).

The σ^P(local) contribution to shielding is particularly marked for transition metal complexes because the summation in eqn.(6) is dominated by the lowest electronic excitation energy, provided the upper state has the correct symmetry properties. Ligand field theory shows this energy difference is often rather small. Furthermore, the greater polarizability of the p and d orbitals and the variety of oxidation states can have a large influence on nuclear shielding of transition metal atoms.

The effects of the various terms of eqn.(6) are clearly illustrated in the ^{43}Ca and ^{113}Cd NMR spectra (Fig.3) of the muscle protein parvalbumin, PA, (MW ~ 11.5 kDa) which contains two high affinity cation binding sites that bind Ca^{2+} and Mg^{2+} with association constants of 10^9 and 10^5 M^{-1} respectively. In these sites, generally termed EF and CD, the Ca^{2+} ions are octahedrally coordinated by carbonyl and carboxylate oxygen atoms (in the EF site one water oxygen coordinates to the Ca^{2+} ion). Only

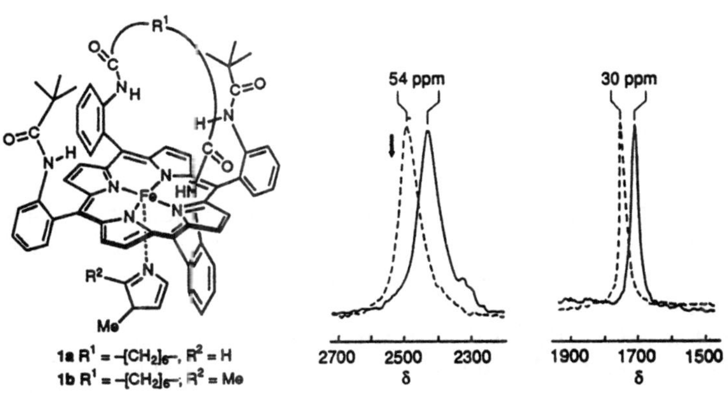

Fig.2 ^{17}O NMR spectra (54.2 MHz) of the oxygenated compounds **1b** (—) and **1a** (----) from use of the normal 90° pulse with a preacquisition delay time $\Delta t \approx 50$ μs and recorded in two steps with the carrier frequency on the absorption resonances. Maximum concentration ≈ 10^{-2}M in toluene solution. Concentration of 1,2-Me_2Im ≈ 5 x 10^{-1} M, T=297 K, number of scans 1080 000 (line-broadening filtering LB = 1 kHz) and 350 000 (LB=500 Hz) for the high- and low-frequency resonances, respectively. Chemical shifts were measured relative to 1,4 dioxane (+0.2 ppm relative to H_2O at 303 K) determined in a separate replacement technique. ↓ denotes the region of the resonance position of the terminal oxygen of the picket-fence porphyrin model, with excess of 1-MeIm, and ether-BHP in which no hydrogen bonding interactions are expected. (Gerothanassis et al 1992).

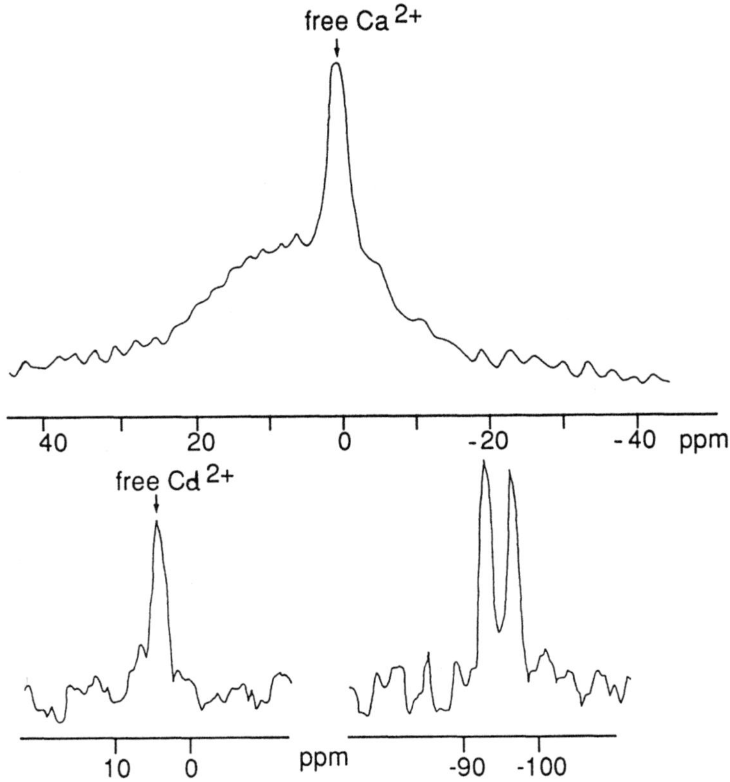

Fig.3 (A) The ^{43}Ca NMR spectrum of a 5.6 mM PA solution, at 23°C and pH 7.3 at a $[Ca^{2+}]/[PA]$ ratio of 2.4. The spectra were recorded by using a 70µs (90°) pulse, a data acquisition time of 20 ms and an instrument dead time of 300 µs. Each spectrum required ca. 2 x 10^6 transients corresponding to ~ 12 h of accumulation (Andersson et al 1982). (B) ^{113}Cd NMR spectrum of $^{113}Cd(II)$-parvalbumin at 30°C. Shifts relative to external 1 M $Cd(ClO_4)_2$. Conditions: 8.8 mM carp parvalbumin, pI 4.25, approx. 21 mM Ca^{2+}, 17.3 mM Cd^{2+} (16.6 mM $^{113}Cd^{2+}$), 0.28 M Na^+, 5.3 mM K^+, 5.3 mM Cl^-, 53 mM Tris-sulphate, 0.45 mM DTT at pH 5.9. Volume 4.74 ml. Total acquisition time 17 min (Drakenberg et al 1978).

a composite ^{43}Ca NMR resonance from Ca^{2+} ions bound to the EF and CD sites is observed which is shifted 10 ppm to high frequency relative to the signal of the free Ca^{2+} (Andersson et al 1982). Contrary to ^{43}Ca, the ^{113}Cd NMR shift spans a wide range from 750 ppm to -200 ppm and the nature of the coordinating ligands and coordination number to Cd^{2+} are strongly reflected in the chemical shift (Fig.4) (Johansson and Drakenberg 1989). Thus the free cations appear as a single signal at

+3.5 ppm while the ^{113}Cd (II) ions bound to the parvalbumin molecule appear as two distinct signals of nearly equal intensity at -93.8 ppm and -97.5 ppm which correspond to the CD and EF sites. The observation of two distinct signals indicates that the two ionic sites of parvalbumin in solution are very similar but not identical. The displacement towards low frequency of the protein-bound ^{113}Cd signals relative to the free ^{113}Cd signal indicates that the cadmium ions are coordinated only to oxygen atoms. It is worth noting that the largest chemical shift reported up to now for ^{113}Cd is for Cd^{2+} bound to the non-catalytic Zn^{2+} site of horse liver alcohol dehydrogenase ($\delta \simeq 751$ ppm). In this case Cd^{2+} is coordinated to four cysteine residues. The ^{113}Cd NMR spectra of bovine insulin displays two signals at 165 and -36 ppm. The resonance at 165 ppm is due to $^{113}Cd^{2+}$ bound to the Zn^{2+} site on insulin where Cd^{2+}

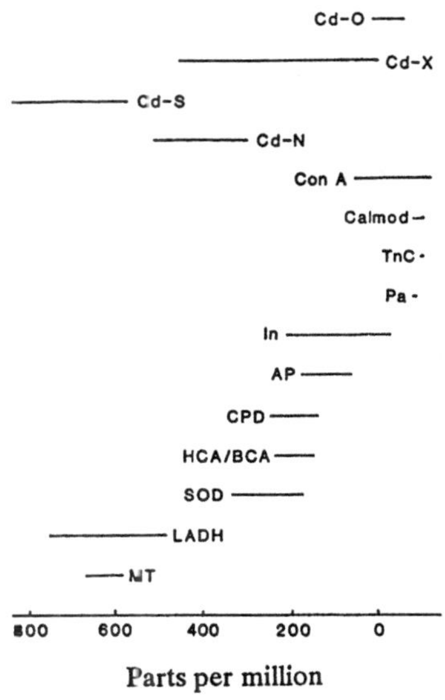

Fig.4 Chemical shift ranges of cadmium metalloproteins compared with chemical shifts determined for cadmium complexes. The shift ranges for the complexes are for pure ligand type. Abbreviations: Con A, Concanavalin A; Pa, Parvalbumin; TnC, Troponin C; Calmod, Calmodulin; In, Insulin; CPD, Carboxypeptidase A; HCA, Human Carbonic Anhydrase; BCA, Bovine Carbonic Anhydrase; AP, Alkaline Phosphatase; SOD, Superoxide Dimutase; LADH, Liver Alcohol Dehydrogenase; MT, Metallothionein (Ellis 1983).

is coordinated to three imidazole nitrogens and three water molecules; the resonance at -36 ppm is due to a Cd^{2+} ion bound to a Cd^{2+} site with only oxygen ligands.

The NMR shieldings of heavy atoms are influenced by two relativistic effects. The first arises from changes in the radial parts of the wave functions of heavier nuclei. The second is due to the breakdown of Russell-Saunders coupling in favour of j-j coupling of angular momenta which results in a spin triplet character of the ground state. Detailed results of relativistic calculations are provided in tables for atoms and ions with closed shell with nuclear charge up to Z=56 and a representative selection of ions with $56 < Z \leq 92$. However, evaluation of relativistic effects for heavy nuclei in molecules still remains to be investigated (Webb 1986; Jameson and Mason 1987).

MAGNETIC ANISOTROPY OF NEIGHBOURING BONDS

The magnetic field B_o will induce magnetic moments in all the electron orbitals of the molecule. Induced moments in a group G of bonding or atomic electrons will affect a nucleus X at a distance (Harris 1983). Clearly, there is such an effect in any given molecular orientation (Fig.5) and it can be shown that, provided the induced moment in G is not isotropic, the effect at X does not average to zero with molecular tumbling. Induced moments are proportional to magnetic susceptibilities; thus, when the magnetic field intensity is along a principal axis (l) of magnetic susceptibility for G, the induced magnetic moment is:

$$\mu_{il}(G) = \chi_l(G)H_l \tag{11}$$

where χ_l is the susceptibility (per molecule of G in direction l), and H_l is the field intensity. It can be shown that if the induced magnetic moments are considered to act as point magnetic dipoles, the contribution to $\Delta\sigma$ from a group G can be shown to reduce to:

$$\Delta\sigma = \frac{1}{3r^3}\left[(\chi_l - \chi_\perp')(3\cos^2\theta_\perp' - 1) + (\chi_l - \chi_\perp'')(3\cos^2\theta_\perp'' - 1)\right]/4\pi \tag{12}$$

where χ_l is the longitudinal component of susceptibility (i.e. along the bond direction),

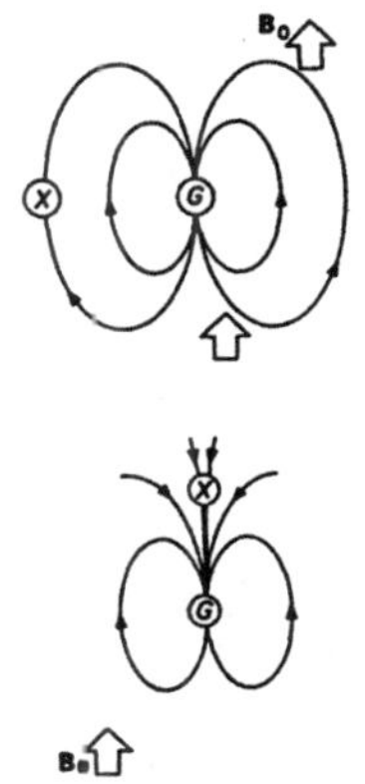

Fig.5 The induced magnetic field produced from an electronic group G will effect a distant nucleus X: (a) $r_{G\text{-}X} \perp B_o$ (X deshielded); (b) $r_{G\text{-}X} \parallel B_o$ (X shielded) (Harris 1983).

$\chi_\perp'$ and $\chi_\perp''$ are the two transverse principal susceptibilities (i.e. perpendicular to the bond), while $\theta_\perp'$ and θ_-'' are the angles between r and the appropriate bond perpendicular. Further simplification occurs if it is considered that the bond is cylindrically symmetrical ($\chi_\perp' = \chi_-''$). This will apply to a good approximation to all single and triple bonds, when the cone axis is the bond direction. In such a case:

$$\Delta\sigma = \frac{1}{3r^3}[(\chi_\parallel - \chi_\perp)(1 - 3\cos^2\vartheta)/4\pi \qquad (13)$$

where θ is the angle between the bond axis and r. The resulting shift will be either to high or to low frequency, depending on the signs of the susceptibility anisotropy and of the (1 - 3 $\cos^2\theta$) term, which changes on either side of θ = 54° 44′. There exist, therefore, regions of space enclosed by a double cone of half-angle 54° 44′ in which the resonance of a nucleus which happens to lie in these regions is shifted in one direction, whereas outside the cone the shift is in the opposite direction. Stereochemical information about the molecule in the neighbourhood of the anisotropic group can thus be deduced.

Of the common groups, C=C and C=O have positive susceptibility anisotropies

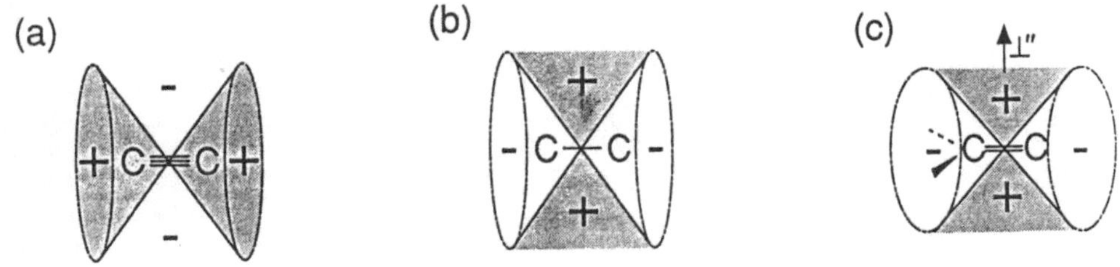

Fig.6 The nodal cones of shielding arising from anisotropy of bond magnetic susceptibility for (a) the C≡C triple bond, (b) the C−C single bond, (c) the carbonyl group. The effective centres of the bond magnetic dipoles are not necessarily midway along the bonds. The + and - signs refer to the effect on the shielding constant, σ, of a distant nucleus in the regions of space designated (Harris 1983).

and their effect is depicted in Fig.6; inside the two cones the contribution to σ is negative and the resonance shifts to low frequency, equation (13). For C≡C the anisotropy is of opposite sign and the effect is reversed. Small shifts of this type can be associated even with C-C and C-H bonds. As expected, the neighbour anisotropy effect falls off rapidly with distance and is observed only in the shifts of nuclei close to the anisotropic group of the molecule. The magnetic anisotropy of the peptide group is likely to contribute significantly to chemical shifts of proteins, but it has been difficult to develop quantitative models for σ^{ani} (Asakura et al 1992). One approach is to treat σ^{ani} as the sum of the magnetic anisotropies arising from the C=O, (O=)C-N and the C^{α}-N bonds. A second approach is based on the concept of atomic anisotropies developed by Pople rather than bond anisotropies.

RING CURRENTS

When electrons are delocalized the nuclei may not hinder the motion. As the circulations are over large areas, the magnetic moment induced by the field can be quite large. In aromatic systems it is assumed that the delocalized π electrons give rise to a ring current when the field is perpendicular to the molecular plane (this concept, however, has been criticized and large basis set ab initio calculations of proton shielding indicated that σ, rather than π, electrons are responsible for ring current effects). Since hydrogen nuclei have a small nuclear shielding range, the shielding changes of a few ppm attributed to ring currents can be of significance in

^{1}H-NMR. In general, ring currents effects are ignored for other nuclei.

There are three main approaches to the problem of calculating ring current effects on proton shieldings. These are the classical dipole model (D) of Pople (1956), the semiclassical current loop model of Johnson and Bovey (JB) (1958) and the quantum mechanical model of Haigh and Mallion (HM) (1972). In the D model the delocalised electrons are represented by an equivalent point dipole at the centre of the ring. In the JB model, the local magnetic field is considered to be produced by two current loops parallel to the plane of the ring and at a given distance above and below this plane. The HM model incorporates London's theory of diamagnetic ring currents.

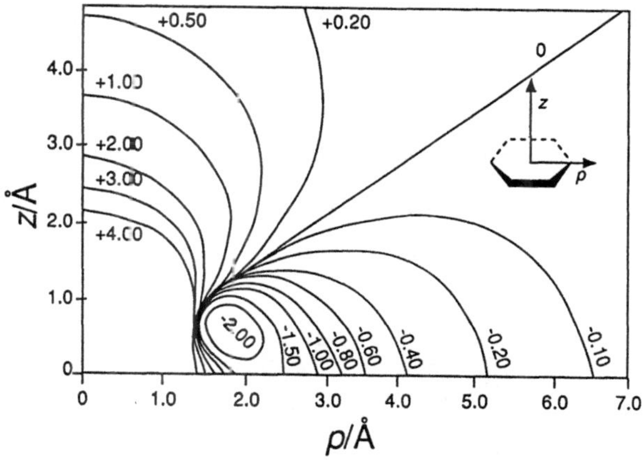

Fig. 7 Isoshielding lines which will be experienced by hydrogens as a result of the ring currents associated with the benzene ring. The *z*-direction is along the hexagonal axis of the benzene ring while ρ is the direction in the plane of the carbon atoms (Johnson and Bovey 1958).

The general form of the ring current equation can be defined by the empirical expression

$$\delta_R \times 10^{-6} = iBG\ (\rho,\ z,\ \phi) \tag{14}$$

where i is the ring current intensity factor which is specific for the aromatic ring type, B is a constant of proportionality and G(ρ, z, ϕ) is a spatial term relating the cylindrical coordinates ρ, z and φ of the proton to each ring plane and its centre. The geometrical term G(ρ, z, ϕ) has particularly large values in the short range vicinity of each aromatic ring, which is exemplified in the case of the Johnson-Bovey equation in Fig. 7. The ring current effects on protons beyond the van der Waals dimensions of the conjugated ring are very similar for the three models, however, they are significantly different for protons in the plane of the ring and close to it (Fig.8).

Many aromatic rings of biochemical interest are heterocyclic, e.g. purine and pyrimidine rings. In such cases the earlier calculations were based on the estimation of the ring current effects following an approach very similar to that of Johnson and Bovey. Recently these calculations have been refined to take into account the atomic diamagnetic anisotropies (Giessner-Prettre and Pullman 1987). Fig.9 shows calculations of the anisotropy of purine and pyrimidine rings. It can be noticed that the observed shieldings depend critically on the orientation of the bases with respect to one another.

The ring-current shifts associated with porphyrin rings are, in general, much larger than those of other aromatic ring systems. The porphyrin ring system has 18 electrons that can circulate around four pyrrole rings and a large "inner" ring. Abraham (1961)

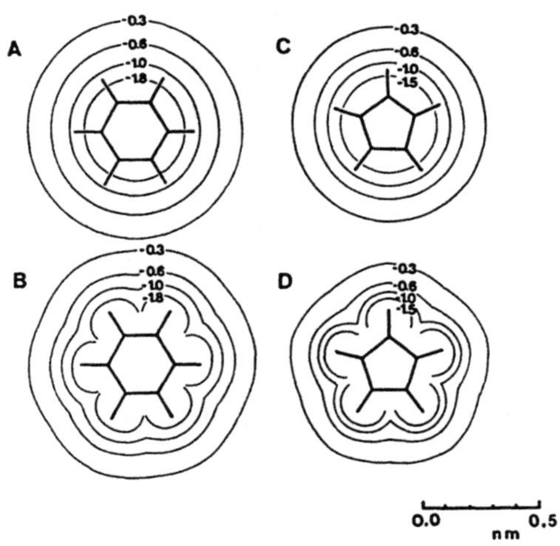

Fig. 8. Contour diagrams to show the magnitude of the high frequency shift in ppm in the plane of the aromatic ring. (A) and (C) show the cylindrical symmetry of the ring-current shift predicted by the Johnson-Bovey equation of the six- and five-membered rings, where there is no dependence of shift on the angle ϕ (eq. 14). (B) and (D) show the hexagonal and pentagonal symmetry of the shift which is predicted from the ϕ dependence of the shift in the Haigh-Mallion equation (Perkins and Dwek 1980).

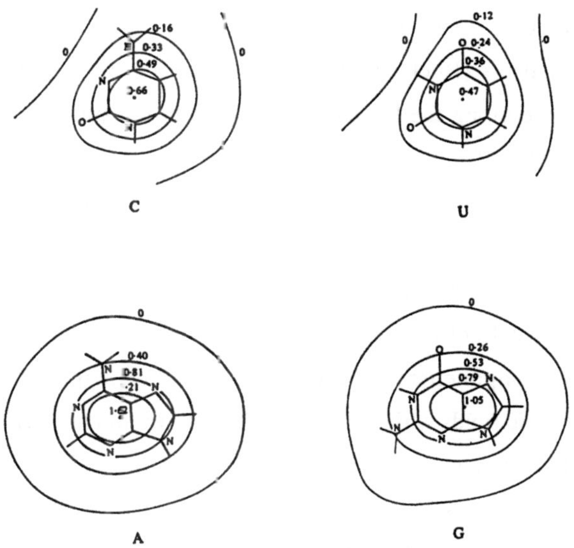

Fig. 9 Contour diagrams to show the magnitude of the anisotropy of purine and pyrimidine rings on the chemical shift of a nucleus at a distance of 3.4 Å above the plane of the base (Giessner-Prettre and Pullman 1987).

tackled the problem using essentially the same theoretical approach as Johnson and Bovey for each of the individual rings. However, the calculated shifts for protons on the porphyrin ring were about 50% larger than the observed shifts. The calculations by Shulman et al (1970), neglecting the spin-orbital coupling, have been quantitatively more accurate.

NEIGHBOURING GROUP ELECTRIC DIPOLES

If a molecule contains polar groups there will be intramolecular electric fields which distort the electron density around nuclei and will therefore affect screening constants (similar effects may be obtained in principle from externally applied electric fields). In the case of a uniform electric field E it can be shown that shielding of e.g. a proton in an axially symmetric situation (Fig.10) there are two terms contributing to σ^e

$$\sigma^e = - A E_z - BE^2 \qquad (15)$$

where z is the direction of the bond of the hydrogen atom and A and B are the electric field shift tensors (Buckingham 1960). The first, or linear, term is due to polarization of charge along the bonds on the atom in question. It is composed of two contributions, the uniform field and field gradient (Batchelor 1975). The second order term is due to orbital distortion, resulting in the mixing of excited states into the molecular wave function and a rise in the ground state energy (actually, if exact electronic wave functions were known there would be no need to invoke an intramolecular electric field effect, since a direct calculation of σ^d and σ^p would suffice). Again, two components can be distinguished (Buckingham 1960), one arising from the steady electric fields which also contribute to the linear term, and one from the fluctuating fields which give rise to Van der Waals forces. The electric field shift tensors A and B differ for various nuclei. For several heteronuclei, e.g., ^{19}F, the second term may be more important than the first.

A simple semiempirical method for estimating the uniform-field linear electric field shift (LEFS) terms is ^{13}C titration shifts has been proposed by Batchelor (1975). This method is based on the hypothesis that a carbon with perfect tetrahedral bond symmetry will show no uniform field LEFS, since the flow of electrons onto the carbon will exactly equal the flow from it. Thus, the difference in titration shift between the carbon of interest and quaternary carbon in an otherwise similar

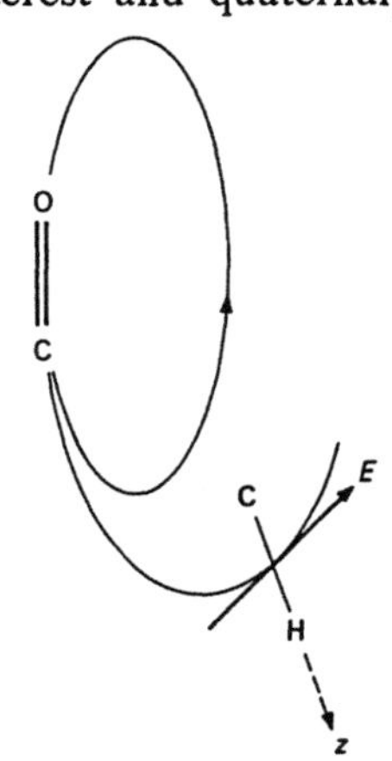

Fig.10 The linear electric field effect on the chemical shift of a C-H proton due to a distant carbonyl group. Note that electrons drift towards the tail of the electric field vector E and therefore, since in this case E_z is negative, the proton is shielded (Harris 1983).

environment is taken as the uniform-field LEFS. Although this component of the electric field shift contribution to ^{13}C titration shifts shows the expected monotonic decrease with increasing distance from the titrating group, the total shift shows more complicated behaviour, as illustrated in Fig.11 (Batchelor et al 1975), because the individual contributions have different signs and different distance dependencies. Recently, the results of ab initio derivative Hartree-Fock calculations of the dipole and quadrupole shielding polarizabilities and hyperpolarizabilities of a number of small molecules were reported, together with estimates of the electric fields and field gradients present in proteins (Augspurger et al 1992). It was suggested that electric field induced shifts may be large ($\approx$ 5 ppm for ^{13}C, $\approx$10 ppm for ^{17}O and ^{19}F) due to the low dielectric constants found in proteins and, in some cases, they may dominate the experimentally observed shifts.

MEDIUM EFFECTS

The effects of solvents on the nuclear shielding of solute molecules may be categorized into specific and non specific classes. The specific interactions include hydrogen bonding because it requires the contact between "specific" sites and the partners, e.g. the proton donor site OH group of a bound molecule of water and the proton acceptor site like the lone pairs of electrons of the peptide oxygen. Specific

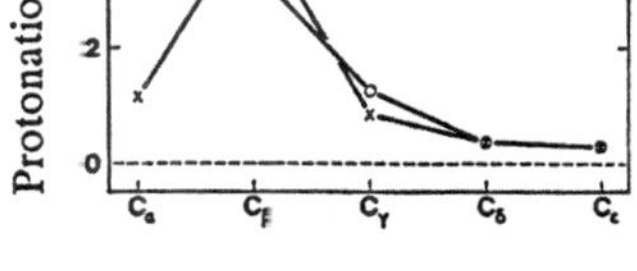

Fig.11 Low frequency shifts of norleucine ^{13}C resonances due to protonation of the amino group. (x) Total protonation shifts. (o) Uniform field LEFS contributions. (Batchelor et al. 1975).

interactions will be examined in the hydrogen bond subsection. Non specific influences can arise from electronic interactions between the dipole moments of the solute and solvent molecules.

Non specific solvent effects on NMR parameters may be discussed on the basis of two models. In one the solvent is taken as a continuum, characterised by its macroscopic dielectric constant, ε. The electrostatic interactions between the solute and solvent molecules are accounted for in terms of the reaction field theory (Bottcher 1973). The second type of model permits the shielding changes to be interpreted in terms of interactions between pairs of molecules. A general pair interaction model includes terms due to van der Waals forces, bulk susceptibility effects, electric moment effects and the magnetic anisotropy of nonspherical solute molecules (Raynes et al 1961).

Precise mathematical expressions are not available for many non-bonded interactions. Consequently, continuum models tend to be more popular for investigating solvent effects on NMR parameters. The most widely used of these, to date, seems to be the solvaton model (Ando and Webb 1981). By means of this model the solute-solvent interaction can be introduced into the paramagnetic screening contribution σ^p(local) and the solvent induced shift is proportional to $(\varepsilon - 1)/2\varepsilon$. Some INDO/S SOS shielding calculations, incorporating the solvaton model, predict the correct trends of shielding and suggest that this model may be suitable for determining the solvent induced changes in heteronuclear shielding that may occur in biomolecules.

HYDROGEN BONDS

Hydrogen bonding plays a very important role in all classes of molecules of biological interest and both intramolecular and intermolecular hydrogen bonding will have an effect on the chemical shift. For example, NMR studies have shown that sharp resonances exist at high frequencies for protons involved in the Watson-Crick base pairing. The deshielding of the proton involved in hydrogen bonding poses some theoretical problems. Pople et al (1959) have pointed out that the decreased shielding arises from two general effects in the case of intermolecular hydrogen bonding between molecules XH and Y to form X−H···Y. First, the stationary magnetic field will induce currents in Y, with the resulting magnetic field caused by those currents tending to shield the hydrogen bonding proton. Second, the electron donor molecule Y will distort the electronic structure in the X−H bond, leading to different diamagnetic currents. The result is a deshielding of the proton. The observed high frequency chemical shifts with hydrogen bonding indicate that effect (b) predominates in most cases; the exceptions involve a few aromatic systems.

Wagner and collaborators (1983) suggested an equation to relate amide proton chemical shifts (corrected for ring-current effects) to the oxygen (acceptor) to proton (donor) distance r:

$$\Delta\delta = 19.2\ r^{-3} - 2.3 \tag{16}$$

where $\Delta\delta$ is the chemical shift difference from random coil values in ppm. The chemical nature of the acceptor group seems not to play an important role and the oxygen atom may belong to a carboxyl group, a carbonyl group or an internal water molecule.

Wishart et al (1991) suggested that, although there is no obvious connection

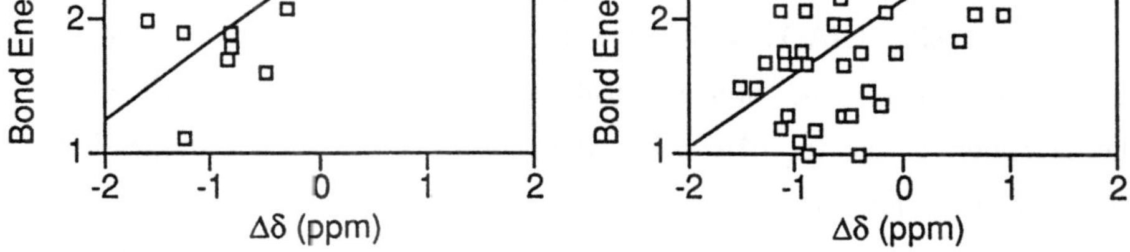

Fig. 12 Plot of $\Delta\delta$ versus hydrogen bond energy in kcal/mol for (a) BPTI and (b) E.coli thioredoxin. $\Delta\delta = \delta_{obs} - \delta_{rc}$ where δ_{rc} is the random coil amide proton chemical shift and δ_{obs} is the observed chemical shift. Only hydrogen bonded pairs in which bond energies exceed 1 kcal/mol were included (Wishart et al 1991).

between proton chemical shifts and solvent exposure, there is a strong dependence on amide proton chemical shifts and hydrogen bond energies (the latter were calculated using an r^{-1} dependency). Fig.12 demonstrates the relationship between uncorrected or raw amide proton chemical shifts and bond energies for residues known to be in hydrogen bonded networks. A relatively good linear correlation for both bovine pancreatic trypsin inhibitor (BPTI) and E.coli thioredoxin is quite evident. From this observation, the following relationship was suggested:

$$\Delta\delta = 19.0\ r^{-1} - 9.7 \qquad (17)$$

Equations (16) and (17) may help to explain the observation that amide protons experience low frequency shift (on average) when placed in helices and a high frequency (on average) when placed in β-strands. This effect is likely due to the distance dependence of amide chemical shifts in hydrogen bonded networks. Studies of highly resolved X-ray structures (Baker and Hubbard 1984) have found that, on average, the hydrogen bonds in β-strands are shorter than the hydrogen bonds in helices (oxygen to proton distances of 1.90Å versus 2.02 Å). By using either equations (16) or (17) this difference in hydrogen bond distance might explain the observed "average" difference in amide chemical shift tendencies (about 0.6 ppm) for both helices and β-strands.

The shieldings of several heteronuclei are quite sensitive to hydrogen bonding interactions. From the theoretical stand point these may be considered by the supermolecule approach. A good example is provided by formamide. Ab initio calculations using GIAO's within the FPT framework have been reported for formamide in the presence of its first hydration shell (Ribas Prado et al 1981). With the exception of the formyl proton, which is not involved in hydrogen-bonding, significant variations in the shielding of the formamide nuclei are produced by hydration. A large increase is observed in the ^{17}O shielding and smaller decreases in that of the ^{13}C and nitrogen nuclei as hydration occurs (Table 1). These predictions are in qualitative agreement with experiment. More specifically the ^{17}O shielding values reported for the complexes FMA-$(H_2O)_2$ show that the shift observed when the oxygen atom is directly involved in the hydrogen bond is much larger than the one

Table 1. Calculated Nuclear Shielding of Hydrated Formamide (Ribas Prado et al 1981).

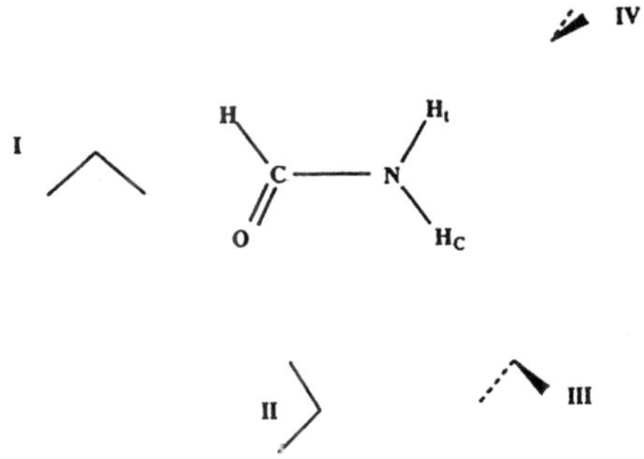

Nucleus	Formamide	Formamide(H_2O) I	II	III	IV	Formamide (H_2O)$_4$
N	235.99	234.51	232.45	233.20	235.24	225.03
C	110.42	107.03	109.25	109.08	108.87	103.05
O	-216.50	-169.88	-169.27	-195.70	-201.71	-88.47
H(c)	23.27	23.16	23.31	23.40	23.21	23.22
$H(N)_c$	27.32	27.24	27.02	25.24	27.61	24.99
$H(N)_t$	27.20	27.01	26.99	27.29	25.01	24.60

occurring when the water molecules are bound to the NH bonds. The variation was calculated to be to low frequency in both cases. This is in agreement with ^{17}O NMR experimental data on formamide and several amides in different solvents. It was subsequently noticed that the shift to low frequency when going from aceton to water (≈ 53 ppm) is significantly smaller than the one calculated between FMA-$(H_2O)_2$ III+IV and FMA-$(H_2O)_4$ (93 ppm). However, a quantitative agreement is obtained when a comparison is made with the extrapolated shielding constants of amides in vacuum (≈ 100 ppm) (Gerothanassis and Vakka submitted).

SECONDARY AND TERTIARY STRUCTURAL EFFECTS ON NMR SHIELDINGS

A number of relationships between chemical shifts and biomolecular structural elements have been demonstrated as a result of the analysis of a large chemical shift/protein structure data from a variety of published sources. Furthermore, this has permitted a re-evaluation of random coil or average chemical shifts for ^1H, ^{13}C, ^{15}N

and ^{31}P nuclei in peptides, proteins and nucleic acids. In particular, it has been found that the ^{1}H chemical shifts of the α-CH proton of all 20 naturally occurring amino acids experience a low frequency shift (with respect to the random coil values) when in a helical configuration and a comparable high frequency shift when in a β-strand extended configuration. An empirical analysis of ^{1}H chemical shifts from 17 proteins whose X-ray structures have been determined indicates that a significant improvement in the estimation of the conformation dependent part of the shift of the α-protons can be made by including the effects of the magnetic anisotropy of the peptide group and of backbone electrostatic contributions. Figure 13(a) shows that the horizontal ellipse observed in the predicted vs experimental shifts for the ring current calculation is significantly tilted towards a correlation with a slope of unity when an estimate of the peptide magnetic anisotropy and electrostatic interactions was included, Fig.13(b).

Due to the introduction of improved isotopic substitution strategies and the development of new methodologies in heteronuclear 2D- and 3D-NMR, ^{13}C and ^{15}N spectroscopy has become increasingly important and an analysis of a small, but rapidly expanding database is now feasible. Spera and Bax (1991) have reported an empirical correlation between protein backbone conformation and $C\alpha$ and $C\beta$ ^{13}C chemical shifts of 442 residues of four proteins for which complete ^{13}C chemical shift assignments are available: BPTI, calmodulin, interleukin-1β and SNase. The X-ray structures of these proteins have been solved and the ϕ and ψ angles are known with good precision. Fig.14 A shows the distribution of secondary shifts in α-helix and β-sheet fragments. All residues for which either the amide or the carbonyl (or both) is involved in an α-helical or antiparallel β-sheet type hydrogen bond are included except residues that have large deviations from the canonical backbone angles ($-90° < (\phi,\psi) < -10°$ for α-helix and $-210° < \phi < -80°$ and $90° < \psi < 200°$). For α-helix (119 residues) the average $C\alpha$ secondary shift is 3.09 ± 1.00 ppm and the average $C\beta$ secondary shift is -0.38 ± 0.85 ppm. For β-sheet (126 residues), $\delta_{C\alpha} = -1.48 \pm 1.23$ and $\delta_{C\beta} = 2.16 \pm 1.91$ ppm. The larger rms spread for residues in β-sheets is largely caused by the fact that the ϕ and ψ angles of these residues show substantial variation ($<\phi> = -114 \pm 32°$; $<\psi> = 142 \pm 20°$) where 32° and 20° are the standard

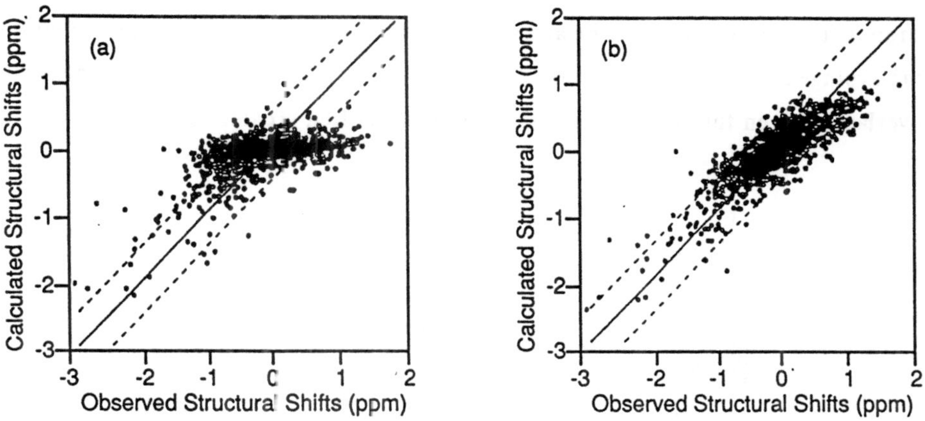

Fig.13 Correlation between observed and calculated structural shifts for 1553 Cα protons: (a) calculation using the original ring currents alone, r = 0.547, rms = 0.468 ppm and (b) calculation which includes the effects of peptide group, magnetic anisotropy and electrostatic interactions, r = 0.849, rms = 0.298 ppm (Osapay and Case 1991).

deviations. In α helices these angles have a much narrower distribution ($<\phi> = -62 \pm 8^\circ$; $<\psi> = -42 \pm 10^\circ$).

The important effect of ϕ and ψ torsion angles on ^{13}C shielding was demonstrated theoretically by De Dios et al (1993) for Cα and Cβ carbons in a series of model Ala molecular fragments. The method used was an ab initio calculation with a G-311G** basis set for all atoms without the formamide hydrogen bond partners. The ~ 5 ppm separations between helical and sheet residues found experimentally were reproduced in the calculation (Fig.14B). Similarly, the theoretical shieldings for the 12 Ala Cα groups in SNase were found to agree well with the experimental values (slope ~ 0.85 and regression coefficient r=0.94). Additional calculations with the incorporation of formamide molecules (to represent hydrogen bond partners) resulted in improved correlation (slope ~ 1.18, r=0.97). Similar results were obtained for Cβ. It was concluded that although ϕ and ψ angles are overwhelmingly dominant for Cα and Cβ carbons, improvements can be obtained by the incorporation of specific hydrogen bond interactions and charge fields (Fig.14C,D).

Conformational dependent shifts are also evident for ^{15}N nuclei. Thus, low frequency shifts were found for residues in helices and high frequency shifts for β-

strands (threonine appears to be an exception to this general rule). However, despite the average shift difference between helices and β-strands being more than 3 ppm the overlap between the two conformational groups is often greater than 3 ppm.

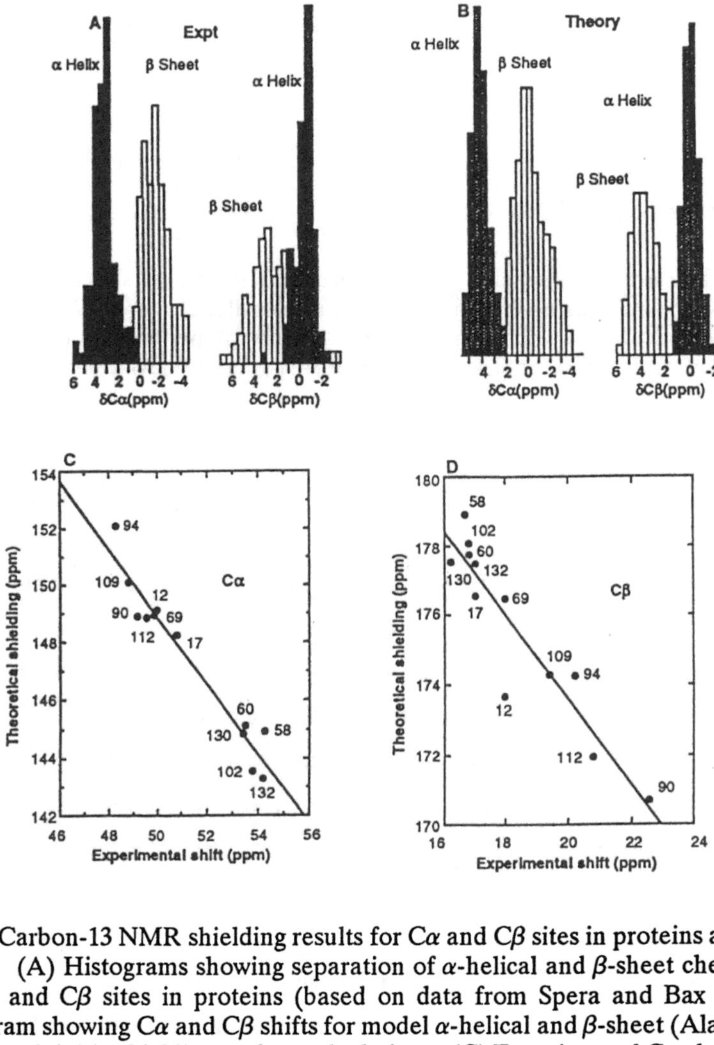

Fig.14 Carbon-13 NMR shielding results for Cα and Cβ sites in proteins and a model system. (A) Histograms showing separation of α-helical and β-sheet chemical shifts for Cα and Cβ sites in proteins (based on data from Spera and Bax 1991). (B) Histogram showing Cα and Cβ shifts for model α-helical and β-sheet (Ala) fragments based on ab initio shielding surface calculations. (C) Experimental Cα chemical shifts for the 12 Ala sites in SNase versus computed shieldings, m = 1.2 and r = 0.97. (D) Experimental Cβ chemical shifts for the 12 Ala sites in SNase versus computed shieldings, m = 1.2 and r = 0.93. The X-ray structure was energy minimized after "protonation", without ligands, for comparison with the solution NMR results (De Dios et al 1993).

Detailed analysis of the backbone ^{15}N chemical shifts of apamin and BPTI indicated a correlation with the torsion angle ψ for β-sheet residues. No other dominant factor was found to account for the large (± 15 ppm) variation from model peptides (Glushka et al 1989).

De Dios et al (1993) have performed ab initio calculations of the Val residues in SNase which show extremely large shielding range (> 30 ppm). Fig.15A shows ^{15}N shielding for a Val fragment calculated as a function of χ^1 torsion angle. The spread of the experimental shifts indicates the importance of additional factors. Direct hydrogen bonding (Fig.15B) and indirect hydrogen bonding of the C=O group (Fig.15C) need to be incorporated in order to obtain close agreement with the experimental results (Fig.16).

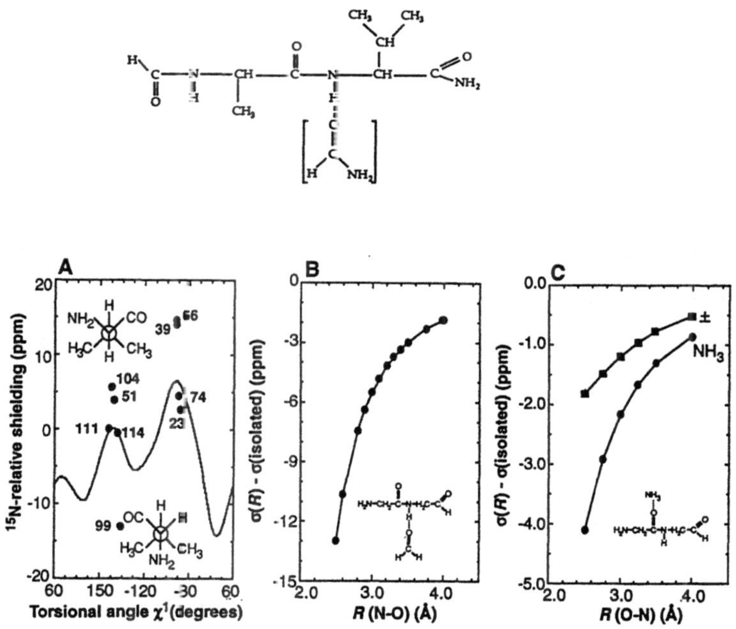

Fig.15 Nitrogen-15 NMR shielding for model systems. (A) Effect of χ^1 torsion angle on ^{15}N shielding in a Val model compound. The experimental shieldings for the Val sites in SNase are indicated. (B) Effects of linear hydrogen bonding to a carbonyl acceptor (formaldehyde) on the ^{15}N shielding of an NH group in a model peptide. (C) Indirect effect of linear hydrogen bonding, modelled by interacting the $H_2NCH_2CONHCH_2CHO$ peptide-carbonyl with NH_3 (●) or a point charge model for NH_3 (■) (De Dios et al 1993).

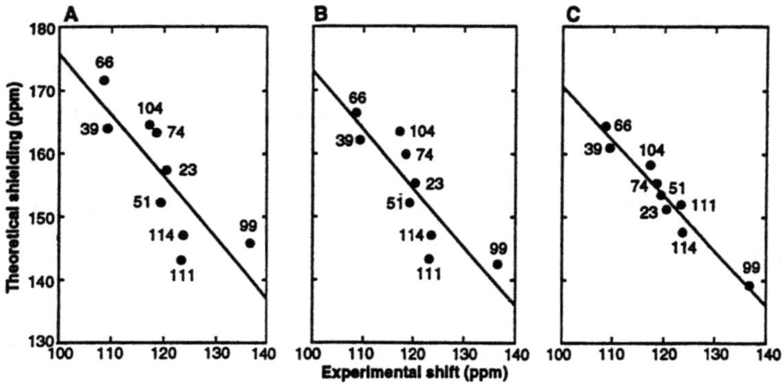

Fig.16 Nitrogen-15 chemical shieldings for the valine sites in SNase. (A) Isolated SNase ^{15}N Val fragments, $m = 0.97$ and $r = 0.81$. (B) As in (A) but with inclusion of a hydrogen-bond partner, $m = 0.92$ and $r = 0.86$. (C) As in (B) but with additional inclusion of CFP, $m = 0.87$ and $r = 0.97$ (De Dios et al 1993).

It is expected that empirical correlations and theoretical calculations of nuclear shielding might play a much more important role in determining secondary and tertiary structures than was previously thought. Interestingly, a technique has already been described for rapidly and quantitatively determining the identity, extent and location of secondary structural elements of proteins based on the α-CH ^1H resonance assignments (Wishart et al 1992). It has been suggested that this chemical shift index protocol might complement the traditional NOE based methods for secondary structure determination.

ACKNOWLEDGEMENTS

Financial support from the Research Committee of the University of Ioannina and the General Secretariat of Research and Technology is gratefully acknowledged. I thank L. Dunphy, Medical Research Institute, Mill Hill, London, for typing and preparing the manuscript.

REFERENCES

Abraham RJ (1961) The proton magnetic resonance spectra of porphyrins. Part II. Ring current effects in the porphyrins. Mol Phys 4: 145-152

Andersson T, Drakenberg T, Forsen S, Thulin E and Sward M (1982) Direct observation of the ^{43}Ca NMR signals from Ca^{2+} ions bound to proteins. J Am Chem Soc 104: 576-580

Ando I and Webb GA (1981) Some quantum chemical aspects of solvent effects on NMR parameters. Org Magn Reson 15: 111-130

Ando I and Webb GA (1983) Theory of NMR parameters. Academic Press / London

Asakura T, Niizawa Y and Williamson MP (1992) Determination of the magnetic anisotropy of the oxygen atom and ^{1}H chemical-shift calculation of proteins. J Magn Reson 98: 646-653

Augspurger J, Pearson JG, Oldfield E, Dykstra CE, Park KD and Schwartz D (1992) Chemical-shift ranges in proteins. J Magn Reson 100: 342-357

Baker EN and Hubbard RE (1984) Hydrogen bonding in globular proteins. Progr Biophys Mol Biol 44: 97-179

Batchelor JG (1975) Theory of linear electric field shifts in carbon-13 nuclear magnetic resonance. J Am Chem Soc 97: 3410-3415

Batchelor JG, Feeney J and Roberts GCK (1975) Carbon-13 NMR protonation shifts of amines, carboxylic acids and amino acids. J Magn Reson 20: 19-38

Böttcher C (1973) Theory of electric polarisation. Elsevier / Amsterdam vol 1

Buckingham AD (1960) Chemical shifts in the nuclear magnetic resonance spectra of molecules containing polar groups. Can J Chem 38: 300-307

De Dios AC, Pearson JG and Oldfield E (1993) Secondary and tertiary structural effects on protein NMR chemical shifts: an ab initio approach. Science 260: 1491-1496

Drakenberg T, Lindman B, Cave A and Parello J (1978) Non-equivalence of the CD and EF sites of muscular parvalbumins. A ^{113}Cd NMR study. FEBS Letters 92: 346-350

Ebraheem KAK and Webb GA (1977) Semi-empirical calculations of the chemical shifts of nuclei other than protons. Progr NMR Spectrosc 11: 149-181

Ellis PD (1983) Cadmium-113 magnetic resonance spectroscopy. Science 221: 1141-1146

Gerothanassis IP, Loock B and Momenteau M (1992) Structural differences of the iron-dioxygen moiety of haemoprotein models with and without an axial hindered base as revealed by ^{17}O NMR and FTIR spectroscopy in solution. J Chem Soc Chem Commun: 598-600

Gerothanassis IP and Vakka C (1993) ^{17}O NMR chemical shifts as a tool to study specific hydration sites of amides and peptides - correlation with the IR amide I stretching vibration. Submitted

Giessner-Prettre C and Pullman B (1987) Quantum mechanical calculations of NMR chemical shifts in nucleic acids. Quart Rev Biophys 20: 113-172

Glushka J, Lee M, Coffin S and Cowburn D (1989) ^{15}N chemical shifts of backbone amides in bovine pancreatic trypsin inhibitor and apamin. J Am Chem Soc 111: 7716-7722

Haigh CW and Mallion RB (1972) New tables of "ring current" shielding in proton

magnetic resonance. Org Magn Reson 4: 203-228
Harris RK (1983) Nuclear magnetic resonance spectroscopy. Pitman / London
Jameson CJ and Gutowsky HS (1964) Calculation of chemical shifts. I. General formulation and the Z dependence. J Chem Phys 40: 1714-1724
Jameson CJ and Mason J (1987) The chemical shift. In: Mason J (ed). Multinuclear NMR. Plenum Press / New York : 51-88
Jardetzky O and Roberts GCK (1981) NMR in molecular biology. Academic Press / New York
Johansson C and Drakenberg T (1989) Metal-ion NMR studies of ion binding. Annual Reports on NMR Spectroscopy 22: 1-59
Johnson CE and Bovey FA (1958) Calculation of nuclear magnetic resonance spectra of aromatic hydrocarbons. J Chem Phys 29: 1012-1014
Osapay K and Case DA (1991) A new analysis of proton chemical shifts in proteins. J Am Chem Soc 113: 9436-9444
Perkins SJ and Dwek RA (1980) Comparisons of ring-current shifts calculated from the crystal structure of egg white lysozyme of hen with the proton nuclear magnetic resonance spectrum of lysozyme in solution. Biochemistry 19: 245-258
Pople JA (1956) Proton magnetic resonance of hydrocarbons. J Chem Phys 24: 1111
Pople JA, Schneider WG and Bernstein HJ (1959) High resolution nuclear magnetic resonance. McGraw-Hill / New York ch. 15
Pople JA (1962) Molecular-orbital theory of diamagnetism. I. An approximate LCAO scheme. J Chem Phys 37: 53-59
Raynes WT, Buckingham AD and Bernstein HJ (1962) Medium effects on proton magnetic resonance. I. Gases. J Chem Phys 36: 3481-3488
Ribas Prado F, Giessner-Prettre C, Pullman A, Hinton JF, Harpool D and Metz KR (1981) Non empirical quantum mechanical calculations of the ^{1}H, ^{13}C ^{15}N and ^{17}O magnetic shielding constants and of the spin-spin coupling constants in formamide, hydrated formamide and N-methylformamide. Theor Chim Acta 59: 55-59
Shulman RG, Wüthrich K, Yamane T, Patel DJ and Blumberg WE (1970) Nuclear magnetic resonance determination of ligand-induced conformational changes in myoglobin. J Mol Biol 53: 143-157
Spera S and Bax A (1991) Empirical correlation between protein backbone conformation and $C\alpha$ and $C\beta$ ^{13}C nuclear magnetic resonance chemical shifts. J Am Chem Soc 113: 5490-5492
Wagner G, Pardi A and Wüthrich K (1983) Hydrogen bond length and ^{1}H NMR chemical shifts in proteins. J Am Chem Soc 105: 5948-5949
Webb GA (1986) Theoretical background to high resolution NMR parameters and nuclear shielding calculations. In: Axenrod T and Ceccarelli G (eds). NMR in living systems. Reidel / Holland Dordrecht: 19-37
Wishart DS, Sykes BD and Richards FM (1991) Relationship between nuclear magnetic resonance chemical shift and protein secondary structure. J Mol Biol 222: 311-333
Wishart DS, Sykes BD and Richards FM (1992) The chemical shift index: a fast and simple method for the assignment of protein secondary structure through NMR spectroscopy. Biochemistry 31: 1647-1651

NMR STUDIES OF PROTEIN LIGAND INTERACTIONS

J Feeney

Laboratory of Molecular Structure
National Institute for Medical Research
Mill Hill
London NW7 1AA

INTRODUCTION

NMR spectroscopy has proved to be a useful technique for studying interactions between proteins and other molecules in solution. These interactions are important in many molecular recognition processes in biology, typical examples being found in complexes of enzymes with substrates and inhibitors, of drugs with their receptors and of transcription factors with DNA duplexes. In this article we will examine how the appropriate NMR data can be analysed to provide detailed information concerning interactions, conformations and dynamic processes within such protein-ligand complexes. We will also discuss several novel applications to other systems taken from the literature in order to illustrate certain aspects of the subject, for example protein-DNA interactions.

Several workers have provided reviews of NMR studies of protein-ligand interactions (Jardetzky and Roberts 1981; Feeney 1990; Handschumacher and Armitage 1990; Sheard and Bradbury 1970; Roberts and Jarkdetzky 1970). In the present article we will concentrate on a relatively small number of systems which illustrate how particular types of information can be obtained from the NMR data. Many of the examples will be taken from work in our own laboratory dealing with the binding of antifolate drugs to dihydrofolate reductase. This enzyme reduces dihydrofolate to tetrahydrofolate using NADPH as a coenzyme and is an essential enzyme in all cells. Studies of complexes of DHFR with antifolate drugs provide examples of drug-receptor complexes where the pharmacological action results from an inhibitor molecule binding selectively to an essential enzyme in an invasive cell. Scheme 1 shows the structures of substrates **1, 2** and the antifolate drugs trimethoprim **3** and methotrexate **4.**

NATO ASI Series, Vol. H 87
NMR of Biological Macromolecules
Edited by C. I. Stassinopoulou

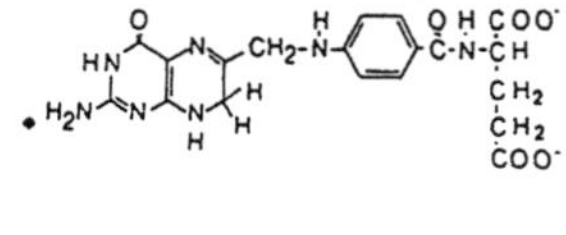

Dihydrofolate 1

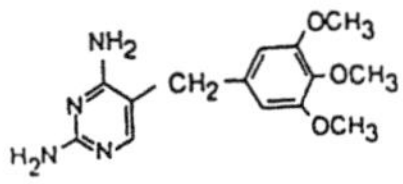

Trimethoprim 3

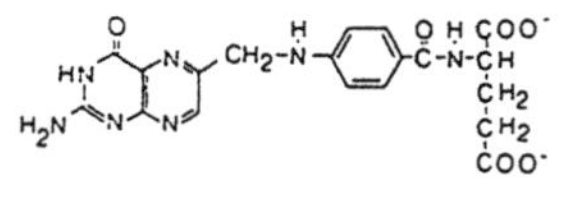

Folate 2

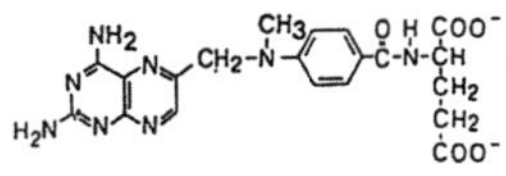

Methotrexate 4

Earlier NMR studies made their major contributions by complementing X-ray structural data by defining specific interactions and identifying ionisation states of interacting groups on the protein and ligand, by monitoring dynamic processes relating to specific interactions within the protein-ligand complexes and, most importantly, by detecting the presence of multiple conformational states in solution. In many of these applications NMR offers considerable advantages over the X-ray method. More recently, NMR methods have been developed for obtaining detailed three-dimensional structural information for small proteins (<20 KDa). A few protein-ligand complexes have now been examined in such detail (for example, the lac repressor headpiece (Kaptein *et al.*, 1989), and homeo-domain proteins (Otting *et al.*, 1990) binding to the appropriate DNA duplexes). One can expect that these structural methods will be applied more generally to protein ligand complexes over the next few years.

EQUILIBRIUM BINDING STUDIES

The binding of reversible inhibitor, I, to an enzyme, E, results in the formation of a non-covalently bound complex, EI, represented by the equilibrium

$$E + I \underset{k_{-1}}{\overset{k_1}{\rightleftharpoons}} EI \qquad (1)$$

where the equilibrium binding constant K_a is given by

$$K_a = \frac{[EI]}{[E][I]} = \frac{k_1}{k_{-1}} \qquad (2)$$

For example such measurements might be made for a series of complexes where the inhibitor (or indeed the protein) is systematically modified in order to measure changes in K_a resulting from the introduction or removal of particular interactions in the complexes. Such investigations must always be accompanied by structural studies on the complexes to ensure that only the predicted effects have taken place and that no major conformational perturbations have occurred in the rest of the system. X-ray crystallography has often been used for such studies and this can provide a very detailed description of the structures. However NMR offers the advantage that it examines the complexes in solution, does not require crystals and furthermore provides a convenient method for monitoring interactions, ionisation effects and dynamic processes in the complexes.

PREPARATION OF THE SAMPLE

NMR experiments for these studies need large quantities of purified protein. The concentrations needed will vary from 10μM for 1D spectra to up to 5mM (in 0.6 ml samples) for some 2D experiments.

Large quantities of protein are best prepared by cloning the appropriate gene into an overexpressing bacterial cell line. Mammalian proteins can also be obtained in this way. To assist in spectral assignment of large proteins it is necessary to prepare proteins uniformly labelled with ^{13}C and ^{15}N by growing the cells on ^{13}C-glucose or ^{15}N-ammonium salts. This requires an organism which grows on minimal media that uses only glucose and ammonium salts as the source of carbon and nitrogen respectively. If one wants to specifically introduce an isotopically labelled amino acid, it is necessary to use an organism which is auxotrophic for that amino acid.

ANALYSIS OF THE NMR SPECTRA

Fast and slow exchange conditions

Before any detailed structural and dynamic information can be obtained from the NMR spectra of the complexes, the spectral resonances need to be assigned to specific nuclei in the ligand and the protein. An important first step is to ascertain whether the bound and free species coexist under conditions of fast or slow exchange on the NMR time scale. If a nucleus has chemical shifts of ω_B and ω_F in the bound and free species

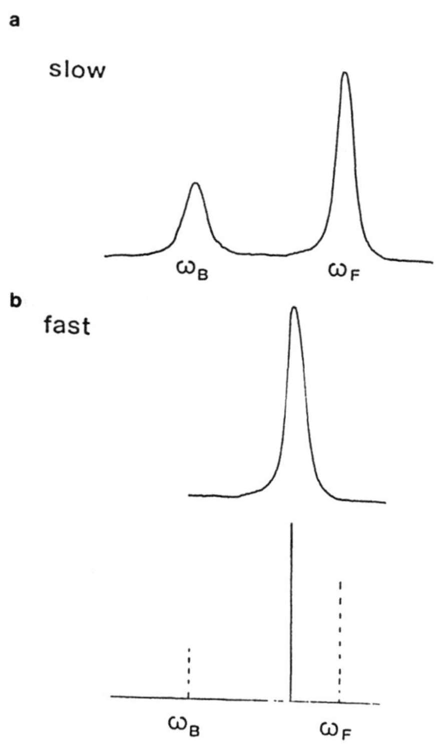

Fig.1. The bound and free species with chemical shifts ω_B and ω_F give rise to (a) the slow exchange spectrum and (b) the fast exchange spectrum depending on the lifetime of the complex and the chemical shift difference.

respectively and the lifetime of the complex is long compared with $(\omega_B - \omega_F)^{-1}$, then separate signals are seen for the bound and free species: this is designated the slow exchange condition (see Figure 1a). If the lifetime of the complex is short compared with $(\omega_B - \omega_F)^{-1}$ then conditions for fast exchange prevail and a single spectrum averaged spectrum is weighted according to the populations and chemical shifts of the two forms (Figure 1b). The frequency of the observed averaged resonance ω_{obs}, is given by

$$\omega_{obs} = \frac{[EI]\omega_B}{I_T} + \frac{[I]\omega_F}{I_T} \tag{4}$$

where I_T is the total ligand concentration. Since $I_T = [I] + [EI]$ equation 4 can be rearranged to give

$$\omega_{obs} = \omega_F + \frac{[EI]}{I_T}(\omega_B - \omega_F) \tag{5}$$

For studies of protein resonances the corresponding equation is

$$\omega_{obs} = \omega_F + \frac{[EI]}{E_T}(\omega_B - \omega_F) \tag{6}$$

where E_T is the total protein concentration.

When the lifetimes are $\sim(\omega_B - \omega_F)^{-1}$ then intermediate exchange conditions exist giving rise to broad complex spectra which are difficult to analyse. Clearly it is important to ascertain which exchange regime obtains before further work can be attempted. With typical ^1H chemical shift differences seen between bound and free species in proton spectra (usually less than 1 ppm), an equilibrium binding constants of 10^3 M^{-1} or less will usually result in spectra in fast exchange while binding constants of 10^6 M^{-1} or greater will almost certainly give rise to spectra in slow exchange. However, in all cases it is important to determine these exchange regimes experimentally.

This can be done by first recording the ^1H spectra of the protein complexes with 0.4, 0.7 and 1.0 equivalents of ligand. One then observes whether the resolved enzyme signals are shifting (fast exchange) or changing area (slow exchange). When more than one equivalent of ligand is added, if no further changes in the protein resonances are observed, this suggests slow exchange. Ligand is then added until signals from the free ligand can clearly be observed. If the ligand chemical shift is constant this indicates slow exchange; if it changes during the titration this corresponds to fast exchange. If the ligand signal sharpens with increasing temperature this also indicates fast exchange; broadening indicates slow exchange.

A typical example of a spectrum showing fast exchange behaviour is provided by the ^1H spectrum of the complex of para-aminobenzoyl-L-glutamate (PABG) with dihydrofolate reductase. PABG is a fragment of the substrate folate and has a K_a value of 0.83 x 10^3 M^{-1}. The chemical shift data for the PABG H_2',H_6', signals plotted as a function of the ligand concentration, are given in Figure 2 (Roberts *et al.*, 1974.) Such

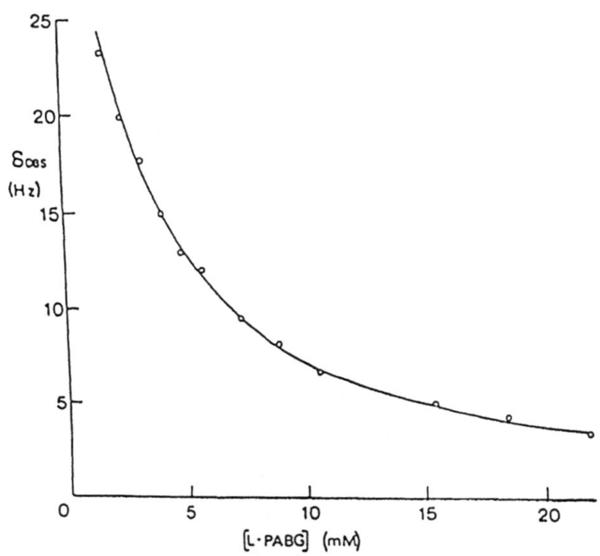

Fig. 2. The dependence of the chemical shift of the H_2',H_6' resonance of para-aminobenzoyl-L-glutamate (PABG) on the concentration of ligand in the presence of 1.4 mM dihydrofolate reductase. The chemical shift is expressed in Hz relative to the free ligand chemical shift (measurements at 100 MHz) (Reproduced with permission from (Roberts *et al.*, 1974).

binding data can be analysed with a non-linear regression program to fit equations 5 and 7 to provide the values of the chemical shifts of the bound ligand and the K_a value.

$$[EI] = \frac{1}{2}\{E_T + I_T + 1/K_a - ((E_T + I_T + 1/K_a)^2 - 4E_T I_T)^{1/2}\} \qquad (7)$$

It is important to examine a range of I_T/E_T ratios sufficient to allow a full binding curve to be obtained.

If the dissociation rate constant is not sufficiently fast to give true fast exchange conditions but fast enough to give coalesced non-Lorentzian signals for free and bound signals in intermediate exchange, the analysis is more difficult; we have estimated the errors in determined values of K_a and the bound chemical shifts which can result if the assumed fast exchange condition is not fully fulfilled. It could be shown that if the observed frequency for the average resonance of the ligand can be measured at very low ligand concentration (for example $[I]_T$ = 0.1mM and $[E]_T$ = 1mM) then a reliable estimate of the bound chemical shift could be obtained (Feeney *et al.*, 1979). However it is usually impossible to see such small ligand signals in ^1H spectra in the presence of the complex spectrum from the protein.

When working under conditions of fast exchange, the assignment of bound signals is usually straightforward since the progressively shifting signals monitored during the titration can be related directly to the assigned signals in the free species. However, such studies necessarily involve complexes formed with weakly binding ligands and are usually of less interest than those on complexes formed with tightly binding ligands.

Slow exchange

Examples of spectra from a complex in slow exchange are shown in Figure 3. These are the ^{31}P signals from the 2′-phosphate group of $NADP^+$ measured as a function of the $NADP^+$ concentration in the presence of dihydrofolate reductase (Hyde *et al.*, 1980). In the presence of 1.3 molar equivalents of $NADP^+$ one observes a signal for bound $NADP^+$ at 3 ppm and a small broad signal at the position of the free $NADP^+$. When increasing amounts of $NADP^+$ are added to the sample, the signal for bound $NADP^+$ remains unchanged while that at the position of free $NADP^+$ increases in intensity and decreases in line width. Clearly the exchange between the bound and free $NADP^+$ is

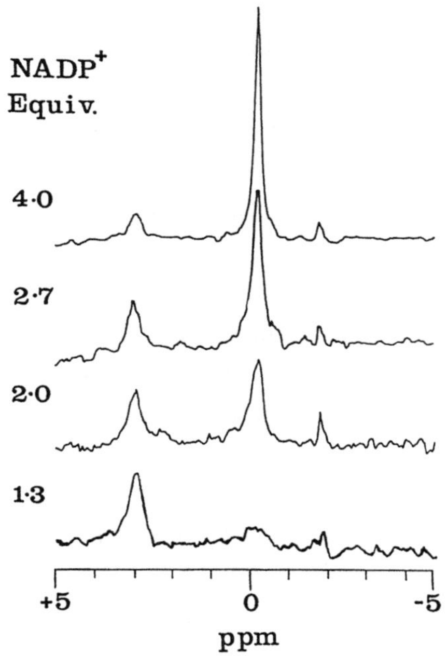

Fig. 3. 2′-Phosphate ^{31}P resonance of $NADP^+$ in the presence of dihydrofolate reductase as a function of $NADP^+$ concentration. Microlitre volumes of concentrated $NADP^+$ solution were added to the enzyme sample and the pH readjusted as necessary. The $NADP^+$ concentration is expressed as moles of $NADP^+$ per mole of enzyme. (Reproduced with permission from Hyde *et al.*, 1980b).

sufficiently slow that ω_B-ω_F > $1/\tau_F + 1/\tau_B$ (where τ_F and τ_B are the lifetimes of the free and bound states, and $1/\tau_B = k_{-1}$). For this bound ligand the line width is given by

$$\pi\Delta\nu_{\frac{1}{2}B} \;=\; 1/T_{2B} + k_{-1} \qquad (8)$$

and for the free ligand

$$\pi\Delta\nu_{\frac{1}{2}F} \;=\; 1/T_{2F} \;+\; 1/\tau_F \;=\; 1/T_{2F} \;+\; p_B\, k_{-1}/p_F \qquad (9)$$

where p_B and p_F are the mole fractions of the ligand in the free and bound states and

$$p_B \;=\; [EI]\,/\,([EI] + [I]) \;=\; 1 - p_F \qquad (10)$$

Thus the line width of the bound ligand signal is independent of the ligand concentration but that of the free ligand signal decreases as the ligand concentration is increased. A plot of $\pi\Delta v_{\frac{1}{2}F}$ against p_B/p_F is a straight line with a gradient of k_{-1}. The p_B/p_F ratio is equal to [EI]/[I] and can be calculated from the total concentration of protein and ligand if K_a is known.

ASSIGNMENT OF PROTEIN AND LIGAND RESONANCES

Before one can obtain detailed information about interactions, conformations, ionisation states and dynamic processes within protein-ligand complexes from the NMR spectra, it is first necessary to assign the protein and ligand resonances. It is also important to ensure that the protein is fully saturated with the bound ligand.

By using 2D and 3D (and even 4D) NMR methods in combination with ^{2}H, ^{13}C and ^{15}N labelled proteins it is now possible to obtain almost complete sequential assignments for proteins of MWt ~ 20 KDa (Clore and Gronenborn, 1991). In our own studies on *L casei* DHFR, backbone (NH and αCH) and side chain resonance assignments have been obtained for almost all 162 amino acids using such multidimensional NMR techniques on a sample of DHFR uniformly labelled with ^{15}N and ^{13}C (Carr *et al.*, 1991; Soteriou *et al.*, 1993). The 3D TOCSY.HMQC, NOESY.HMQC (Ikura *et al.*, 1990), HMQC.NOESY.HMQC (16) and HNCA (Ikura *et al.*, 1990; Kay *et al.*, 1990) were the most useful experiments for backbone resonance assignments while the 3D HCCH.COSY and HCCH.TOCSY (Bax *et al.*, 1990; Fesik and Zuiderweg, 1990) experiments provided most of the side-chain resonance assignments. Once the assignments had been made for one complex it proved relatively easy to transfer the assignments to related complexes because of the similarities in the connectivity patterns seen in the 2D spectra of the different complexes. Assigned DHFR resonances have been used to monitor ionisation changes, to characterise conformational mixtures and to provide conformational information from NOE measurements for the various complexes of ligands with DHFR (Feeney, 1990).

For large proteins where sequential assignment methods might be difficult to apply assignments can sometimes be made by using NOE measurements to correlate resonances from nuclei which are known to be close in the crystal structure of the complex. Many of the early assignments for DHFR were made in this way and they have all proved to be in complete agreement with the sequential assignments made subsequently. Obviously for this approach to succeed the solution and crystal structures need to be similar. Frequently this appears to be the case.

Assignments of ligand resonances are particularly important because ligand nuclei, of necessity, are well placed to provide direct information about the binding site in the complex. We have seen that for complexes with weakly binding ligands ($K_a < 10^3 M^{-1}$) fast exchange behaviour is usually observed and the bound chemical shifts can be calculated from the analysis of the binding curves (see Figure 2). The signals from very tightly binding ligands ($K_a > 10^9$ M^{-1}) are more difficult to assign. The usual method is to examine isotopically labelled analogues (^2H,^3H,^{15}N,^{13}C) in combination with various experimental procedures. For example, deuterated ligands can assist ^1H assignments by producing differences between ^1H spectra of complexes formed with deuteriated and non-deuteriated ligands. We have used this approach to assign the nicotinamide ring proton resonances in complexes of DHFR.NADP$^+$ formed with selectively deuterated NADP$^+$ analogues (Birdsall *et al.*, 1982). Another method is to selectively label the ligand with tritium (^3H) and then to observe the ^3H spectra of the complexes: for example, by examining [^3H-7,3′,5′-) folic acid in its complex with DHFR and NADP$^+$ using ^3H NMR it was possible to confirm the presence of three different conformational states by monitoring the assigned tritium resonances for the 7,3′,5′-tritium nuclei (the ^3H chemical shifts are directly related to the ^1H chemical shifts) (Evans *et al.*, 1985). For ^{15}N or ^{13}C labelled ligands their complexes can be examined directly by using ^{15}N or ^{13}C NMR: only the signals from nuclei at the enriched positions are observed and thus the assignment problem is usually trivial. Where the protons are directly attached to ^{15}N or ^{13}C, the opportunity arises for using editing pulse sequences. Heteronuclear multiple quantum coherence (HMQC) experiments allow the attached protons to be detected selectively (they are then characterised by both the ^1H and ^{15}N (or ^{13}C) frequencies, the X nuclei being detected indirectly). Fesik *et al.*, (1990) have shown how these experiments can be used to study protein-ligand complexes. A powerful extension of these experiments is the 3D NOESY.HMQC experiment : this allows selective detection of the NOEs from the ligand protons (on ^{13}C or ^{15}N) to neighbouring protons in the protein and the observed NOESY cross peaks to be dispersed over the X-chemical shift frequency range. This considerably simplifies the NOESY spectrum at any particular X-frequency and is very important for examining complexes involving large proteins since the normal NOESY spectrum containing all the ligand-protein and protein-protein NOEs is often too complex to analyse (Ikura *et al.*, 1990).

For complexes formed using less tightly binding ligands ($K_a \sim 10^6\ M^{-1}$) the spectra can show slow exchange behaviour, giving separate signals for bound and free species, but still have dissociation rate constants sufficiently large to allow transfer of saturation or 2D exchange methods to be used to connect the resonance of the bound and free species. Because the assignments in the free ligand are usually known with certainty, this allows the assignments of the connected signals from the bound species to be made. This method can also be used to measure the dissociation rate constants (see Section on Dynamic Processes in Protein-ligand Complexes).

Complexes formed with ligands containing phosphorus can be examined by ^{31}P NMR. Even in systems where only a few different ^{31}P nuclei are involved, the assignment is not trivial: for example in ^{31}P spectra of NADPH bound to a protein the two pyrophosphate ^{31}P nuclei can only be assigned unequivocally by resorting to isotopic labelling procedures or to heteronuclear correlation experiments (Georthanassis *et al.*, 1992). For the complex formed by NADPH with *L.casei* dihydrofolate reductase, the ^{31}P studies provided detailed information about the ionisation states and conformations in the bound state (Feeney *et al.*, 1975).

When fluorine containing ligands or protein are available, ^{19}F NMR measurements can be used to examine their complexes. The assignment of ^{19}F signals from the ligand is often straightforward since usually only one site or two sites are labelled. The simple spectra are ideal for monitoring multiple conformations and dynamic processes in the complexes (Birdsall *et al.*, 1990). ^{19}F signals from fluorine-containing proteins have been assigned by observing the ^{19}F spectra following systematic removal of each fluorine-containing amino acid by site-directed mutagenesis.

Once the detailed assignments of the resonances are known, the NMR parameters for the various nuclei can be measured and compared for free and bound species in a protein-ligand complex or series of related complexes. Changes in chemical shifts, spin-coupling constants and relaxation behaviour (including nuclear Overhauser effects) have all been monitored to provide information about interactions, conformations and dynamic processes in the complexes. Chemical shift changes can best be used for determining ionisation states or hydrogen bonding states where there is a well-documented difference in chemical shift between the two states. Studies of nuclei other than ^1H can be very useful in this regard: for example the ^{15}N chemical shift of a pyrimidine N_1 ring nitrogen changes by 80 ppm on protonation (Bevan *et al.*, 1985).

Measurements of bound coupling constants can provide localised conformational information since there are well-established relationships between three bond vicinal coupling constants and dihedral torsion angles. Surprising few examples of such studies have been reported (Feeney *et al.*, 1975; Rodgers and Roberts. 1973) probably because of the difficulties of obtaining good coupling constant data from the broad-lines measured for most protein-ligand complexes. By far the most direct method of obtaining conformational information is by measuring protein-ligand NOEs from 2D NOESY spectra of the complex (see earlier Chapters). Using this approach spatial information within the binding site can be obtained and details of the specific interactions deduced.

Conformational information can sometimes be obtained by introducing paramagnetic probes at defined sites into the complex (transition or lanthanide metal ions or stable free radicals in the form of nitroxide spin labels) and observing their shielding and relaxation effects on assigned signals from neighbouring protons. There are many problems associated with such approaches (for example the difficulty of allowing for weak non-specific binding in the case of the paramagnetic ions) but useful conformational information can be obtained when the methods are applied with care (Mildvan, 1989).

SPECIFICITY OF INTERACTIONS

Detailed information about the groups on the protein and ligand which are involved in specific interactions can be obtained by determining the three dimensional structure of the complex. Most structures of protein-ligand complexes have been obtained from X-ray diffraction studies on crystals but an increasing number of structures are being determined in solution by NMR methods.

Determination of protein ionisation states

NMR has proved to be particularly effective for studying electrostatic interactions involving charged residues on the protein or ligand. Changing the charge state of an ionisable group in a protein or ligand is usually accompanied by large, characteristic changes in shielding of nuclei close to the ionisable group. NMR can thus be used to report on the ionisation states of specific groups in proteins and ligands, to measure their pK values and to detect the changes in pK values which accompany protein-ligand complex formation (Markley, 1975; Wüthrich, 1976; Jardetzky and Roberts, 1981). X-ray crystallography is unable to provide such direct information about ionisation states and clearly this is an area where NMR can make a unique contribution to understanding interactions within the binding site.

The ionisation behaviour of histidine residues have been intensely studied even from the earliest days of biological NMR. The pK values of histidines in proteins are typically in the range 5.5 to 8.5 and they can easily be studied by carrying out pH titrations over a pH range accessible for many proteins.

In D_2O solutions the ^1H signals from the C2 protons of the imidazole rings of histidines are relatively easy to detect even in the 1D spectra of complex proteins. The signals of the $C\epsilon_1$(C2) and $C\delta_2$(C4) protons of the imidazole ring are shifted downfield on protonation by 1.0 (7.7 to 8.7) and 0.4 (7.0 to 7.4) ppm respectively.

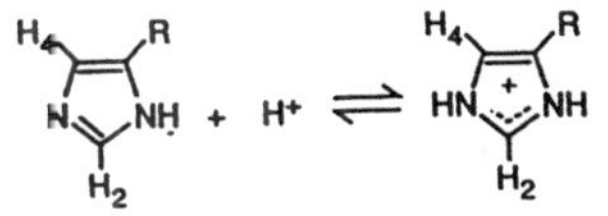

Because there is fast exchange on the chemical shift timescale between the protonated and unprotonated species, the ^1H signals from the ring protons have averaged chemical shifts weighted according to the populations of the two states. A plot of the chemical shifts against the pH values can be fitted to the Henderson-Hassalbalch equation and thus provide values for the pK value and the chemical shifts of the protonated and non-protonated species.

For some proteins, the His titration curves deviate from a perfect Henderson-Hassalbalch curve and this indicates the presence of interacting ionisable groups. This was seen, for example, for two of the histidines in the c-AMP receptor protein from *E. coli* and the data could be analysed in terms of two neighbouring interacting His residues with pK values of 6.21 and 7.47 (Clore and Groenborn, 1982) by using equations developed by Schrager and coworkers (Schrager *et al.*, 1972).

L. casei dihydrofolate reductase has seven histidine residues and these have been extensively studied by NMR. Five of the histidines are influenced by substrate or inhibitor binding (Birdsall *et al.*, 1977). One of the most useful findings from the histidine titration studies is that substrates and inhibitors which contain the L-glutamic acid moiety (for example methotrexate) cause the pK value of His 28 to be increased from 6.8 to ~7.8. From the crystal structure (see Figure 4a) it is seen that the γ-carboxylate of the L-glutamic acid moiety is near to His 28 and the α-carboxylate is near

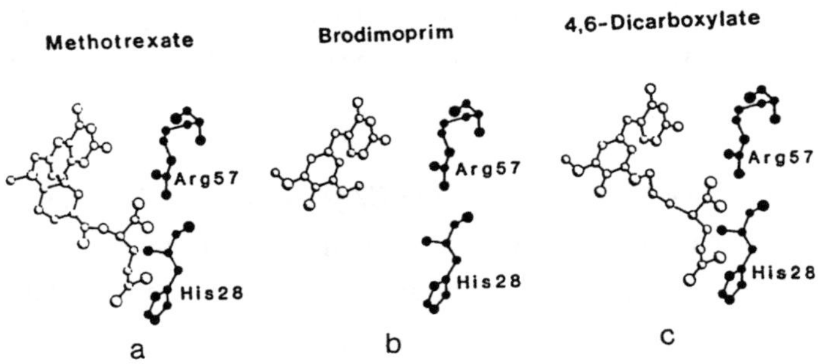

Fig. 4. Detail from the X-ray crystal structure of *L.casei* dihydrofolate reductase complex with methotrexate and NADPH (Bolin *et al.*, 1982) showing (a) the proximity of the α and γ carboxyl groups of methotrexate to histidine 28 and arginine 57. Panel (b) shows brodimoprim **7** in its bound conformation making no contacts with these residues. Panel (c) shows a model of the mode of binding of the 4,6-dicarboxylate analogue of brodimoprim **8** (from crystal structure data of Matthews and coworkers). (Reproduced with permission from Birdsall *et al.*, 1984b).

to the conserved Arg 57 residue (Bolin *et al.*, 1982). The NMR determined pK behaviour of His 28 provides a very useful probe for determining whether or not these interactions are present in various complexes. NMR studies of complexes of DHFR with methotrexate **4** and its α-amide **5** and γ-amide **6** analogues have provided interesting results.

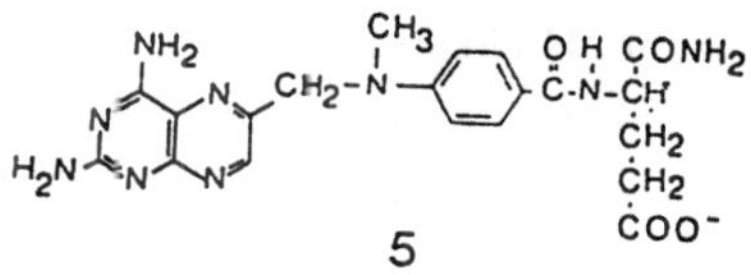

5

α-amide

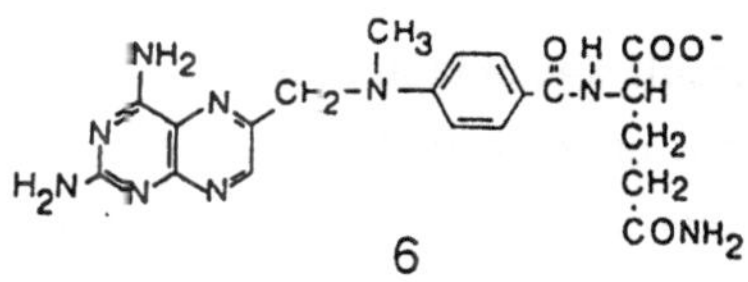

6

Υ-amide

The binding of the α-amide to DHFR is a 100 fold weaker than that of methotrexate whereas the γ-amide binds only 9 times more weakly than the parent drug (Antonjuk *et al.*, 1984). From the histidine titration curves of the complexes it is clear that, in the complex of the γ-amide with DHFR, His 28 does not show a large perturbation in pK compared with free DHFR (see Figure 5). This result is as expected since the γ-amide is unable to form the ionic interactions with the histidine imidazole

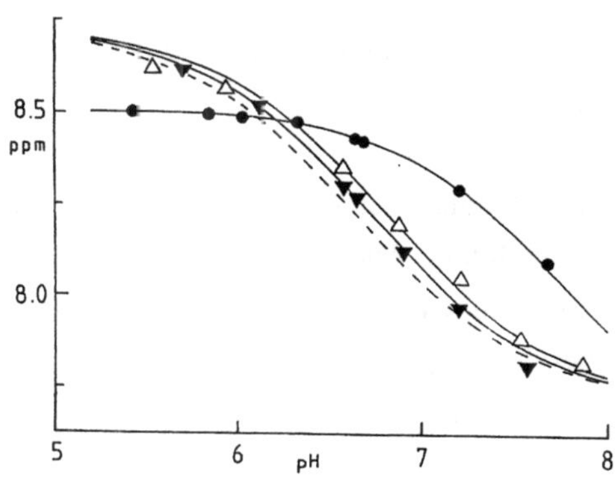

Fig. 5. The pH dependence of the chemical shift of histidine 28 in several dihydrofolate reductase complexes; (●) methotrexate, (▼) methotrexate γ- amide, (Δ) methotrexate α-amide (---) enzyme alone. Chemical shift scale is in ppm and referenced from DSS. (Reproduced with permission from Antonjuk *et al.*, 1984).

group. More detailed studies of the ^{1}H chemical shifts in the various complexes using assigned COSY 2D spectra, revealed that the γ-amide complex has a very similar spectrum to the methotrexate complex (except for His 28) suggesting that this modification causes only a local perturbation involving the removal of the γ-carboxylate - His 28 interaction. For the complex with the α-amide analogue, where the γ-carboxylate is still present, an unexpected result was obtained; the histidine titration curves for this complex clearly indicate that once again His 28 does not increase its pK. This indicates that disruption of the interaction between the α-carboxylate group and Arg 57 has caused a conformational change which prevents the γ-carboxylate from interacting with His 28. The ^{1}H COSY spectrum of the α-amide complex shows chemical shift changes for the protons of Leu 27, Phe 49, Leu 54 and Leu 19 all of which are close to the benzoyl ring of the methotrexate and expected to be influenced by its ring current shielding : this indicates that the disruption of structure caused by introducing the α-amide modification also influences the position of the benzoyl ring in the complex. The NMR findings, indicating that both the α- and γ-carboxylate interactions are affected in the complex with the α-amide offer an explanation for why the decrease in binding caused by the α-amide modification is greater than that caused by the γ-amide modification. From the observed changes in binding constant in the different complexes and knowing that both the α- and γ-carboxylate interactions are absent in the α-amide complex, it could be estimated that each interaction is making a similar contribution to the binding energy (~ 5.6 KJ $mole^{-1}$).

One can also use the pK behaviour of the His 28 to monitor interactions in DHFR complexes with novel inhibitors. The antibacterial drug brodimoprim 7 binds to DHFR with its 2,4-diaminopyrimidine ring in the same binding site as that used by the corresponding part of the pteridine ring of methotrexate. However, there are no direct interactions of brodimoprim with Arg 57 and His 28 (see Figure 4b). Several analogues

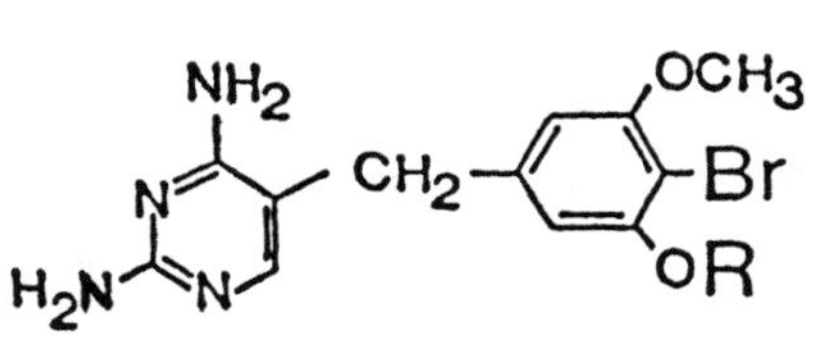

7 R= CH_3 Brodimoprim

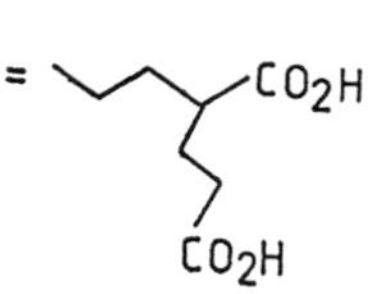

8 R=

of brodimoprim have been prepared with modifications at the 3'-position of the benzoyl ring introducing carboxylate groups aimed at making interactions with Arg 57 and His 28. From molecular modelling studies, based on the X-ray structural data (Bolin *et al.*, 1982), the 4,6-dicarboxylate derivative **8** was predicted to bind to both Arg 57 and His 28 (Birdsall *et al.*, 1984b). In fact, this analogue was found to bind 1000 times more tightly than does brodimoprim and furthermore the selectivity of binding to the bacterial enzyme compared with the mammalian enzyme is retained. NMR could then be used to prove that the inhibitor is binding in the predicted manner since the pH dependence of the C2 proton signal of His 28 shows an increase of one unit in the pK value on complex formation.

Determination of ligand ionisation states

NMR can also monitor the ionisation states in the bound ligand. For the case where the ionisation state is a protonated species, it is sometimes possible to use NMR to detect directly the proton involved in this protonation. If the protonation is on a nitrogen atom then the use of selective ^{15}N labelling can provide an unambiguous method of assigning the bonded proton. Such proton signals can sometimes be detected directly, but in other cases it is necessary to use 2D or 3D HMQC based experiments to detect the ^{15}NH proton selectively. When inhibitors containing a 2,4-diamino pyrimidine ring (such as antibacterial drug trimethoprim) bind to DHFR, NMR evidence clearly shows that the N_1 position of the pyrimidine ring is protonated. In a ^1H NMR study examining ^{15}N enriched trimethoprim in its complex with DHFR in H_2O solution, a doublet was observed at 14.79 ppm in the spectrum which could be assigned to the N_1H proton (see Figure 6a): the doublet splitting (~90Hz) is characteristic of one bond ^{15}N - ^1H spin-coupling and furthermore, the splitting disappears in the ^1H spectrum of the DHFR complex formed with non-labelled trimethoprim (Bevan *et al.*, 1985) (Figure 6b). The ^{15}N chemical shift of the N_1 nitrogen also has a value characteristic of the protonated species (80 ppm different from the non-protonated species). Earlier studies using [^{13}C-2]-trimethoprim had already shown that the N_1 position is protonated in the bound state and had indicated that the pK value for this protonation was displaced by at least 2 units as a result of complex formation (Roberts *et al.*, 1981). A consideration of the X-ray data (Bolin *et al.*, 1982) strongly suggested that the protonated N_1 group is interacting with the carboxylate anion of the conserved Asp 26 residue (see Figure 4a).

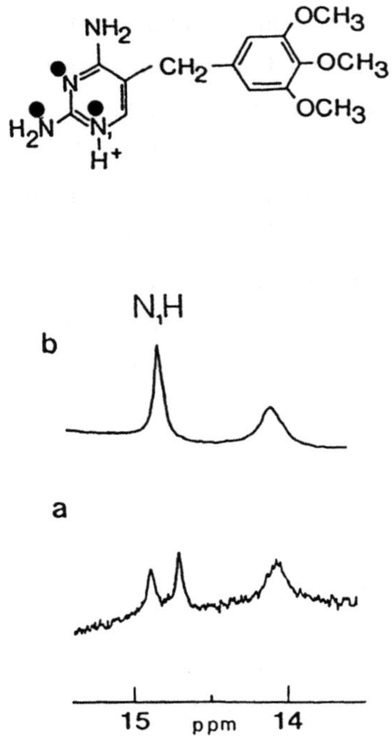

Fig. 6. Comparison of the low field region of the ^1H NMR spectra in 90% H_2O/10% ^2H_2O of the complexes of dihydrofolate reductase with (a) [$^{15}N_3$]-trimethoprim and (b) unlabelled trimethoprim (Reproduced with permission from Bevan *et al.*, 1985).

Ionisation states of phosphate groups can be monitored using ^{31}P NMR and we have used this approach in studies of coenzyme (NADPH or $NADP^+$) binding to DHFR. Figure 7 shows the plots of ^{31}P chemical shifts of the 2′-phosphate group as a function

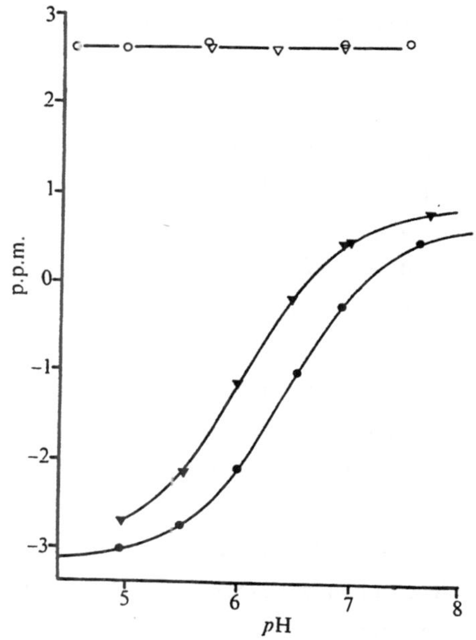

Fig. 7. The pH dependence of the chemical shift of the 2′-phosphate ^{31}P resonance of NADPH (∇,▼) and $NADP^+$ (○,●). The open symbols refer to coenzyme bound to dihydrofolate reductase and the solid symbols to free coenzymes. (Reproduced with permission from Feeney *et al.*, 1975).

of pH for complexes of DHFR with oxidised and reduced coenzyme together with the same data for the free coenzymes. While the data for the free coenzymes clearly indicate pK values 6.1 (NADPH) and 6.4 ($NADP^+$), the ^{31}P chemical shift of the 2′-phosphate group of the coenzymes in the DHFR complexes does not change with pH over the range 4.5 to 7.5. The bound chemical shifts indicate that the 2′-phosphate is binding in exactly the same manner in complexes with the oxidised and the reduced coenzyme. In each case the phosphate group is binding as the dianionic form with its pK perturbed by at least three units. These studies are relatively straightforward because one is dealing with tightly bound ligands, where the bound and free species exist under conditions of slow exchange.

DETERMINATION OF CONFORMATIONS OF PROTEIN-LIGAND COMPLEXES

It is only relatively recently that NMR has been able to provide the full three dimensional structure for protein-ligand complexes in solution. Fesik (1993) has reviewed this area from the perspective of its value in drug design and refers to several studies of ligand receptor for complexes where NMR has provided structural information to assist in drug design (Michnik *et al.*, 1991; Moore *et al.*, 1991; Ikura *et al.*, 1992; Theriault *et al.*, 1992; Meadows *et al.*, 1993).

The general methods for biomolecular structure determination using NMR have been discussed in earlier Chapters. First the ^1H resonance assignments are made and then interproton distance constraints are obtained from NOE measurements and dihedral angle constraints from vicinal coupling constants. Families of structures are then calculated from geometric and covalent structure constraints using distance geometry or restrained molecular dynamics (MD) calculations. The final structures are refined using restrained energy minimisation or molecular dynamics calculations.

Ideally, one would like to determine the structures of the unbound species as well as the complex. This has been achieved in a few cases particularly for protein-DNA complexes. Kaptein and coworkers (1989) have used NOE information to determine the structures of complexes of the lac repressor headpiece (HP56) with various DNA duplexes. About 25 protein-DNA NOEs could be defined. The lac repressor headpiece and the DNA do not change conformation appreciably on complex formation and the complex was modelled first by using an ellipsoid algorithm to achieve the docking followed by a restrained molecular dynamics refinement to produce a structure of the complex which satisfied all the observed intermolecular NOEs. The derived structure provided the important result that the second helix of the helix-turn-helix motif (ie. the recognition helix which makes most of the contacts with the major groove) is oriented with respect to the dyad axis of the operator in a manner opposite to that found in X-ray structures of several other repressor-operator complexes : Kaptein and coworkers (1989) have postulated that there are two classes of helix-turn-helix proteins differing with respect to the orientation of the recognition helix with respect to the dyad axis. Amino acids on the recognition helix were found to form hydrogen bond interactions with protons of the bases within the major groove of the DNA. Some small distortion of the recognition helix on binding was also noted in these studies.

For the complex of the Antennapedia homeodomain protein binding to a DNA 14-mer (Otting *et al.*, 1990; Neri *et al.*, 1990), the intramolecular NOEs were found to be similar in the free and bound states indicating that the free and bound conformations are very similar. Intermolecular NOEs between nine sets of protons on the DNA and protein were measured for the complex and this was sufficient to obtain a good characterisation of the conformation of the complex. This was achieved by docking the protein and DNA in the fixed conformations previously determined by NMR using the measured intermolecular NOEs (translated into distance constraints of 5Å) in conjunction with an ellipsoid algorithm calculation.

DETERMINATION OF THE BOUND LIGAND CONFORMATION

Many ligands are flexible in solution but adopt a single conformation when bound to their receptor protein. From the aspect of drug design it is important to know the bound conformation since this could provide the basis for designing more rigid and effective inhibitors. Several methods based on determining intramolecular NOEs in the bound ligand have been proposed.

Use of transferred NOEs

The transferred NOE (TRNOE) technique is a useful method for obtaining conformational information about the bound ligand (Albrand *et al.*, 1979; Cayley *et al.*, 1979; Feeney *et al.*, 1983; Clore and Gronenborn, 1987; Gronenborn and Clore, 1990). In this method, cross-relaxation (NOE) between two protons in the bound ligand is transferred to the free molecule by chemical exchange between bound and free species (see Figure 8). The negative NOEs from the bound state can thus be detected by carrying out NOE experiments on the free ligand (in slow exchange), or on the averaged signals for free and bound ligand (in fast exchange). Thus, if one selectively irradiates either the free or the averaged signal of proton S in a 1D experiment then a negative transferred NOE will be observed on the free or averaged signal of proton I and the magnitude of the transferred NOE is proportional to the cross-relaxation rate σ^{B}_{IS} between I and S in the bound state. The method works when

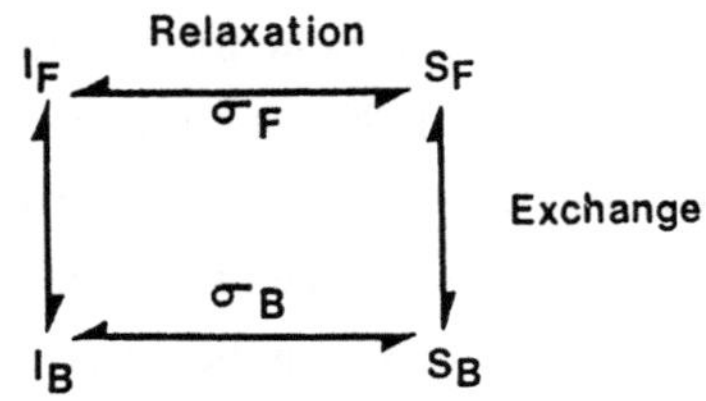

Fig. 8. Schematic indication of processes relating to the magnetisation of two nuclei I and S in a ligand binding to a protein. (Reproduced with permission from Albrand *et al.*, 1979).

i) the chemical exchange rate k is much faster than the relaxation rate of the free proton, ρ^F_I

$$k \geq 10\, \rho^F_I$$

ii) and $| (1-p_F)\ \sigma^B_{IS} | \gg | p_F\, \sigma^F_{IS} |$

where p_F is the mole fraction of free ligand and σ^B_{IS} and σ^F_{IS} are the cross-relaxation rates between I and S in the bound and free states respectively. If the bound cross-relaxation rates for two pairs of spins (IS and IT) can be measured then Equation 11 can be used to estimate the ratio of their internuclear separations. If one of the distances is known, the other can be calculated.

$$\frac{\sigma^B_{IS}}{\sigma^B_{IT}} = \left(\frac{r_{IT}}{r_{IS}}\right)^6 \tag{11}$$

In practice, the method works best for large proteins bound to ligands in fast exchange but it is also effective for slowly exchanging ligands if the above conditions can be met. The first observation of such transferred NOE effects for slowly exchanging ligands was in the complex of trimethoprim with a selectively deuterated DHFR sample (Cayley *et al.*, 1979). Irradiation of the H6 signal from free ligand was found to cause a decrease in intensity at the H_2' H_6' signal of the free ligand due to a transferred NOE effect. This result indicated that H6 is close to H_2' (or H_6') in the folded conformation of bound trimethoprim. The method was also shown to be useful for determining the glycoside bond orientations for bound nucleotides. For the complex $NADP^+$.DHFR a transferred negative NOE was observed between the nicotinamide ribose 1'-proton of free $NADP^+$ and the nicotinamide ring 2-proton of free $NADP^+$ implying an anti-conformation for the glycoside bond in bound $NADP^+$. Gronenborn and Clore (1982) have developed the theory for the transferred NOE effect and several workers have exploited the technique particularly for ligands such as nucleotides binding to proteins under conditions of fast exchange (for a review see Gronenborn and Clore, 1990). Transferred NOE effects can also be detected in 2D NOESY spectra and Clore *et al.*, 1986 have used this approach to obtain a set of 23 distance constraints between pairs of ligand protons in the tetrapeptide, acetyl-Pro-Ala-Pro-Tyr-NH_2, bound to porcine pancreatic elastase. They were able to use these constraints in combination with molecular dynamics calculations to define the conformation of the bound ligand. By combining this with some limited crystallographic data they could locate the bound ligand in its protein binding site where it was found to have the reverse orientation to that found in other serine protease-inhibitor complexes.

In the absence of exchange between bound and free ligand, it is not possible to use the transferred NOE method to determine the bound conformation. In such cases it is necessary to measure the intramolecular NOEs within the ligand directly. The main problem is one of detecting the relevant NOEs and the presence of a large number of overlapping cross-peaks from the protein. Various strategies involving isotopic labelling have been adopted to overcome this problem.

Use of perdeuteriated proteins

Hsu and Armitage (1992) have used a very direct strategy involving measuring intramolecular ^1H-^1H NOEs for ligands bound to perdeuteriated proteins. Because only the bound ligand ^1H signals are detected, the 2D COSY and NOESY spectra are relatively simple. This approach has been used to examine cyclosporin A (CsA) in its complex with perdeuterated cyclophilin (CyP) (the latter was prepared by expressing the human CyP gene in *E.coli* grown on deuteriated algal hydrolysate in 98% D_2O). The NOE data indicated that the bound cyclosporin A exists in an all trans conformation with no detectable secondary structure and no intramolecular hydrogen bonds.

Isotope editing using labelled ligands

Fesik and coworkers (1988, 1990) have proposed an elegant technique for measuring intra- and intermolecular NOEs in protein ligand complexes based on using isotopically labelled (^{13}C or ^{15}N) ligands and examining their complexes by using NMR isotope-editing procedures. NOEs involving ligand protons directly attached to ^{13}C or ^{15}N can thus be selectively detected. They demonstrated the method in an ^{15}N-edited 2D NOE experiment on a pepsin/inhibitor (1:1) complex formed with ^{15}N-labelled inhibitors (Fesik *et al.*, 1988). The spectra are simple in that they show NOE cross-peaks only between the amide protons attached to ^{15}N and their neighbouring protons in the protein. The measured NOEs could be interpreted by assuming that the inhibitor has an extended backbone conformation when complexed with the porcine pepsin (Fesik *et al.*, 1988). Isotope-editing methods (Otting *et al.*, 1986; Weber *et al.*, 1991) have been applied to study ^{13}C- and ^{15}N-labelled cyclosporin A bound to cyclophilin. Intramolecular NOE information for free and bound cyclosporin A was obtained and it could be shown that the bound ligand has all its peptide bonds in the trans form in contrast to free cyclosporin which has a cis peptide bond between the methyl leucine residues at positions 9 and 10.

DETECTION OF MULTIPLE CONFORMATIONS

One of the most useful applications of NMR spectroscopy in the study of protein-ligand complexes is its ability to detect different conformational states in solution. In some cases the different conformations are in slow exchange such that a separate NMR spectrum is observed for each conformation. It is important to detect and characterise the presence of different conformations of the bound ligand since each conformation offers a new starting point for the design of improved inhibitors. Such measurements of conformational mixtures are also important when one is considering structure activity relationships. It is worth noting that NMR is the only method which can provide quantitative information about such conformational equilibria in solution.

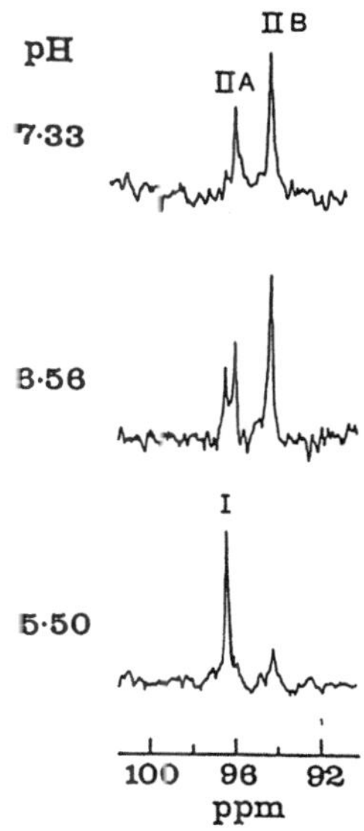

Fig. 9. Carboxamide carbon region of the 50.3 MHz ^{13}C spectrum of the complex of dihydrofolate reductase with folate and [3-carboxamide-^{13}C]NADP$^+$ at three pH values. the resonances from bound NADP$^+$ in conformations I, IIa and IIb are indicated. (Reproduced with permission from Birdsall *et al.*, 1982).

We have observed several examples of multiple conformations in our NMR studies of complexes of DHFR (Birdsall *et al.*, 1981, 1984a; Gronenborn *et al.*, 1981; Birdsall *et al.*, 1989; Birdsall *et al.*, 1990). In most cases the different conformations correspond to a flexible ligand occupying essentially the same binding site but in different conformational states. For example, such multiple conformational states have been detected in NMR spectra of complexes of the substrate folate with DHFR (Birdsall *et al.*, 1989). NMR studies of the folate.DHFR and folate.NADP$^+$.DHFR complexes have shown that each exists as a mixture of three slowly interconverting conformations in solution (Forms I, IIa and IIb). Each form gives rise to a separate spectrum and it is possible to monitor changes in populations as a function of pH (see Figure 9) (Birdsall *et al.*, 1982). Most of the spectral differences are in signals from the ligands or from protein residues in the binding site.

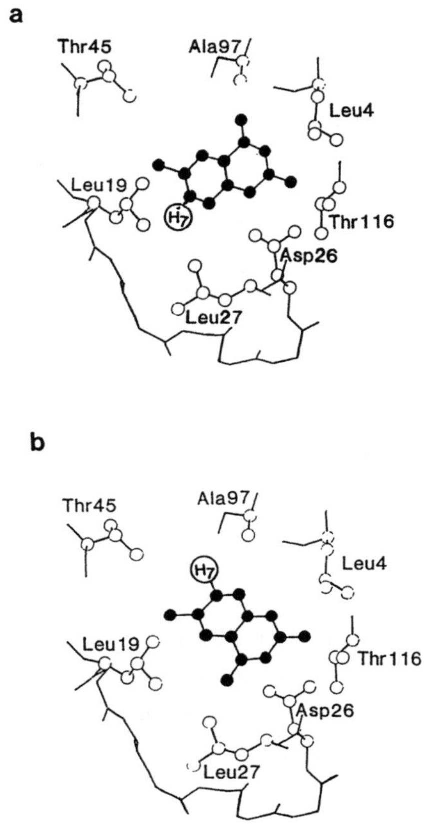

Fig. 10. (a) Conformation of the methotrexate pteridine ring in its binding site in DHFR (from the data of Matthews and coworkers). (b) Proposed conformation of pteridine ring in the 'productive' conformation of the folate.DHFR complex.

COSY spectra for all three forms are very similar to that recorded for the DHFR.MTX.NADP$^+$ complex indicating that any differences in the global protein conformation between the different complexes must be small. Recent ^{13}C NMR studies have indicated that Form IIb exists in the keto form whereas Forms I and IIa exist as enolic forms (Cheung *et al.*, 1993). Irradiation of the folate H7 proton signal in the low

pH Form I and in Form IIa (but not IIb) gives rise to NOE connections to protons in the methyl groups of Leu 19 and 27. Similar NOE effects were seen previously in the methotrexate. DHFR complex (which is known to have a 'turned over' orientation of its pteridine ring, 180° different from that in the catalytically active folate complex (see Figure 10) (Charlton *et al.*, 1985; Birdsall *et al.*, 1989; Cheung *et al.*, 1993)). For Form IIb, an NOE connection from the folate H7 proton to the Ala 97 methyl protons was detected as would be expected for the 'turned over' conformation (Figure 10b).

Thus Forms I and IIa have the same pteridine ring orientation as bound methotrexate and in their enolic forms could thus bind in a very similar way to the pteridine ring in methotrexate. Form IIb has its folate pteridine ring turned over, ~ 180° about an axis approximately coincident with the C2-NH_2 bond (Figure 10b) and this orientation would allow the catalytic reduction to proceed in a manner which would give the correct stereochemistry for the reduction.

We have observed multiple conformations in several other complexes of *L. casei* DHFR (for example, with $NADP^+$ and trimethoprim (Gronenborn *et al.*, 1981) and with substituted pyrimethamines (Birdsall *et al.*, 1990)), and other workers have seen multiple conformations in complexes with *S. faecium* DHFR (London *et al.*, 1979) and *E.coli* DHFR (Falzone *et al.*, 1991): it seems likely that many other protein-ligand complexes exist as mixtures of conformations. It should be remembered that such conformations would be difficult to detect directly if the conformations are in fast exchange.

DYNAMIC PROCESSES IN PROTEIN-LIGAND COMPLEXES

NMR measurements can characterise many of the dynamic processes occurring within complexes: this dynamic information complements the static structural information and provides a more complete description of the complex. Studies using NMR relaxation, line-shape analysis and transfer of magnetisation have provided a wide range of dynamic information relating to protein-ligand complexes. The NMR accessible motions range from fast ($>10^9$ s^{-1}) small amplitude oscillations of fragments of the complex (available from ^{13}C relaxation measurements) to slow motions (1 to 10^3 s^{-1}) involved in the rate of dissociation of the complex, rates of breaking and reforming protein-ligand interactions and rates of flipping of aromatic rings in the bound ligands. The dynamic processes characterised by NMR studies in several complexes of dihydrofolate reductase will be discussed.

Dissociation rate constants from transfer of saturation studies.

Consider a proton existing in two magnetically distinct environments, for example, a ligand free in solution and bound to the protein under conditions of slow exchange where separate signals are seen for the two forms. If the resonance of the bound proton is selectively saturated, then this saturation will be transferred to the signal of the free proton via the exchange process and the intensity of the free proton signal will decrease in intensity. The rate of decrease of the magnetization in the free state as a function of the time, *t*, for which the bound proton is saturated (assuming saturation occurs in a time short compared to *t*) is given by Forsén & Hoffman (1963) in Equation 12, where M_{z0}^{F} is the equilibrium magnetization of the free proton in the absence of irradiation and τ_{1F}

$$\frac{dM_z^F}{dt} = \frac{M_{z0}^F}{T_{1F}} - \frac{M_z^F}{\tau_{1F}} \qquad (12)$$

is the lifetime of the proton spin state defined by Equation 13, where T_{1F} is the spin-lattice relaxation time and τ_F is the residence time of the proton in the free state.

$$\frac{1}{\tau_{1F}} = \frac{1}{T_{1F}} + \frac{1}{\tau_F} \qquad (13)$$

The solution of Equation 12 is given by Equation 14.

$$M_z^F(t) = M_{z0}^F \left[\frac{\tau_{1F}}{\tau_F} \exp(-t/\tau_{1F}) + \frac{\tau_{1F}}{T_{1F}} \right] \qquad (14)$$

The free proton signal is seen to show an exponential decay, with a time constant τ_{1F}, to a new equilibrium value (Equation 15).

$$M_z^F(\infty) = M_{z0}^F \frac{\tau_{1F}}{T^{1F}} \qquad (15)$$

The value of τ_{1F} can be determined by measuring the signal height as a function of *t* and fitting these data (in the form of a plot of $M_z^F(t)$ vs. *t*) to Equation 14 by nonlinear regression. Combination of this value with a measurement of $M_z^F(\infty)/M_{z0}^F$ (cf Equation 15) allows both τ_F and T_{1F} to be calculated. The dissociation rate constant can then be calculated from Equation 16, where p_F and p_B are the fractions of the total ligand in the free and bound states, respectively.

$$k_{off} = \frac{p_F}{p_B \tau_F} \tag{16}$$

This method has been used to measure the dissociation rate constant for the $NADP^+$.DHFR complex in the presence of an excess of free $NADP^+$: the signal height of the A2 proton resonance in the free $NADP^+$ was measured as a function of irradiation time, t, at the frequency of the corresponding bound proton. Analysis of this curve, according to Equations 14 and 15, allows the lifetime of the complex to be determined and, from this, the dissociation rate constant (20 s^{-1} at 284K) (Hyde *et al.*, 1980a). Similar measurements on the trimethoprim.DHFR complex yielded a value for the dissociation rate constant of 6 s^{-1} at 313K (Searle *et al.*, 1988).

Rapid segmental motions in protein-ligand complexes

Information about rapid molecular motions ($>10^9$ s^{-1}) can often be obtained by measuring ^{13}C relaxation times. We have measured the ^{13}C relaxation behaviour of the two enriched carbons in [7,4'-OMe-^{13}C]-trimethoprim bound to DHFR and found that the calculated correlation time (assuming isotropic motion) is 15.4 ± 1.5 ns at 295 K which is very close to that expected for the overall trumbling of the protein (Searle *et al.*, 1988). The relaxation behaviour of the 4'-OCH_3 is quite different having a 6-fold slower relaxation rate. Analysis of this data by procedures outlined by Lipari and Szabo (1982) provides information about rapid motions about the C7-C1', C4'-O and O-CH_3 bonds. The results provide an effective correlation time for the internal motions affecting the 4'-OCH_3 (4.3 (± 0.8) ps) and information on an order parameter for the additional motion about the three bonds linking the OCH_3 group to C7 (Searle *et al.*, 1988). The observed order parameter could not be explained simply in terms of one additional motion corresponding to the methyl group rapidly spinning about its axis and there must be additional motion on the sub-nanosecond time scale about at least one of the other two bonds.

Rates of ring flipping

Aromatic rings of Phe and Tyr residues sometimes take up fixed orientations in a protein such that nuclei at the 2′,6′ positions (and the 3′,5′ positions) are non-equivalent due to the asymmetric nature of their environments in the protein. More usually, Phe and Tyr aromatic rings are rapidly flipping, probably by 180° jumps, such that averaged NMR signals are seen for nuclei at the 2′,6′- (and 3′5′-) positions (Wüthrich, 1976). Slow and fast ring-flipping has also been observed and characterised in ligands bound to proteins, for example in complexes of methotrexate (Clore *et al.*, 1984) and trimethoprim analogues (Cayley *et al.*, 1979; Searle *et al.*, 1988) with DHFR. For example, we have used ^{19}F NMR spectroscopy to make such measurments for the complex of 3′,5′-difluoromethotrexate bound to the enzyme over the temperature range 273 to 308° (Clore *et al.*, 1984). At 308°K, a single resonance is observed for the two ^{19}F nuclei but as the temperature is progressively lowered the signal broadens and eventually separates into two signals of equal intensity at 273° (see Figure 11). The spectra are

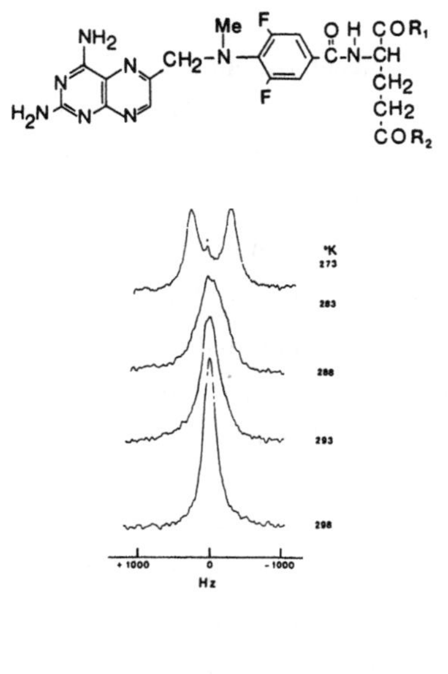

Fig. 11. 188.2 MHz ^{19}F NMR spectra of 3′,5′-difluoromethotrexate complexed with *L. casei* dihydrofolate reductase examined at a series of temperatures. (Reproduced with permission from Clore *et al.*, 1984).

characteristic of nuclei undergoing exchange between two equally populated non-equivalent sites. At low temperature the benzoyl ring is in a fixed orientation on the protein which leads to non-equivalence of the 3′- and 5′-fluorine nuclei in the asymmetric binding site. The simplest exchange process leading to equivalence of the nuclei at higher temperatures would be a 'flipping' motion about the symmetry axis of the benzoyl ring (Wüthrich, 1976). Line shape analyses of the data for the 3′,5′-difluoromethotrexate.enzyme complex and its ternary complexes with $NADP^+$ and NADPH indicate that the rate of ring-flipping increases in the ternary complexes even though the 3′,5′-difluoromethotrexate is bound more tightly (k_{298} = 7.27 x 10^3 s^{-1} (binary); 1.88 x 10^4 s^{-1} (ternary with $NADP^+$); 2.04 x 10^4 s^{-1} (ternary with NADPH). These results show that the coenzyme binding results in a conformational change which allows faster ring-flipping to take place. Examination of the crystal structure of the enzyme.methotrexate.NADPH complex reveals that steric interactions would forbid ring flipping if the rigid binding site implied by the crystal structure were maintained: however, theoretical studies (Karplus *et al.*, 1981) on the flipping rates of aromatic rings in proteins show that the low barriers for ring flipping can be accounted for by transient displacements of the protein atoms near the rings producing a relaxed protein structure where the ring-flipping is possible.

The availability of ^{13}C-labelled ligands can also facilitate such studies. For example, ^{13}C line-shape analysis on the signals from the enriched carbons in [m-methoxy-^{13}C]-brodimoprim bound to DHFR has been used to measure the 'flipping' rates of the benzyl ring in the bound ligand (Searle *et al.*, 1988). Again, we have analysed the ^{13}C line shapes on the basis of McConnell's modification of the Bloch equations and obtained estimates of the rates of 'flipping' and the activation parameters (at 298 K trimethoprim, 250 s^{-1}; E_a 42 KJ/mol;brodimoprim 140 s^{-1}, E_a 59 KJ/mol). In both cases these rates are greater than the dissociation rates of the compexes and thus the flipping takes place many times during the lifetime of the intact complex. A consideration of the structure of bound trimethoprim and its analogues (based on X-ray and NMR findings) revealed that in its bound conformation, trimethoprim would find it impossible to undergo a 180° flip of the benzyl ring because of severe steric interactions between the atoms at the pyrimidine C6 and benzyl C2′,6′ positions. In fact, ring 'flipping' can only be achieved by rotating θ_1 by at least 30° (see Figure 12) and this in turn requires a substantial conformational change of the protein. Thus the measured rate of flipping is indirectly monitoring transient fluctuations in the conformation of the enzyme structure which are required to allow the flipping to proceed.

Hydrogen exchange rates with solvent

Extensive NMR measurements of exchange rates between solvent and labile protons on protein or ligand have been reported. These are usually based on line shape analysis or transfer of magnetisation methods.

We have made such measurements for complexes of trimethoprim analogues with DHFR (Searle *et al.*, 1988). In Figure 9 we saw how the N_1 proton of bound trimethoprim could be assigned unambiguously using ^{15}N-labelled trimethoprim. On changing the temperature the line shape of the N_1 proton signal varies due to changes in the exchange rate of this proton with the H_2O solvent. This line-width data can be analysed according to Equation 8 in order to estimate the exchange rate. In a protein-ligand complex such exchange can be considered as a two step process : in the first step the structure opens to allow access of the solvent and in the second step the chemical exchange process takes place (see Leroy *et al.*, 1988). Adding imidazole, an effective catalyst for N-H exchange, to the enzyme-trimethoprim complex, did not perturb the observed exchange process, which indicates that the 'opening' rate is determining the rate of exchange. In this case, the N_1H forms a hydrogen bond with the carboxylate group of Asp26 (Figure 12) and the measured exchange rate (34 s^{-1} at 298K) is thus the rate of breaking and reforming this hydrogen bonding interaction. Once again a very

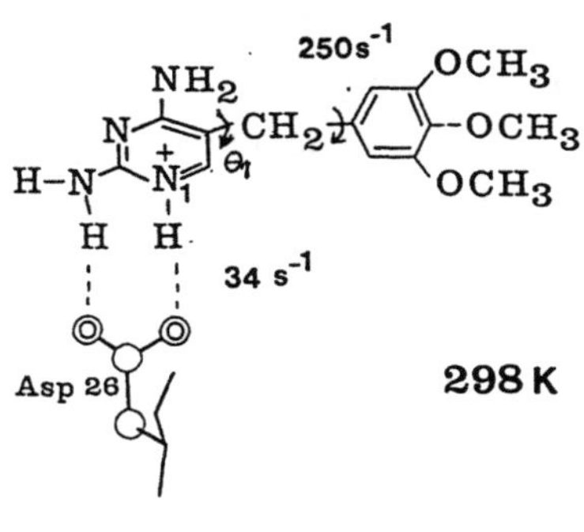

Fig. 12. Dynamic processes in complex of trimethoprim with *L. casei* dihydrofolate reductase (Reproduced with permission from Searle *et al.*, 1988)

important interaction in the complex is breaking and reforming at a rate much faster than the dissociation rate. Thus both the pyrimidine ring and benzyl ring are involved in transient fluctuations in their interactions with the protein during the lifetime of the complex : the rates are summarised on Figure 12. When these structural fluctuations take place in close succession they probably form part of a sequence of events leading to dissociation of the complex.

CONCLUSION

It is clear that advances in NMR methodology, particularly in multidimensional NMR experiments used in conjunction with isotopically labelled molecules will provide even more detailed information about protein-ligand complexes in solution. These applications should become more widespread once the required NMR facilities become more generally available.

ACKNOWLEDGEMENT

I would like to acknowledge the collaboration of many colleagues in the work on dihydrofolate reductase, in particular Berry Birdsall and Gordon Roberts. I would like to thank Linda Dunphy for processing this manuscript.

References

Albrand, J.P., Birdsall, B., Feeney, J., Roberts, G.C.K. and Burgen, A.S.V. (1979). The use of transferred nuclear Overhauser effects in the study of the conformations of small molecules bound to proteins. Internat. J. Biol. Macromolecules **1**, 37-41.

Antonjuk, D.J., Birdsall, B., Burgen, A.S.V., Cheung, H.T.A., Clore, G.M., Feeney, J., Gronenborn, A., Roberts, G.C.K. and Tran, W. (1984). A ^{1}H NMR study of the role of the glutamate moiety in the binding of methotrexate to dihydrofolate reductase. Brit. J. Pharmacol. **81**, 309-315.

Bevan, A.W., Roberts, G.C.K., Feeney, J. and Kuyper, I. (1985). ^{1}H and ^{15}N NMR studies of protonation and hydrogen-bonding in the binding of trimethoprim to dihydrofolate reductase. Eur. Biophys. J. **11**, 211-218.

Birdsall, B., Griffiths, D.V., Roberts, G.C.K., Feeney, J. and Burgen, A.S.V. (1977). ^{1}H Nuclear magnetic resonance studies of *Lactobacillus casei* dihydrofolate reductase: effects of substrate and inhibitor binding on the histidine residues. Proc. Roy. Soc. Lond.B. **196**, 251-265.

Birdsall, B., Roberts, G.C.K., Feeney, J. and Burgen, A.S.V. (1977). ^{31}P NMR studies of the binding of adenosine-2'-phosphate to *L. casei* dihydrofolate reductase. FEBS Letters **80**, 313-316.

Birdsall, B., Hyde, E.I., Burgen, A.S.V., Roberts, G.C.K. and Feeney, J. (1981). Negative cooperativity between folinic acid and coenzyme in their binding to *L. casei* dihydrofolate reductase. Biochemistry. **20**, 7186-7195.

Birdsall, B., Bevan, A.W., Pascual, C., Roberts, G.C.K., Feeney, J., Gronenborn, A. and Clore, G.M. (1984a). Multinuclear NMR characterisation of two coexisting conformational states of *Lactobacillus casei* dihydrofolate reductase trimethoprim-NADP+ complex. Biochemistry. **23**, 4733-4742.

Birdsall, B., Feeney, J., Pascual, C., Roberts, G.C.K., Kompis, I., Then, R.L., Müllеs, K. and Krohn, A. (1984b). A ^{1}H NMR study of the interactions and conformations of rationally designed brodimoprim analogues in complexes with *L.casei* DHFR. J. Med. Chem. **27**, 1672-1676.

Birdsall, B., Feeney, J., Tendler, S.J.B., Hammond, S.J. and Roberts, G.C.K. (1989). Dihydrofolate reductase: Multiple conformations and alternative modes of substrate binding. Biochemistry. **28**, 2297-2305.

Birdsall, B., Gronenborn, A., Hyde, E.I., Clore, G.M., Roberts, G.C.K., Feeney, J. and Burgen, A.S.V. (1982). ^{1}H, ^{13}C and ^{31}P NMR studies of dihydrofolate reductase-NADP+-folate complex: characterisation of three co-existing conformational states. Biochemistry. **21**, 5831-5838.

Birdsall, B., Tendler, S.J.B., Arnold, J.R.P., Feeney, J., Griffin, R.J., Carr, M.D., Thomas, J.A., Roberts, G.C.K. and Stevens, M.F.G. (1990). NMR studies of multiple conformations in complexes of *L. casei* dihydrofolate reductase with analogues of pyrimethamine. Biochemistry. **29**, 9660-9667.

Bolin, J.T., Filman, D.J., Matthews, D.A., Hamlin, R.C. and Kraut, J. (1982). Crystal structures of *E. coli* and *L. casei* dihydrofolate reductase refined to 1.7 Å resolution. I. General features and binding of methotrexate. J. Biol. Chem. **257**, 13650-13662.

Carr, M.D., Birdsall, B., Jimenez-Barbero, J., Polshakov, V.I., Bauer, C.J., Frenkiel, T.A., Roberts, G.C.K. and Feeney, J. (1991). Dihydrofolate reductase: Sequential resonance assignments using 2D and 3D NMR and secondary structure determination in solution. Biochemistry **30**, 6330-6341.

Cayley, P.J., Albrand, J.P., Feeney, J., Roberts, G.C.K., Piper, E.A. and Burgen, A.S.V. (1979). NMR studies of the binding of trimethoprim to dihydrofolate reductase. Biochemistry **18**, 3886-3895.

Cheung, H.T.A., Birdsall, B., Frenkiel, T.A., Chau, D.D. and Feeney, J. (1993) ^{13}C NMR determination of the tautomeric and ionisation states of folate in its complexes with *L.casei* dihydrofolate reductase. Biochemistry **32**, 6846-6854.

Charlton, P.A., Young, D.W., Birdsall, B., Feeney, J. and Roberts, G.C.K. (1985). Stereochemistry of reduction of vitamin folic acid by dihydrofolate reductase. J. Chem. Soc. Perkin II, 1349-1353.

Clore, G.M. and Gronenborn, A.M. (1982). Proton nuclear magnetic resonance study of the histidine residues of the *Escherichia coli* adenosine cyclic 3',5'-phosphate receptor protein. pH titration behavior, deuterium exchange, and partial assignments. Biochemistry. **21**, 4048-4053.

Clore G.M. and Gronenborn, A.M. (1982). Theory and applications of the transferred nuclear Overhauser effect to the study of the conformations of small ligands bound to proteins. J. Magn. Reson. **48**, 402-417.

Clore, G.M., Gronenborn, A.M., Birdsall, B., Feeney, J. and Roberts, G.C.K. (1984). NMR studies of 3',5'-difluoromethotrexate binding to *Lactobacillus casei* dihydrofolate reductase. Molecular motion and coenzyme induced conformational change. Biochem. J. **217**, 659-666.

Clore, G.M., Gronenborn, A.M., Carlsson, G. and Meyer, E.F. (1986). Sterochemistry of binding of the tetrapeptide Aceyl-Pro-Ala-Pro-Tyr-NH_2 to porcine pancreatic elastase. J. Mol. Biol. **190**, 259-267.

Clore, G.M., Gronenborn, A.M. (1991). Applications of three- and four-dimensional heteronuclear NMR spectroscopy to protein structure determination. Progress in NMR Spectroscopy **23**, 43.

Evans, E.A., Warrell, D.C., Elvidge, J.A. and Jones, J.R. (1985). Handbook of tritium-NMR spectroscopy and applications. J. Wiley and Sons, Chichester.

Falzone, C.J., Wright, P.E. and Benkovic, S.J. (1991). Evidence for two interconverting protein isomers in the methotrexate complex of DHFR from *E.coli*. Biochemistry **30**, 2184-2191.

Feeney, J. (1990). NMR studies of interactions with ligand with dihydrofolate reductase. Biochem. Pharmacol. **40**, 141-152.

Feeney, J., Batchelor, J.G., Albrand, J.P., Roberts, G.C.K. (1979). The effect of intermediate exchange processes on the estimation of equilibrium constants by NMR spectroscopy. J. Magn. Resonance **33**, 519-529.

Feeney, J., Birdsall, B., Roberts, G.C.K. and Burgen, A.S.V. (1975). ^{31}P NMR studies of NADPH and $NADP^+$ binding to *L. casei* dihydrofolate reductase. Nature. **257**, 564-566.

Feeney, J., Birdsall, B., Roberts, G.C.K. and Burgen, A.S.V. (1983). The use of transferred nuclear Overhauser effect measurements to compare the binding of coenzyme analogues to dihydrofolate reductase. Biochemistry. **22**, 628-633.

Fesik, S.W., Luly, J.R., Erickson, J.W. and Abad-Zapatero, C. (1988). Isotope-edited proton NMR study on the structure of a pepsin/inhibitor complex. Biochemistry **27**, 8297-8301.

Fesik, S.W., Zuiderweg, E.R.P., Olejniczak, E.T. and Gampe, R.T. (1990). NMR methods for determining the structures of enzyme/inhibitor complexes as an aid in drug design. Biochem. Pharmacol. **40**, 161-167.

Fesik, S.W., Gampe, R.T. Jr., Holzman, T.F., Egan, D.A., Edalji, R., Luly, J.R., Simmer, R., Helfrich, R., Kishore, V. and Rich, D.H. (1990). Isotope-edited NMR of cyclosporin A bound to cyclophilin: Evidence for a *Trans* 9,10 amide bond. Science **250**, 1406-1409..

Forsén, S. and Hoffman, R. (1963). J. Chem. Phys. **39**, 2892.

Gronenborn, A., Birdsall, B., Hyde, E.I., Roberts, G.C.K., Feeney, J. and Burgen, A.S.V. (1981). ^{1}H and ^{31}P NMR characterisation of two conformations of the trimethoprim-$NADP^{+}$-dihydrofolate reductase complex. Molecular Pharm. **20**, 145-153.

Gronenborn, A.M. and Clore, (1990). Determination of ligand conformation in macromolecular complexes using the transferred nuclear Overhauser effect. Biochem. Pharmacol., **40**, 115-119.

Handschumacher, R.E. and Armitage, I.M. (ed.) (1990). NMR methods for elucidating macromolecule-ligand interactions: an approach to drug design. Biochemical Pharmacology **40**, 1-174.

Hsu, V.L. and Armitage, I.M. (1992). Solution structure of cyclosporin A and a nonimmunosuppressive analog bound to fully deuterated cyclophilin. Biochemistry **31**, 12778-12784.

Hyde, E.I., Birdsall, B., Roberts, G.C.K., Feeney, J. and Burgen, A.S.V. (1980a). Proton NMR saturation transfer studies of coenzyme binding to *Lactobacillus casei* dihydrofolate reductase. Biochemistry **19**, 3728-3746.

Hyde, E.I., Birdsall, B., Roberts, G.C.K., Feeney, J., Burgen, A.S.V. (1980b). ^{31}P NMR studies of the binding of oxidised coenzymes to *Lactobacillus casei* dihydrofolate reductase. Biochemistry. **19**, 3746-3754.

Ikura, M., Clore, G.M., Gronenborn, A.M., Zhu, G., Klee, C.B. and Bax. A. (1992). Solution structure of a calmodulin-target peptide by multidimensional NMR. Science **256**, 632-638.

Jardetzky, O. and Roberts, G.C.K. (1981). In: NMR in molecular biology, Academic Press, London.

Kaptein, R., Boelens, R. and Lamerichs, R.M.J.N. (1989). Top. Mol. Struct. Structural Bio. **10**, 35.

Karplus, M. and McCammon, J.A. (1981). The internal dynamics of globular proteins. Critical Rev. in Biochem. **9**, 293-349.

Leroy, J.L., Kochoyan, M., Hyn-Dinh, T. and Guéron, M. (1988). Characterization of base-pair opening in deoxynucleotide duplexes using catalyzed exchange of the imino proton. J. Mol. Biol. **200**, 223-238.

Lipari, G. and Szabo, A. (1982). Model-free approach to the interpretation of nuclear magnetic resonance relaxation in macromolecules. 1. Theory and range of validity. J. Am. Chem. Soc. **104**, 4546-4559.

London, R.E., Groff, G.P. and Blakley, R.L. (1979). ^{13}C NMR evidence for the slow exchange of tryptophans in dihydrofolate reductase between stable conformations. Biochem. Biophys. Res. Commun. **86**, 779.

Markley, J.L. (1975). Observation of histidine residues in proteins by means of nuclear magnetic resonance spectroscopy. Acc. Chem. Res. **8**, 70-80.

Meadows, R.P., Nettesheim, D.G., Xu, R.X., Olejniczak, E.T., Petros, A.M., Holzman, T.F., Severin, J., Gubbins, E., Smith, H. and Fesik, S.W. (1993). Three-dimensional structure of the FK506 binding protein/ascomycin complex in solution by heteronuclear three- and four-dimensional NMR. Biochemistry **32**, 754-765.

Michnick, S.W., Rosen, M.K., Wandless, T.J., Karplus, M. and Schreiber, S.L. (1991). Solution structure of FKBP, a rotamase enzyme and receptor for FK506 and rapamycin. Science **252**, 836-839.

Mildvan, A.S. (1989). NMR studies of the interactions of substrates with enzymes and their peptide fragments. FASEB. J. **3**, 1705-1714.

Moore, J.M., Peattie, D.A., Fitzgibbon, M.J. and Thomson, J.A. (1991). Solution structure of the major binding protein for the immunosuppressant FK506. Nature **351**, 248-250.

Muller, L. (1979). Sensitivity enhanced detection of weak nuclei using heteronuclear multiple quantum coherence. J. Am. Chem. Soc. **101**, 4481-4484.

Neri, D., Otting, G. and Wüthrich, K. (1990). New nuclear magnetic resonance experiment for measurements of the vicinal coupling constants $^3J_{HB\alpha}$ in proteins. J. Am. Chem. Soc. **112**, 3663-3665.

Otting, G., Senn, H. Wagner, G. and Wüthrich, K. (1986). Editing of 2D ^1H NMR spectra using X half-filters. Combined use with residue-selective ^{15}N labeling of proteins. J. Magn. Reson., **70**, 500-505.

Otting, G., Qian, Y.Q., Billeter, M., Müller, M., Affolter, M., Gehring, W.J. and Wüthrich, K. (1990). Protein-DNA contacts in the structure of a homeodomain - DNA complex determined by nuclear magnetic resonance spectroscopy in solution. EMBO J. **9**, 3085-3092.

Roberts, G.C.K. and Jardetzky, O. (1970). Nuclear magnetic resonance spectroscopy of amino acids, peptides, and proteins. Adv. Protein Chem. **24**, 447-545.

Roberts, G.C.K., Feeney, J., Burgen, A.S.V., Yuferov, V., Dann, J.and Bjur, R. (1974). NMR studies of the binding of substrate analogues and coenzyme to dihydrofolate reductase from *L.casei*. Biochemistry. **13**, 5351-5357.

Roberts, G.C.K., Feeney, J., Burgen, A.S.V. and Daluge, S. (1981). The charge state of trimethoprim bound to *L. casei* dihydrofolate reductase. FEBS Letters **131**, 85-88.

Rodgers, P. and Roberts, G.C.K. (1973). ^{13}C NMR studies of the conformation of *N*-Acetyl-L-Tryptophan in its complex with chymotrypsin. FEBS Lett. **36**, 330-333.

Searle, M.S., Forster, M.J., Birdsall, B., Roberts, G.C.K., Feeney, J., Cheung, H.T.A., Kompis, I. and Geddes, A.J. (1988). The dynamics of trimethoprim bound to dihydrofolate reductase. Proc. Natl. Acad. Sci. (USA) **85**, 3787-3791.

Schrager, R.I., Cohen, J.S., Heller, S.R., Sachs, D.H. and Schecter, A.N. (1972). Mathematical models for interacting groups in nuclear magnetic resonance titration curves. Biochemistry **11**, 541-547.

Sheard, B. and Bradbury, E.M. (1970). Nuclear magnetic resonance in the study of biopolymers and their interaction with ions and small molecules. Prog. Biophys. Mol. Biol. **20**, 187-246.

Theriault, Y., Logan, T.M., Meadows, R., Yu, L., Olejniczak, E.T., Holzman, T.F., Simmer, R.L. and Fesik, S.W. (1992). Solution structure of the cyclosporin A/cyclophilin complex by NMR. Nature **361**, 88-91.

Weber, C., Wider, G., von Freyberg, B., Traber, R., Braun, W., Widmer, H. and Wüthrich, K. (1991). The NMR structure of cyclosporin A bound to cyclophilin in aqueous solution. Biochemistry **30**, 6563-6574.

Wüthrich, K. (1976). NMR in biological research: peptides and proteins. North Holland Publishing Co.

NMR STUDIES OF LIGAND-PROTEIN INTERACTIONS INVOLVING QUADRUPOLAR NUCLEI

Ioannis P. Gerothanassis
University of Ioannina, Department of Chemistry, Ioannina GR-45110, Greece

INTRODUCTION

A significant number of biologically important elements have isotopes that possess nuclear magnetic spin quantum numbers greater than half and thereby possess nuclear quadrupole moments which shorten relaxation times and broaden lines. This renders the nuclei less suitable, in general, for biological NMR applications than nuclei with I=½. During the last decade or so there has, however, been considerable methodological progress in this particular area and NMR studies of quadrupolar nuclei can often provide unique and valuable biochemical information (Forsen et al 1987; Sanders and Tsai 1989).

EXPERIMENTAL

It has been shown that at constant magnetic field, B_o, signal strengths are proportional to $\gamma^3NI(I+1)$, where N is the natural abundance of the isotopic species concerned. This quantity called the receptivity, D, of the nucleus can be taken as a crude guide to the ease of obtaining a signal for a given concentration of the relevant atoms in the solution at a constant magnetic field (Harris 1983). According to this definition several quadrupolar nuclei (including the biologically important nuclei ^{7}Li, ^{23}Na,^{25}Mg, ^{51}Mn and ^{59}Co) have receptivities greater than that of ^{13}C and ^{31}P (Table 1). For the nuclei ^{2}H, ^{17}O, ^{32}S, ^{43}Ca, ^{57}Zn which have very low receptivities, a tenfold or higher D^x values can in these cases be achieved through isotope enrichment. However, such comparisons between quadrupolar and spin-½ nuclei are not very significant because of the linewidth problems for the former. Diamagnetic biomolecular complexes in solution exhibit quadrupolar linewidths ranging from

NATO ASI Series, Vol. H 87
NMR of Biological Macromolecules
Edited by C. I. Stassinopoulou

TABLE 1 A partial list of the spin properties of selected quadrupolar nuclei of biological interest

Isotope[a]	Spin	Natural Abundance N/%	Magnetogyric ratio $\gamma/10^7$ rad $T^{-1} s^{-1}$	Quadrupole moment[b] $Q/10^{-28}$ m^2	Relative receptivity[c] D^P	D^c
$^{2}H^{d}$	1	0.015	4.1064	2.73×10^{-3}	1.45×10^{-6}	8.21×10^{-3}
^{6}Li	1	7.42	3.9366	-8×10^{-4}	6.31×10^{-4}	3.58
^{7}Li	$^3/_2$	92.58	10.396	-4.5×10^{-2}	0.272	1.54×10^{3}
$^{14}N^{d}$	1	99.63	1.9324	1.6×10^{-2}	1.00×10^{-3}	5.69
^{17}O	$^5/_2$	0.037	-3.6266	-2.6×10^{-2}	1.08×10^{-5}	6.11×10^{-2}
^{23}Na	$^3/_2$	100	7.0760	0.12	9.25×10^{-2}	5.25×10^{2}
^{25}Mg	$^5/_2$	10.13	-1.6370	0.22	2.71×10^{-4}	1.54
^{33}S	$^3/_2$	0.76	2.0517	-6.4×10^{-2}	1.71×10^{-5}	9.73×10^{-2}
^{35}Cl	$^3/_2$	75.53	2.6212	-7.89×10^{-2}	3.55×10^{-3}	20.02
^{37}Cl	$^3/_2$	24.47	2.182	-6.21×10^{-2}	6.64×10^{-4}	3.77
^{39}K	$^3/_2$	93.1	1.2484	5.5×10^{-2}	4.73×10^{-4}	2.69
(^{41}K)	$^3/_2$	6.88	0.6852	6.7×10^{-2}	5.78×10^{-6}	3.28×10^{-2}
^{43}Ca	$^7/_2$	0.145	-1.7999	0.2 ± 0.1	9.27×10^{-6}	5.27×10^{-2}
^{51}V	$^7/_2$	99.76	7.032	0.3	0.381	2.16×10^{3}
^{55}Mn	$^5/_2$	100	6.598	0.55	0.175	9.94×10^{2}
^{59}Co	$^7/_2$	100	6.3171	0.40	0.277	1.57×10^{3}
^{63}Cu	$^3/_2$	69.09	7.0904	-0.16	6.43×10^{-2}	3.65×10^{2}
^{65}Cu	$^3/_2$	30.91	7.5958	-0.15	3.54×10^{-2}	2.01×10^{2}
^{67}Zn	$^5/_2$	4.11	1.6726	0.15	1.17×10^{-4}	0.665
(^{79}Br)	$^3/_2$	50.54	6.7021	0.33	3.97×10^{-2}	2.26×10^{2}
^{81}Br	$^3/_2$	49.46	7.2245	0.28	4.87×10^{-2}	2.77×10^{2}
^{95}Mo	$^5/_2$	15.72	1.743	0.12	5.07×10^{-4}	2.88
(^{97}Mo)	$^5/_2$	9.46	-1.780	1.1	3.25×10^{-4}	1.84
^{127}I	$^5/_2$	100	5.3521	-0.69	9.34×10^{-2}	5.30×10^{2}
^{133}Cs	$^7/_2$	100	3.5089	-3×10^{-3}	4.74×10^{-2}	2.69×10^{2}

[a] Nuclei in brackets are considered to be not the most favourable for the element concerned. [b] The sign of *Q* relates to the distribution of charge within the nucleus. It should be noted that reported values of *Q* may be in error by as much as 20-30%. [c] D^P is the receptivity relative to that of ^1H whereas D^C is relative to ^{13}C. [d] A useful isotope of $I = ½$ exists (Harris 1983).

several Hz up to several KHz. Relatively short relaxation times thus permit rapid acquisition and short pulse repetition times (~ $4T_1$). Observation of very broad peaks, however, is a potential problem due to baseline distortions. The most severe origin of baseline artifacts is the transient response of the NMR probe, often referred to as "acoustic ringing", caused by the generations of ultrasonic waves from the action of the rf pulse. The most commonly used method of avoiding baseline distortions has been the insertion of a sufficiently long delay time, Δt, between the end of the pulse and the beginning of data acquisition. Although this procedure can reduce the baseline artifacts, it inevitably results in a considerable reduction in the S/N ratio by a factor $\exp(-\Delta t/T_2)$ and integrated peak areas are not consistent. Further, a phase shift is apparent across the spectrum.

In the last decade several pulse sequences for the suppression of acoustic ringing have been reported and readers should consult a relevant review article for details (Gerothanassis 1987). For experiments in solution, the extended spin-echo sequence was found to be particularly useful since it utilizes minimum pulse and preacquisition delay times. The basis of the technique is a spin-echo and the inversion of a spin-echo:

$$180^\circ_{\phi_1}-\tau_1-\overbrace{\theta^\circ_{\phi_2}-\tau-180^\circ_{\phi_3}-\tau-\mathrm{FID}_{R_\phi}}^{\text{spin-echo}}-$$

$$\overbrace{-\theta^\circ_{\phi_1}-\tau-180^\circ_{\phi_3}-\tau-\mathrm{FID}_{R_\phi}}^{\text{spin-echo}}$$

ϕ_1:	x	x	x	x	$-x$	$-x$	$-x$	$-x$	y	y	y	y	$-y$	$-y$	$-y$	$-y$
ϕ_2:	$-x$	$-x$	$-x$	$-x$	x	x	x	x	$-y$	$-y$	$-y$	$-y$	y	y	y	y
ϕ_3:	y	y	x	x	y	y	x	x	$-x$	x	$-y$	y	$-x$	x	$-y$	y
R_ϕ:	R_y	R_y	R_{-y}	R_{-y}	R_{-y}	R_{-y}	R_y	R_y	R_{-x}	R_{-x}	R_y	R_y	R_x	R_x	R_{-x}	R_{-x}

where θ°_Φ is a particular flip angle (usually 90°) with transmitter phase Φ, and R_Φ refers to the phase of the receiver (16 step phase cycling). Assuming that the acoustic response of each individual pulse is a linear phenomenon and independent of any preceding or following pulses, the distortion resulting from the $(\theta^\circ)_{\Phi1,\Phi2}$ pulse is cancelled out after the completion of a sixteen pulse transmitter and receiver cycle. Similarly, the acoustic distortion of the $180^\circ_{\Phi3}$ is cancelled for every x,-x or y, -y receiver phase sequence. Thus a theoretical, overall cancellation of the baseline distortion is achieved.

It has been demonstrated that pulse sequences demanding a series of 90° and 180° pulses can be applied only to a limited spectral window because of the finite rf power and the low γ values. This is a serious disadvantage when recording resonances with large frequency separations. In this case it is necessary to record the spectra in several steps with the carrier frequency on the various resonances. An alternative way is to utilize composite pulses. These are clusters of closely spaced short r.f. pulses which achieve the same nominal flip angle as a conventional single pulse but are less sensitive to resonance offset effects. It is anticipated, however, that the relaxation during the pulse (a very serious problem whenever recording very broad resonances) cannot be avoided.

BASIC PRINCIPLES

If a molecule or ion containing a magnetic nucleus is rapidly exchanging between states, "free" in solution and bound to a macromolecule, the observed chemical shift δ_{obs}, is given by:

$$\delta_{obs} = P_F \delta_F + P_B \delta_B \tag{1}$$

where P_F and P_B are the mole fractions of free and bound nuclei with the chemical shifts δ_F and δ_B, respectively. In many cases a similar relation is valid for the relaxation rates:

$$R_{(1,2)obs} = P_F R_{(1,2)F} + P_B R_{(1,2)B} \tag{2}$$

Whereas Eq. (2) is generally valid for R_1 under situations of fast chemical exchange (which may be defined by the condition $\tau_B^{-1} > R_{(1,2)B}$, where τ_B is the mean lifetime of the bound ion and $R_{(1,2)B}$ is the relaxation rate at the binding site), it is strictly valid for R_2 only when the difference in chemical shift between the free and the bound state, $\delta\omega_{FB}$, is less than $R_{2,B}$. For more general expressions see Dwek (1973). When two or more binding sites are present, additional terms must be added on the right-hand side of Eqs. (1) and (2).

For quadrupolar nuclei changes in relaxation rates upon binding are often more pronounced than changes in chemical shifts. Quadrupolar relaxation is caused by the interaction of a nuclear electric quadrupole moment, eQ, with fluctuating electric field gradients, *eq*, at the position of the nucleus. Electric field gradients are system-dependent properties and may be regarded as either intramolecular or intermolecular in origin. In isotropic systems, when the conditions of extreme narrowing are fulfilled and in the absence of chemical exchange, the T_1 and T_2 relaxation times are equal and are given by:

$$\frac{1}{T_1} = \frac{1}{T_2} = \frac{3\pi^2}{10} \frac{2I+3}{I^2(2I-1)} (1 + \frac{\eta^2}{3}) (\chi)^2 f(\omega,D) \tag{3}$$

where χ is the nuclear quadrupolar coupling constant. An appropriate reference axis system can be chosen to reduce this interaction to two values:

$$\chi = e^2q_{zz}Q/h \text{ (the main component)} \tag{4}$$

and the asymmetry parameter (dimensionless):

$$\eta = (e^2q_{xx}Q - e^2q_{yy}Q)/e^2q_{zz}Q \tag{5}$$

where

$$|e^2q_{zz}Q| \geq |e^2q_{yy}Q| \geq |e^2q_{xx}Q|. \tag{6}$$

The asymmetry parameter η varies from 0 to 1 and describes the deviation of the electric field gradient from axial symmetry ($\eta = 0$ for axial symmetry). Because even in the rare instance that $\eta = 1$ the term $1 + \eta^2/3$ is only 1.33, the asymmetry parameter can usually be neglected. $f(\omega,D)$ is the correlation function depending on the rotational diffusion constant D and the relative orientation of the field gradient components with respect to the diffusion constants. When isotropic reorientation is assumed, $f(\omega,D)$ reduces to the single correlation time τ_c. If the resonance absorption is a Lorentzian lineshape the measurement of the linewidth at half-height, $\Delta\nu_{1/2}$, is a direct measure of the relaxation time:

$$\Delta\nu_{1/2} = \frac{1}{\pi T_2} \tag{7}$$

provided that unresolved indirect spin coupling is not present and chemical exchange phenomena do not take place.

In biological systems the extreme narrowing condition is usually encountered only in the case of "free" ligands in solution or in the case of low molecular weight complexes. More common is the situation when $\omega^2_o\tau^2_c > 1$ which will be discussed below.

(A) Integer Spins

In this case, the longitudinal and transverse components are different and each is the sum of I decaying exponential terms. For I=1 the expression for the longitudinal and transverse relaxation rates, $1/T_1$ and $1/T_2$, are given by Abragam (1961) as:

$$\frac{1}{T_1} = \frac{3\pi^2}{10} \frac{2I+3}{I^2(2I-1)} \left(1+\frac{\eta^2}{3}\right) (\chi)^2 \{ J(\omega_o) + 4J(2\omega_o) \} \tag{8}$$

$$\frac{1}{T_2} = \frac{3\pi^2}{10} \frac{2I+3}{I^2(2I-1)} \left(1+\frac{\eta^2}{3}\right) (\chi)^2 \{ 9J(0) + 15J(\omega_o) + 6J(2\omega_o) \} \tag{9}$$

where $J(\omega)$ is the Fourier transform of the correlation time describing the motion

$$J(\omega) = \frac{2\tau_c}{1+\omega_o^2\tau_c^2} \tag{10}$$

For I=2, the longitudical and transverse relaxations are the sums of two exponentials. Since there are no stable nuclei with spin of two, the results of this case are of no importance.

(B) Half-integer Spins

(a) Relaxation properties.

In this case the decay of the longitudinal (M_1) and transverse (M_2)

magnetizations is no longer exponential but may be written as a weighted sum of $I+½$ exponentials (see Hubbard 1970):

$$M_1(t) = M_1(\infty)\left\{1 - k\left[\sum_{i=1}^{I+½} C_{1,i} \exp(-R_{1,i}\, t)\right]\right\}, \tag{11}$$

$$M_2(t) = M_2(0)\left[\sum_{i=1}^{I+½} C_{2i} \exp(-R_{2,1}\, t)\right], \tag{12}$$

where $\sum_{i=1}^{I+½} C_{ji} = 1$ and k =2 for an inversion recovery experiment. As will be discussed below (cf. 'second-order dynamic frequency shifts') the resonance frequencies of the various components will not in general be the same. Furthermore, although equation (11) is valid for the total magnetization, the decays of the individual components will normally not be exponential and will also depend on the length of the monitoring pulse (Werbelow and Marshall 1981).

Eqns (11) and (12) cannot be solved analytically but numerically for each $\omega_o\tau_c$ value. Such numerical calculations have been performed by Bull et al (1979). Fig. 1 illustrates the longitudinal and transverse relaxation of spin I=5/2 nuclei in the absence of chemical exchange. The longitudinal relaxation may be regarded as a single exponential, within the limits of experimental error, for all correlation times. On the contrary, the transverse relaxation is governed by three contributions and the weights of the different contributions have similar magnitudes for the longer correlation times. For values of $\omega_o^2\tau_c^2 << 1$ (extreme narrowing condition) the transverse relaxation is approximately a single exponential. In the slow motion limit the rapid decay of the component 3 at longer correlation times may allow analysis in terms of just two exponentials.

(b) Second-order dynamic frequency shifts

In addition to non-exponential relaxation for quadrupolar nuclei under conditions of non-extreme narrowing the NMR signal will also be modified by shift changes that have been termed 'second-order dynamic frequency shifts'. Although long recognized theoretically (cf. Abragam 1961) the experimental demonstration of

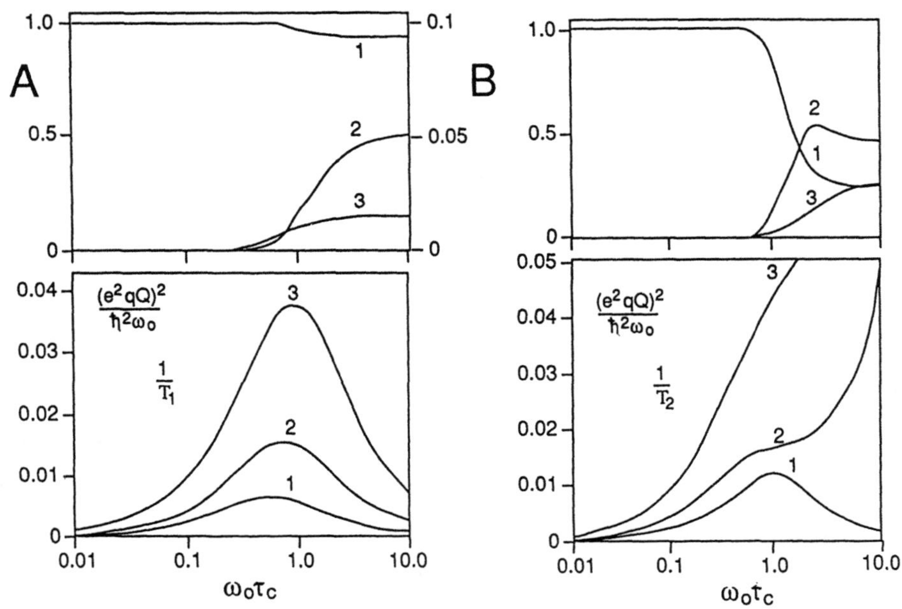

Fig. 1 Longitudinal relaxation (A) and transverse relaxation (B) of spin 5/2 nuclei in the absence of chemical exchange. The upper plot shows the normalized amplitudes $C_{1,i}$ and $C_{2,i}$ of the three exponential components as a function of $\omega_o\tau_c$ (the coefficients 2 and 3 are plotted on the expanded scale given to the right). The lower plot shows their relaxation rates as a function of the same parameter (Bull et al 1979).

such a phenomenon is not straightforward since the symmetry of the composite peak is in practice difficult to distinguish from a small phase misadjustment and thus is likely overlooked (Fig. 2). Furthermore, strongly relaxed transitions may decay so quickly as to be undetected if the instrumental deadtime is too long. The effects of second-order frequency shifts for $I={}^3/_2$ nuclei has been discussed by Werbelow and coworkers (Werbelow and Marshall 1981; Werbelow and Pouzard 1981). The magnitude of the second order effects are dependent on the ratio χ^2/ω_o and will thus be particularly important in cases of large quadrupole coupling constants and low observing frequency. Westlund and Wennerström (1982) have derived expressions for the NMR band shape for $I = {}^5/_2$ and $I = {}^7/_2$ nuclei both in the presence and absence of chemical exchange. The results of this work are schematically summarized in Fig. 3. In the limit $\omega_o\tau_c \gg 1$, the $m = ½ \rightarrow m = -½$ component will dominate the spectrum because it is much narrower than the other components. The intensity of this component relative to the total intensity will be 40% for $I={}^3/_2$, 26% for $I = {}^5/_2$ and 19% for $I = {}^7/_2$. This narrow signal will be shifted towards lower frequency with a value proportional to χ^2/ω_o. Between these two limits the resonance will be more or less asymmetric with an apparent shift either to high or low frequencies depending on the values of ω_o, τ_c and χ. The ratio of the dynamic shift to line width, $\Delta\omega_d/\Delta\omega_{½}$, will in this region depend only on ω_o and τ_c.

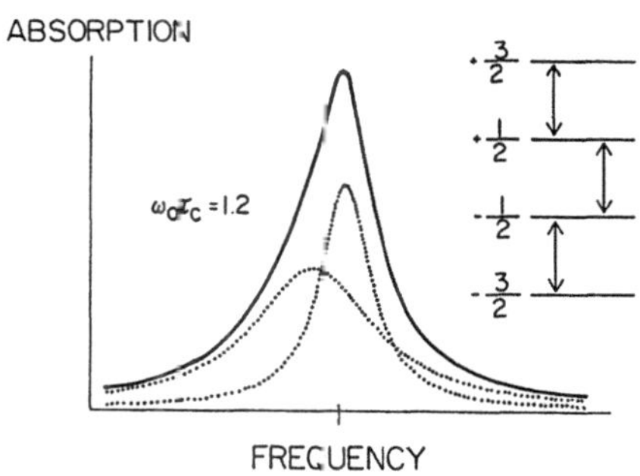

Fig. 2 Energy-level diagram (right) and single-quantum NMR spectrum (left) for a spin ${}^3/_2$ nucleus with rotational correlation time $\tau_c = 1.2/\omega_o \simeq 15$ ns for ^{23}Na at 7.0 Telsa. The narrow component line arises from the -½ to −½ transition, and the broad component from the +½ to $+{}^3/_2$ and $-{}^3/_2$ to -½ transitions. Note the distinct chemical shift difference between the broad and narrow transitions (Marshall et al 1982).

(c) Effects of relaxation mechanisms other than quadrupolar

From Fig. 1 it is evident that the linewidth of the narrow component decreases as τ_c increases in the slow-motion limit, while those of the intermediate and broad components increase. However, Lerner and Torchia (1986) have shown that the linewidth of the narrow component does not necessarily decreases as the τ_c increases if the chemical shift anisotropy (CSA) is large. Oldfield et al (1989) observed such behaviour from temperature and magnetic field strength dependence studies of the

^{17}O NMR spectra of the $MbC^{17}O$ samples. This suggests that the CSA contribution to the linewidth is not negligible.

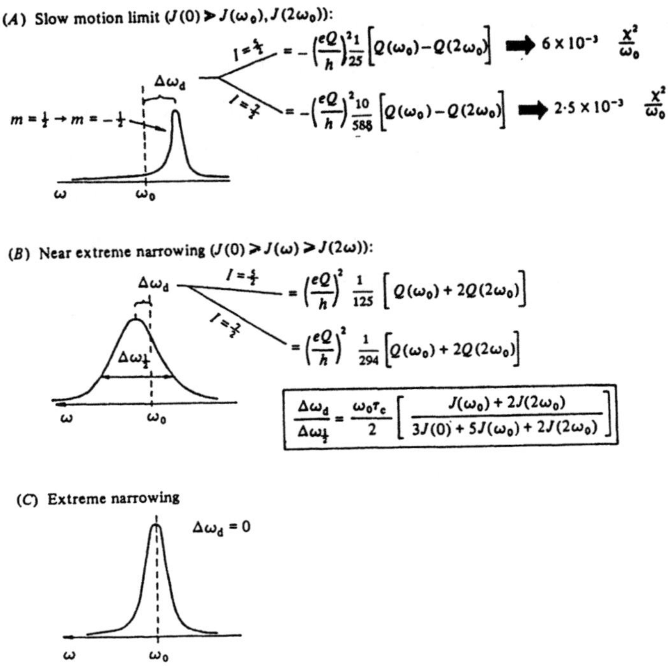

Fig. 3 Some general characteristics of dynamic frequency shifts for I = $^5/_2$ and I = $^7/_2$ nuclei (Forsen et al 1987).

The relaxation rates for the CSA mechanism outside the extreme narrowing condition are given by:

$$\frac{1}{T_1} = \frac{1}{15}\,\gamma^2 B_o^2\,(\sigma_{\parallel} - \sigma_{\perp})^2 \left\{ \frac{2\tau_c}{1+\omega_o^2\tau_c^2} \right\} \tag{13}$$

$$\frac{1}{T_2} = \frac{1}{90}\,\gamma^2 B_o^2\,(\sigma_{\parallel} - \sigma_{\perp})^2 \left\{ \frac{6\tau_c}{1+\omega_o^2\tau_c^2} + 8\tau_c \right\} \tag{14}$$

where the symbols have their usual meaning. For $\tau_c \simeq 14$ ns and $|\sigma_{\|} - \sigma_{\perp}| = 800$ ppm (which is the value for the CO-picket fence porphyrin obtained from NMR experiments in the solid state) a CSA contribution to the linewidth ~ 100 Hz was obtained for the narrow components. This is comparable to the pure quadrupolar contribution to the linewidth (~ 96 Hz). The CSA contribution to T_1 was found to be negligible (~ 170 ms).

SELECTED APPLICATIONS

(A) Metal Ion Binding to Proteins

NMR studies of the binding of Ca^{2+} to parvalbumin, troponin C and calmodulin are representative examples illustrating the procedures, problems and interpretations of the experiments (Forsen et al 1987). In all three cases only one broad resonance is observed for the bound Ca^{2+} though there are two strong sites with slow calcium exchange in each protein. This is due to the small shift between free and bound $^{43}Ca^{2+}$ ions (ca. 10 ppm) compared to the broad resonances from the bound calcium ($\Delta\nu_{1/2}$ ~ 500-800 Hz i.e. 30 to 50 ppm for the magnetic field used). This demonstrates that in this case the relaxation rates are more useful parameters than the chemical shifts. The relaxation rates R_1 and R_2 are not equal (the extreme narrowing condition is not valid) therefore R_1 and R_2 can be used to calculate both the correlation time and the nuclear quadruple coupling constant. Interestingly χ ~ 1.1 MHz and is about twice as large as in the Ca-EDTA complex which has a much lower symmetry.

Additional information concerning the binding of Ca^{2+} and Mg^{2+} to calmodulin, CaM, have been obtained by the use of ^{25}Mg NMR in titration and temperature dependence experiments. The affinity of Mg^{2+} to CaM is much weaker than that of Ca^{2+}, therefore, it is not possible to observe fully bound Mg^{2+}. In the case of slow exchange, increasing temperature will increase the exchange rate (i.e. increase $1/\tau_{ex}$) and cause an increase in the observed $\Delta\nu_{1/2}$ if the increase due to exchange rate is large enough to outweigh the decrease in the intrinsic linewidth (due to increased molecular motion). In the case of fast exchange, the intrinsic linewidths of both free and bound Mg^{2+} decrease due to decreasing τ_c, which will result in a decrease in the observed $\Delta\nu_{1/2}$ upon increasing temperature. It was demonstrated that the $Mg^{2+}/CaM{\cdot}Ca_2^{2+}$ system is in slow exchange at 25° and in the intermediate region at 30-50°C. When such a complete set of data can be obtained, the data can be analyzed by "total line shape analysis" which yielded the following information: K_a = 2000 M^{-1} (assuming two equivalent sites), k_{off} ($\approx k_{ex}$) = 2300 sec^{-1} and $\Delta\nu_{1/2}$ = 3.5 kHz for bound Mg^{2+}. In this case, τ_c and χ cannot be obtained simultaneously. However, if τ_c of Mg^{2+} is assumed to be the same to that of Ca^{2+} when bound to CaM, χ = 1.6 MHz can be obtained. The result unequivocally establishes that Mg^{2+} binds to sites III and IV of CaM, as described in detail by Tsai et al (1987). Complete line shape analysis by the above method is not possible when the quantity of the sample is limited or the protein is sensitive to temperature changes. Further, even if the protein is stable enough for a variable temperature study, the system may remain in slow (or fast) exchange throughout the entire temperature range.

(B) Halide Ion Binding to Proteins

A number of parameters may be obtained from the halide ion NMR studies (Drakenberg and Forsen 1983):

(i) stoichiometry and binding constants of halide ions;

(ii) stoichiometry and accessibility of metal reactive groups on the protein;

(iii) rotation correlations time(s) of the macromolecule;

(iv) quadrupole coupling constant(s) of the halide ion(s) in different macromolecular binding site(s);

(v) rates and activation parameters of halide ion exchange, and

(vi) binding constants and stoichiometry of other ions or molecules through competition experiments.

Haemoglobin and Cl^- transport proteins have been investigated in detail using both ^{35}Cl and ^{37}Cl isotopes. In fact, the $^{35}Cl/^{37}Cl$ line broadening ratio can be used to characterize a site as slowly or rapidly exchanging on the appropriate NMR time scale, and once the characterization has been made the maximum information on chloride exchange rates can be extracted from the line broadening. For the slowly exchanging site, the line broadening specifies the chloride binding and dissociation rate constants and the chloride dissociation constant. For a site in the rapid exchange regime lower limits can be placed on these rate constants.

(C) Substrate Binding to Enzymes

(a) Internal motions of bound substrates

It is widely accepted that enzyme-bound substrates possess a certain extent of internal rotational freedom relative to the enzyme. However, it is difficult to describe the dynamics of bound substrates quantitatively because a single measurement can yield only the effective τ_c, which may include contributions from several motional components. Further, relaxation studies are often complicated by exchange processes as described previously herein. A representative study of this category has been a ^2H NMR investigation of the interaction of deuterated substrates and inhibitors in the presence of the enzyme lipoxygenase-1 which is non-haem iron protein of MW ~ 102 kDa (Viswanathan and Cushley 1981). The absence of a single value for all deuterons indicates the presence of internal motions in the enzyme-bound inhibitors. Any internal motions, however, should be highly restricted since a linewidth of 370 Hz for the bound [12,12,12-^2H] dodecanoic acid was observed relative to the free acid line width of 17 Hz. Furthermore, the degree of mobility increases as the distance of the methylene (or methyl) group from the carboxyl head group increases. The relaxation rates of the deuterons at the double bonds and the doubly allylic position C-H of the substrate are 26% to 29% slower than that of the α-position C-2 of inhibitor dodecanoic acid and considerably faster than the average C^2H_2 position further along the inhibitor acyl chain. It was concluded that while the olefinic region of the substrate is more tightly bound to the enzyme than the same region of the chain in the inhibitor, there is still significant internal motion.

(b) Detection of multiple conformations

External ligands occupy the 6th coordination position of the haem iron in many haemoproteins. Carbon monoxide is one such external ligand and has been extensively used in many spectroscopic studies. Recently, analysis of both ^{17}O relaxation and chemical shift data of $C^{17}O$-ligated haemoproteins has provided information on both the structural and dynamic properties of the haem group (Lee et al 1988, 1989). Peroxidases are haem proteins with MW ~ 40-42 kDa that catalyze oxidations by hydrogen peroxide. They contain iron-protoporphyrin IX as a noncovalently bound prosthetic group. The ^{17}O NMR spectra of CO-horseradish peroxidase isozyme C (Fig.4) indicate a single haemoprotein $C^{17}O$ signal (at 358.3 ppm) below pH 10 and two separate signals at pH 10.5. The new signal, which was assigned to the alkaline form of CO-horseradish peroxidase isozyme C, is broader and approximately 7 ppm to high frequency relative to that of the acidic form. A similar pH-dependent relationship of the chemical shift and linewidth was observed for CO-horseradish peroxidase isozyme A and CO-horseradish peroxidase isozyme C. Furthermore, the new signal decreases in intensity upon lowering the pH, while the chemical shift remains constant. It was concluded that CO-horseradish peroxidase isozyme C undergoes a transition between the acidic and alkaline forms. However, contrary to the case of CO-horseradish peroxidase isozyme A, the transition is slow on the NMR time scale. The pK_{CO} value was estimated to be ~ 10.7. Computer simulation of the pH 10.5 CO-horseradish peroxidase isozyme C spectrum indicates that both signals are broadened by the acid-base transition. In particular, the resonance linewidth of the acidic form at pH 10.5 is about 60 Hz greater than that of the pH 4.5 or 7 which presents the purely acidic form of the enzyme. Assuming that a slow exchange limit is valid the exchange rate between the acidic and alkaline forms of CO-horseradish peroxidase isozyme was estimated to be ~ 1.88×10^2 s^{-1}.

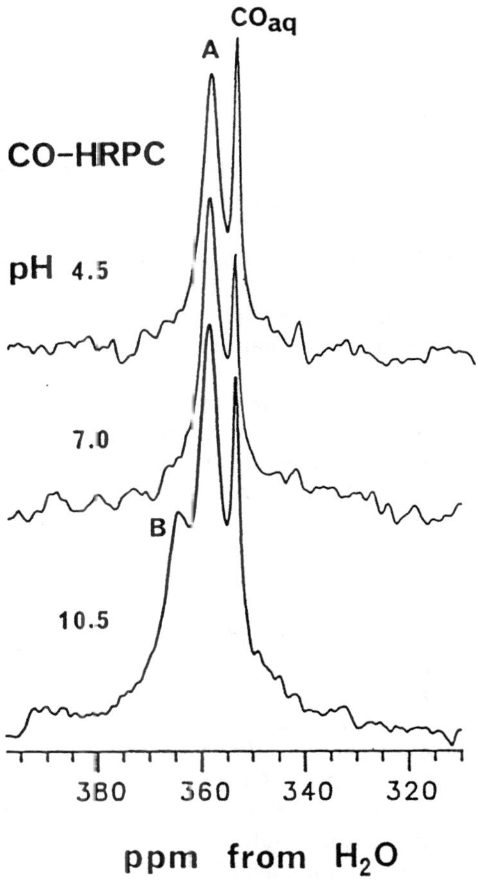

Fig.4 Oxygen-17 NMR spectra of CO-horseradish peroxidase isozyme C (CO-HRPC), concentration 1 mM in 50 mM phosphate buffer at 11.7 Tesla, number of scans 50000-250000, recycle time 100-150 ms, pulse width 38 μs, line broadening 100 Hz. A, B and CO_{aq} denote the acidic form of CO-HRPC, the alkaline form of CO-HRPC and free CO in solution, respectively (Lee et al 1988).

ACKNOWLEDGEMENTS

Financial support from the Research Committee of the University of Ioannina and the General Secretariat of Research and Technology is gratefully acknowledged. I thank L. Dunphy, Medical Research Institute, Mill Hill, London, for typing and preparing the manuscript.

REFERENCES

Abragam A (1961) The principles of nuclear magnetism. Oxford University Press, London

Bull TE, Forsen S and Turner DL (1979) Nuclear magnetic relaxation of spin $^5/_2$ and spin $^7/_2$ nuclei including the effects of chemical exchange. J Chem Phys 70: 3106-3111

Drakenberg T and Forsen S (1983) The halogens - Chloride, bromide, and iodine. In: Lambert JB and Riddell FG (eds). The multinuclear approach to NMR spectroscopy. Reidel Dordrecht: 405-444

Dwek RA (1973) Nuclear magnetic resonance in biochemistry. Oxford Univ. Press (Clarendon), London

Forsen S, Drakenberg T and Wennerstrom H (1987) NMR studies of ion binding in biological systems. Quart Rev Biophys 19: 83-114

Gerothanassis IP (1987) Methods of avoiding the effects of acoustic ringing in pulsed Fourier transform nuclear magnetic resonance spectroscopy. Progr NMR Spectrosc 19: 267-329

Harris RK (1983) Nuclear Magnetic Resonance Spectroscopy. Pitman, London

Hubbard PS (1970) Nonexponential nuclear magnetic relaxation by quadrupolar interactions. J Chem Phys 53: 985-987

Lee HC, Cummings K, Hall K, Hager LP and Oldfield E (1988) Oxygen-17 nuclear magnetic resonance spectroscopic studies of carbonmonoxyperoxidases. J Biol Chem 263: 16118-16124

Lerner L and Torchia DA (1986) An analysis of non-Lorentzian ^{23}Na line shapes in two model systems. J Am Chem Soc 108: 4264-4268

Marshall AG, Wang T-CL, Cottrell CE and Werbelow LG (1982) First experimental demonstration of NMR dynamic frequency shifts : dispersion vs absorption (DISPA) line shape analysis of sodium-23 in aqueous sodium laurate/lauric acid solution. J Am Chem Soc 104: 7665-7666

Sanders II CR and Tsai M-D (1989) Ligand-protein interactions via nuclear magnetic resonance of quadrupolar nuclei. Methods in Enzymology 117: 317-333

Tsai M-D, Drakenberg T, Thulin E and Forsen S (1987) Is the binding of magnesium (II) to calmodulin significant? An investigation by magnesium-25 nuclear magnetic resonance. Biochemistry 26: 3635-3643

Viswanathan TS and Cushley RJ (1981) Deuterium nuclear magnetic resonance study of the interaction of substrates and inhibitors with soybean lipoxygenase. J Biol Chem 256: 7155-7160

Werbelow LG and Marshall AG (1981) The NMR of spin - $^3/_2$ nuclei: the effect of second-order dynamic frequency shifts. J Magn Reson 4: 443-448

Werbelow LG and Pouzard G (1981) Quadrupolar relaxation. The multiquantum coherences. J Phys Chem 85: 3887-3891

Westlund PO and Wennerstrom H (1982) NHR lineshapes of $I=^5/_2$ and $I=^7/_2$ nuclei. Chemical exchange effects and dynamic shifts. J Magn Reson 50: 451-466

Multidimensional NMR Spectroscopy of Peptides

Horst Kessler, Matthias Eberstadt, Wolfgang Schmitt
Technische Universität München
Institute of Organic Chemistry and Biochemistry
Lichtenbergstr. 4, D - 85747 Garching
Germany

Peptides and proteins are of extreme importance in biological systems as carriers of function (enzymes, receptors, inhibitors, transport systems) and regulation of informations (hormones, neurotransmitters, and many others). With exception of some antibiotics these linear biopolymers are ribosomally synthesized and later on specifically cleaved to obtain the biologically active specific peptide/protein. For the understanding of their function knowledge of the spatial structure and molecular mobility is required. A key step in the function of peptides and proteins is the formation of multimeric complexes based on molecular recognition. The 3D structure of these complexes is of great interest. In unique cases this information can be obtained by NMR spectroscopy. In the following we will treat specifically problems that occur with oligopeptides.

On first glance, peptides seem to be more easily to study than proteins, because they are smaller and contain less amino acid residues. However, there are special problems which make NMR of peptides very tricky.

1. **Flexibility**. Small to medium sized peptides are rather flexible molecules. For linear peptides up to a size of 15-25 amino acids there exists no strongly prefered conformation (Kessler 1982). However, there is often a beginning of folding, that shifts the mean NMR parameters to become closer to these typical for specific secondary structures (turn, helix, ß-sheet) (Dyson et al 1988).

NATO ASI Series, Vol. H 87
NMR of Biological Macromolecules
Edited by C. I. Stassinopoulou

2. **Large surface to core ratio**. Mainly 3D-structure determination is based on distances between protons within the molecule (Wüthrich 1986). In very large proteins many of the nuclei are situated in the core of the protein. Hence, the position of such buried protons are well defined by a number of distances. On the other hand protons at the surface have molecular neighbors only on one side, the other being exposed to the solvent. In fact, in small molecules like small peptides all protons may be more or less situated at the surface and the number of structure determinating distances is strongly decreased. Often it might occur in these small molecules that the system is underdetermined, especially if the possibility of multiple conformations in fast equilibrium is taken into account. In the literature it is often found that a conformation is presented based on an insufficient amount of experimental data. Obviously the "structures" derived under these circumstances are far from reality, and most of the time have no relation what so ever to the *bioactive conformation*.

   Another problem, associated with the large surface/core ratio is the sensitivity of conformations to the surrounding environment (crystal, solvent, vacuum, membrane, complexation with ions or receptors) (Mierke et al 1993). Often the conformation or the population between families of distinct conformations changes dramatically when the solvent is changed. A typical example is found in cyclosporin, where the backbone conformation is very similar in lipophilic media ($CDCl_3$ (Kessler et al 1990a), benzene (Kessler et al 1990a), THF (Kessler et al 1990b) and in the crystal (Loosli et al 1985), but changes dramatically when LiCl is added to the THF solution (Köck et al 1992) and when it is bound to its natural receptor cyclophilin (Weber et al 1991; Fesik et al 1991). In polar solvents (MeOH (Ko and Dalvit 1992), DMSO (Kessler et al 1985a), H_2O) four to seven slowly exchanging conformations are populated in comparable amounts. Of course, NMR spectroscopy in the lipophilic medium identifies already several different *side chain conformations* for some of the residues in fast equilibrium (Kessler et al 1990a), whereas in the crystal one of these conformations is frozen out.

3. **Required or expected quality of a structure**. In proteins, at least in large proteins, where solution or crystal conformation is usually the

bioactive conformation, one might be already satisfied when secondary structure elements are identified and a rough folding model often is a great success in understanding the protein function. On the other hand only the identification of a ß-turn in a small peptide is not what we expect as a *reliable* structure. It is normally required to define the type of the turn (ßI, ßI', ßII ...) (Rose et al 1985) and to determine the rotamer population of side chains as well as to have information of specific backbone mobility (Kessler et al 1988) etc.

Hence, the following facts and topics have to be considered when attempting 3D structure determination of peptides by NMR.

Cyclization and "rigidity"

One has to use cyclic peptides, especially small cyclic or bicyclic peptides if one wants to correlate solution conformations to the bioactive conformation (Kessler 1982). As pointed out above, the most stable conformation of a linear peptide in solution can change dramatically in the binding process. Reduced flexibility ("rigidity") of a cyclic peptide on the other hand, is indicated by the following criteria:

- large chemical shift dispersion of NH-protons and of the diastereotopic protons of glycine or side chains (preferred side chain conformations in general parallels prefered backbone conformation)
- strong differentiation in NH temperature gradients
- chemical shifts and coupling constants far apart from random coil values
- small rms deviations in the molecular dynamics refined structure and relative to the observed NMR parameters to be sure that the system is well defined by NOEs and Js
- the resulting conformation should be independent on the starting structure of the calculation
- establishing the same structure in several different solvents (if solubility allows, one should determine the conformation in a very lipophilic solvent ($CDCl_3$) and in water or DMSO).

If one has obtained such a cyclic peptide then normally an increase in biological activity is observed upon cyclization accompanied by high selectivity (in activating different receptor subtypes) and good resistance to enzymatic degradation in vivo.

Since a "structure" is normally represented as a mean over a molecular dynamics run which fulfill all experimental parameters it is important to use many different parameters (NOE, J) which average differently (Mierke et al 1994).

Hence, inclusion of homo- and heteronuclear J-coupling constants can improve the reliability of the "structure". (Mierke and Kessler 1992, Eberstadt et al 1992). As mentioned above, the number of NOE restraints may not be large enough to find the global minimum. The use of NOE for structure determination has been established in numerous publications in the last decade. Here we will emphasize J-coupling constants which have a renaissance because new ways have been found recently to determine coupling constants and to use them in calculations.

Determination of J-coupling constants

For reasons of digital resolution it is advisable to use the lowest dimension for the extraction of coupling constants. However, with increasing size of the molecule and with increasing contribution of different couplings to specific signals in the NMR spectrum problems occur due to overlap of multiplet patterns which forces one to use 2D and 3D techniques. There are now several new methods to determine coupling constants in 2D and 3D NMR spectroscopy.

Four different principles to extract coupling constants from multidimensional NMR spectra can be applied: a) measurement of cross-peak intensities in correlated spectra, b) line shape fitting of cross peak multiplet patterns, c) combination of multiplets of one nucleus with different in-phase/antiphase patterns and d) analyzing of so called E.COSY patterns.

a) Coherence transfer in J-correlated spectra (INEPT or DEPT transfer) results in antiphase patterns of the active coupling in each individual cross peak. The intensities of the cross peaks depend on the size of the J-coupling constants of the involved nuclei. For example, in the COLOC-experiment (Fig. 1), when applied to carbonyl carbons, the cross peak intensity depends on the following transfer function:

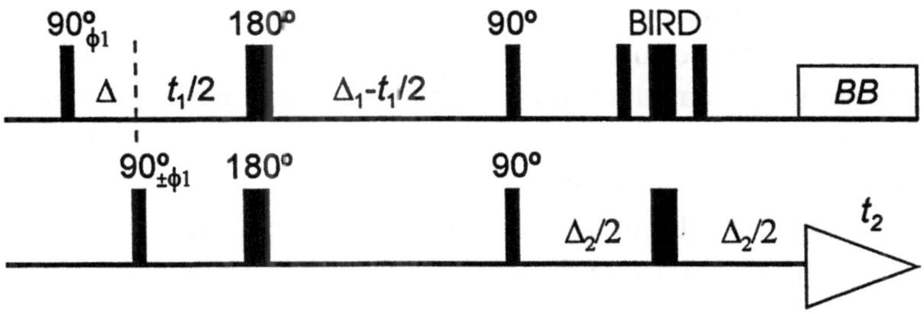

Fig. 1: *CT-Correlation via long-range-coupling (COLOC) with low-pass J-filter and BIRD decoupling to suppress signals via direct coupling.*

$$I \sim \sin\pi{}^{n}J_{CH}\Delta_1 \Pi\cos\pi{}^{n}J_{H,H}\Delta_1 \sin\pi{}^{n}J_{CH}\Delta_2 \Pi\cos\pi{}^{n}J'_{CH} \exp(-\Delta_1/T_{2H})\exp(-\Delta_2/T_{2C})$$

(J'_{CH} = other protons coupled to ^{13}C, Π is the product over all relevant couplings)

If the delays Δ_1 and Δ_2 are small in the first approximation only the sine contributions determine the cross peak intensities. Therefore the relative size of the coupling constant can be measured by integration of the crosspeaks.

$$I_{apparent} \sim \sin\pi\,{}^{n}J_{CH}\,\Delta_1 \cdot \sin\pi\,{}^{n}J_{CH}\,\Delta_2$$

This method has been used to assign diastereotopic ß-protons via qualitative determination of $^3J_{CO,H\beta}$ and $^3J_{N,H\beta}$ (Kessler et al 1987). Together with exact values of homonuclear proton coupling constants between H_α and H_β an unequivocal assignment of $H_\beta{}^{proR}$ and $H_\beta{}^{proS}$ is possible. This information is needed for the calculation of rotamer populations about χ_1, and also to vastly improve the quality of structures via restrained MD calculations. The signal intensities of cross peaks in inverse detected experiments where no refocussing is used, such as in the HMBC, depend only on a single sine contribution. Hence, differences in cross peak intensities are less distinct.

Recently a number of new techniques have been developed that use the same principle as explained above and could be applied to isotopically labeled proteins (Billeter et al 1992, Bax et al 1992, Vuister et al 1993).

b) A direct extraction of J-coupling constants in AX-spin systems from the splitting of antiphase components in COSY spectra is difficult when the J coupling is not large compared to the signal line width. Due to the antiphase pattern of the cross peaks intensity cancellation occurs in the overlapping region and the peak maxima are shifted apart. Hence, the apparent coupling constant is too large (Neuhaus et al 1985). Line shape analysis, or a simplified procedure (Kim and Prestegard 1989) has to be used. The latter method uses a trace from the cross peak phased in two different ways: one to pure absorption and the other to pure dispersion (Fig. 2). The related coupling constant can easily be calculated from the two distances of maxima ν_d and ν_a in Fig. 2.

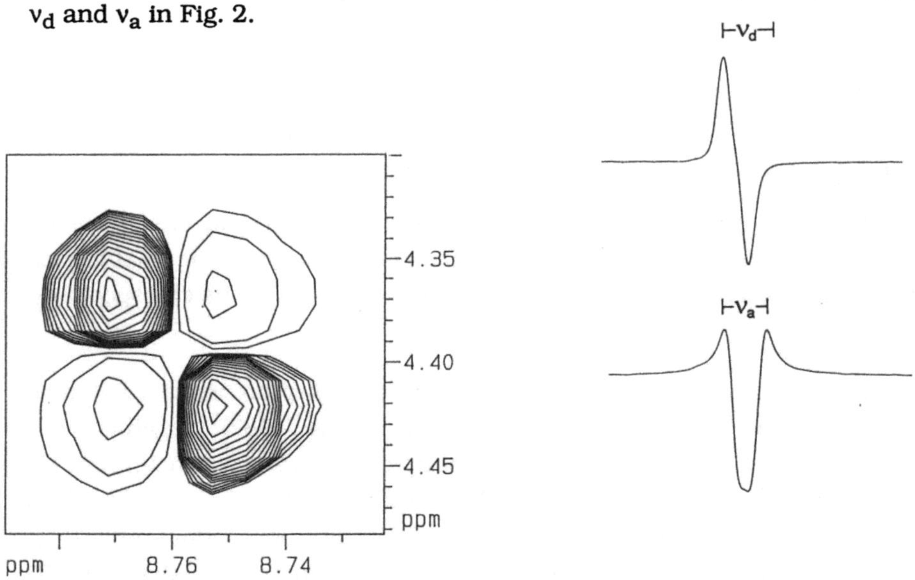

Fig. 2: *Determination of the NH-H$_\alpha$ coupling constant from a COSY cross peak*

c) Each cross peak to a single nucleus in a certain dimension of a COSY spectrum is characterized by the anti-phase pattern of the active coupling whereas all other passive couplings appear in-phase. Hence, a positive (or negative) combination of different cross peak multiplets results in half the number of lines with the sum (or difference) of both active coupling constants in anti-phase (DISCO technique) (Kessler et al 1985b, Kessler and Oschkinat 1985). As the intensity of different cross peaks vary they have to be adjusted accordingly in the DISCO-procedure.

A special case of simplification can be achieved by a combination of cross-peak patterns with pure in-phase signals of the same nucleus. Then the active coupling disappears and a decoupled spectrum results which is shifted by πJ to lower or higher field depending on a positive or negative combination. If the signal of the nucleus in the 1D-spectrum is overlapped by other signals, a z-filtered TOCSY spectrum (pulse sequence, see Fig. 3)

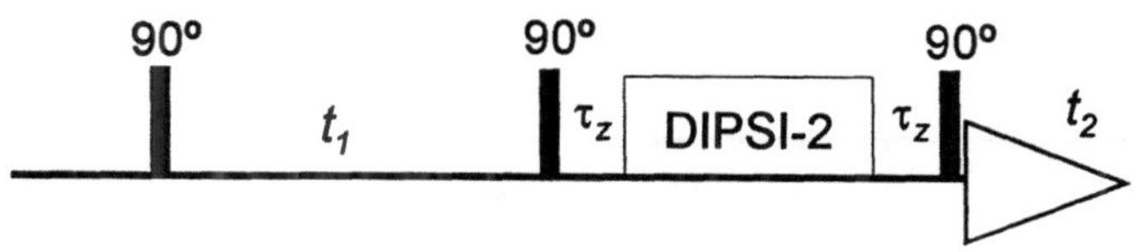

Fig. 3: *z-Filtered TOCSY to obtain pure in-phase signals in overcrowded proton spectra*

which yields pure in-phase patterns of the cross-peaks can be applied for DISCO. Keeler et al. have written a computer program to match intensities and extract the J-coupling constants (Titman et al 1990).

This procedure is especially important for the determination of heteronuclear long range coupling constants to heteronuclei that have no proton attached. A slightly modified HMBC experiment (Fig. 4) is combined with a homo-

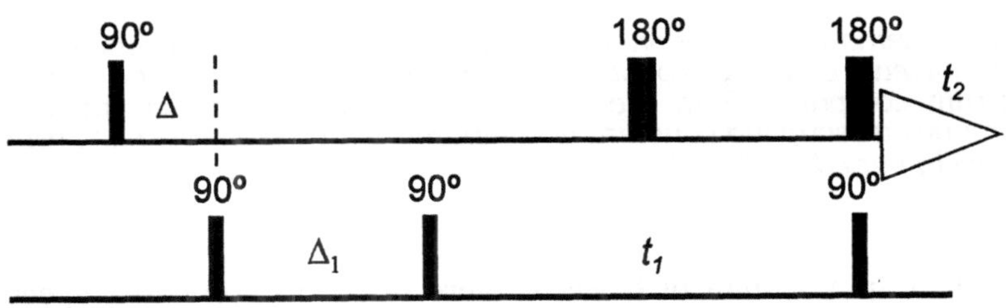

Fig. 4: *Pulse sequence of the modified HMBC experiment with low-pass J-filter*

nuclear proton spectrum which yields signals in pure absorption mode (for instance a z-TOCSY (Fig. 3)). The homonuclear cross peak contains all homonuclear proton couplings, whereas the referring cross peak in the HMBC is additionally modulated by the heteronuclear long range coupling. Using an iterative procedure the desired heteronuclear coupling constant can be calculated by comparing the simulated multiplet to that of the HMBC (Titman et al 1989) (Fig. 5).

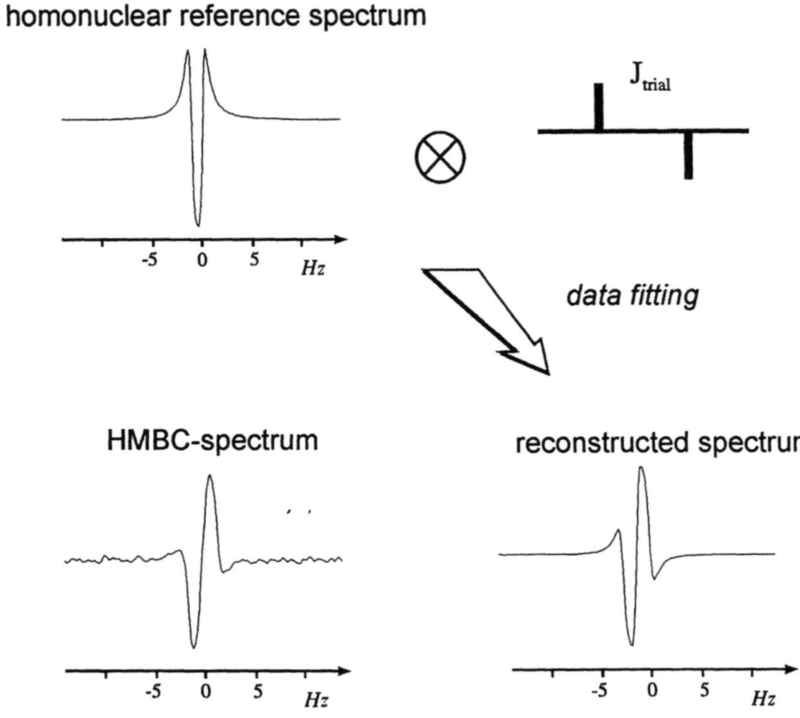

Fig. 5: *Schematic description of the Titman-Keeler method for the evaluation of heteronuclear coupling constants from HMBC spectra. The reconstructed spectrum computed from the homonuclear reference spectrum and a trial heteronuclear coupling constant is compared with the HMBC spectrum.* J_{trial} *is iteratively varied until a fit is obtained.*

d) A spin system of three or more nuclei in correlated spectra often exhibits multiplicities in cross-peaks caused by all possible transitions in the term scheme. However, it is also possible to run spectra which show exclusively transitions which are connected in the term scheme ("E.COSY pattern"), for instance E.COSY, P.E.COSY or HETLOC. The principle of the evaluation of an E.COSY pattern is explained in Fig. 6.

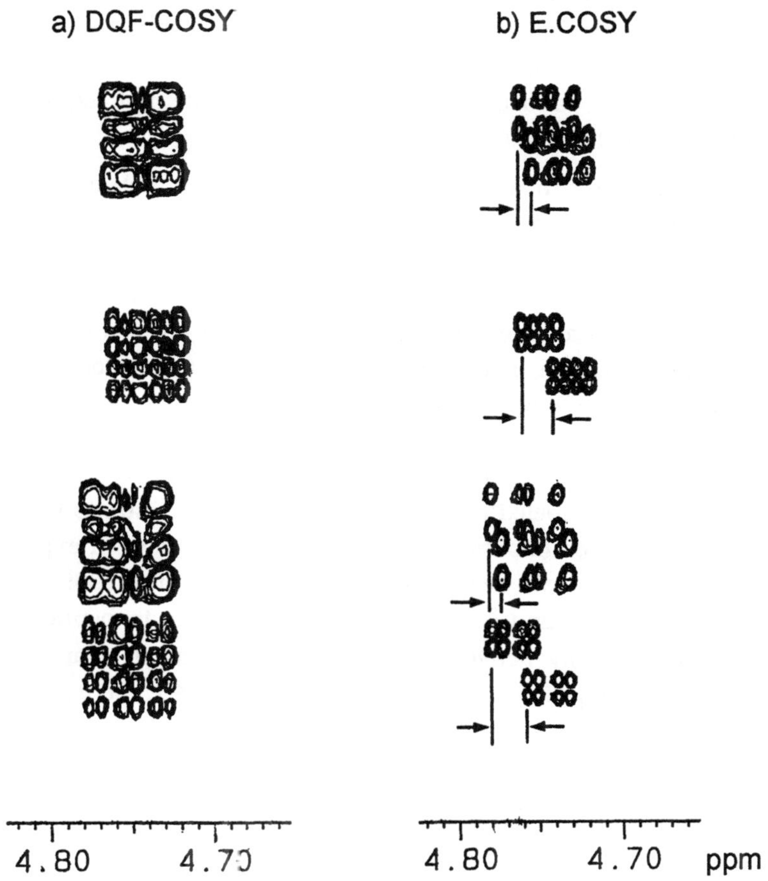

Fig. 6: *Comparison of DQF-COSY (a) and E.COSY (b). Part of the α,β-region of two phenylalanines in a cyclic hexapeptide. The passive coupling* $^3J_{H\alpha,H\beta}$ *and* $^3J_{H\alpha,H\beta'}$ *(indicated by the arrows) can easily be measured with high precision.*

E.COSY patterns are obtained in any two-dimensional correlation (COSY, TOCSY, NOESY) between two nuclei A and B, when a third nucleus C is coupled to them. The requirement to observe only connected transitions in the term scheme ("exclusive" COSY) is that the spin state of the nucleus is not affected in the pulse sequence (C remains α or β in both dimensions in which A and B evolve their shifts). The passive coupling expresses itself by a shift of the A-B multiplet pattern which would be obtained when no coupling to C would occure. The shift vectors of the two multiplets yield the desired passive coupling constants. For the determination of homonuclear $^3J_{H\alpha,H\beta}$ coupling constants (see Fig. 6) this cross peak pattern can be obtained with an E.COSY experiment (Griesinger et al 1985), which results from a combination of a DQF- and a TQF-COSY. A further method requiring less measurement time is the P.E.COSY (Müller 1987).

If a proton is attached to a heteronucleus, the heteronuclear long range couplings can be obtained in the same manner as discribed above using special designed pulse sequences. In such pulse sequences the direct heteronuclear one bond coupling constant (140 Hz for aliphatic carbons or 92 Hz for NH) provides a large splitting of the multiplet pattern in F_1. This is sufficient for a separation of the signals even if the resolution in the F_1-domain is low due to limitations by the incrementation in the indirect dimension. Using such methods J-coupling constants even smaller than the line width can be determined. This technique was first observed in TOCSY or NOESY spectra of ^{15}N labeled proteins where coupling constants from H_β to ^{15}N could be determined (Montelione et al 1989a). Later several experiments based on the same principle were developed (Montelione et al 1989b). For the determination of $^3J_{N,H\beta}$ and $^3J_{H^N,C\beta}$ coupling constants in peptides which contain the magnetically active heteroisotopes in natural abundance the HETLOC sequence (ω_1-filtered TOCSY) (Kurz et al 1991, Schmieder et al 1991) (Fig. 7) was introduced. To filter all proton signals which are not attached to ^{13}C or ^{15}N, respectively, the hetero filter developed previously (Otting et al 1986) was applied.

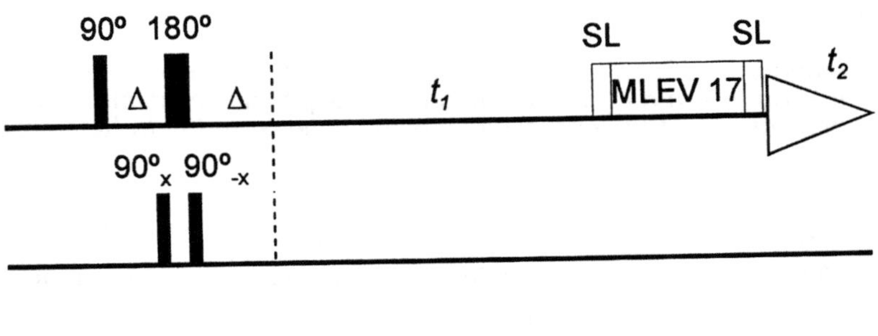

Fig. 7: *Pulse sequence for the HETLOC experiment*

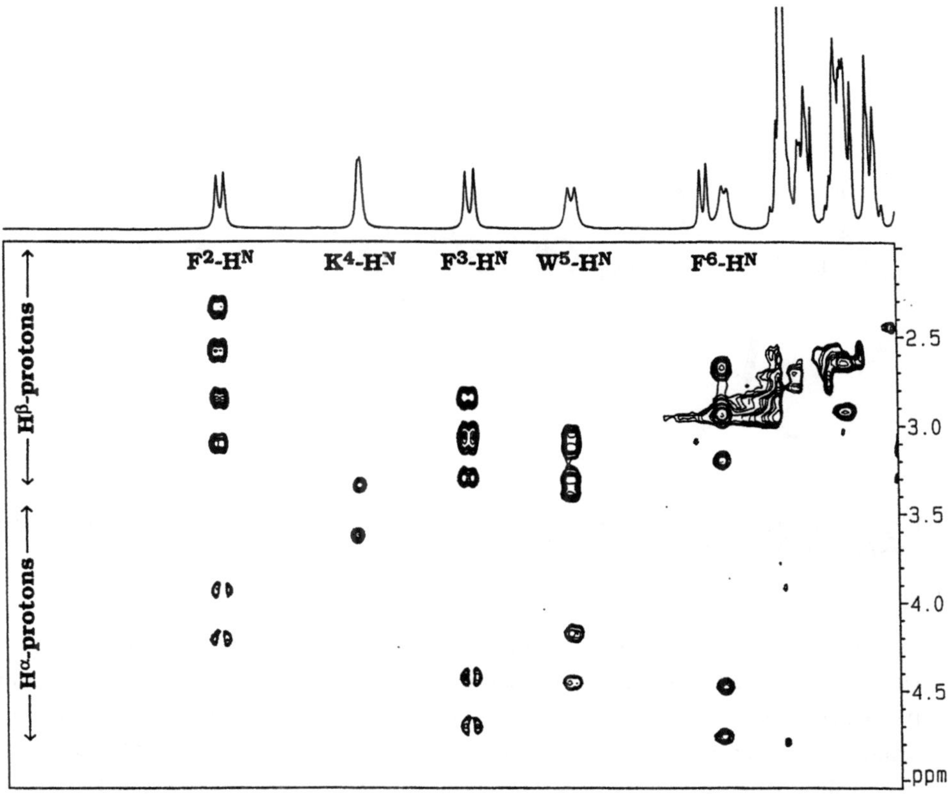

Fig. 8: *HETLOC spectrum of cyclo(-D-Pro-Phe-Phe-Lys-Trp-Phe-) with carbon heteronuclear filtering. From the shift of the doublets at each ß-proton in F2 the* $^3J_{H^N,C\beta}$ *coupling constant can be obtained.*

An example of a ^{13}C filtered TOCSY spectrum is shown in Fig. 8.

The peptide backbone is determined by the three angles ϕ (CO-N-C$^\alpha$-CO), ψ (N-C$^\alpha$-CO-N) and ω (C$^\alpha$-CO-N-C$^\alpha$) (Fig. 9). The peptide bond angle ω in most cases prefers trans-orientation (ω=180°) (see Fig. 9).

There are several homo- and hetero-nuclear ^3J-coupling constants which determine the ϕ angle which are accessible even in peptides with heteroatoms in natural abundance ($^3J_{H^N,H\alpha}$, $^3J_{H^N,C\beta}$, $^3J_{H^N,CO}$, $^3J_{H\alpha,CO}$) (Schmieder and Kessler 1991). In contrast there are difficulties in finding suitable coupling constants for ψ. The $^3J_{N,H\alpha}$ coupling is very small

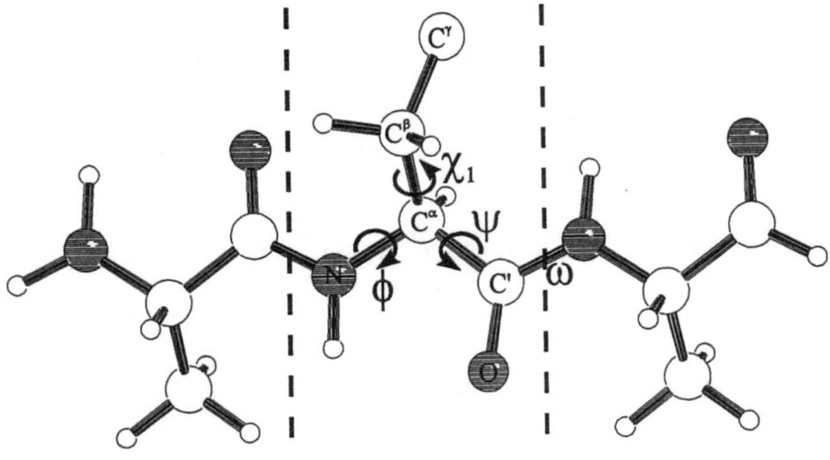

Fig. 9: *A peptide fragment in the extended conformation. The backbone torsione angle* ϕ, ψ, ω *and the side chain dihedral* χ_1 *are indicated.*

for conformational angles normally prefered in peptides (Bystrov 1976). However, it was found long time ago (Egli and von Philipsborn 1981) but not used up to recently that the $^1J_{C\alpha,H\alpha}$ coupling constant depends on ϕ *and* ψ. These couplings can easily be obtained by simply recording a HMQC without decoupling during acquisition. Our group (Mierke and Kessler 1992, Mierke et al 1992) and others (Vuister et al 1993) determined empirically the parameters for the angular dependence of $^1J_{C\alpha,H\alpha}$:

$$^1J = A \cos \phi + B \cos \psi + C$$

This equation can be used in restrained molecular dynamics calculations for structure refinement as will be discussed below.

Application of J-coupling constants for conformational analysis of peptides

Conformational analysis of peptide structures in solution by NMR was up to recently based mainly on atom-atom distance information obtained from NOESY spectra. However, there has been an increased interest in the use of coupling constants as additional restraints for structure determination in order to improve the quality of the structures. The problem with Karplus type curves is that up to four solutions are possible for each coupling constant. This has limited the use of coupling constants and they are often employed as a secondary check only. One method to address the wide range of allowed values is the measurement of additional coupling constants to remove ambiguities. Using this approach, only dihedral angles that are in agreement with all of the couplings are restrained. Usually this allows one to remove some of the possible solutions from further consideration.

Coupling constants were often used to compare their value with the mean structure obtained from a MD trajectory (followed by energy minimization). This is certainly wrong as NOE values and J-coupling constants average differently (Jardetzky 1980). In situations where conformational averaging takes place, the NOEs are averaged as a function of the sixth power of the distance between the two atoms (e.g. r^{-6}). On the other hand, coupling constants are averaged as a function of the cosine series of the dihedral angle subtended by the coupled atoms. In this case NOE restraints and coupling constants will produce different conformations, a fact which is a clear indication that the molecule is undergoing conformational averaging fast on the NMR time scale. A better procedure is to calculate J-couplings for each structure of the trajectory first and average these values.

One of the first applications of coupling constants directly in structure calculations was to restrain the torsion to a range of acceptable dihedral angles. But this method often refered to as *dihedral angle restraining*, was limited only to *extreme large* (> 8 Hz) and *small* (< 6 Hz) $^3J_{H^N,H\alpha}$-coupling constants. For large values a ϕ angle of -120° is assumed, while for small coupling constants ϕ is centered around -60°. However, this method is restricted to protein structures where no strong deviation from the Ramachandran-allowed conformations are found.

A more general method introduced recently, uses the whole Karplus curve for structural refinement calculations (Kim and Prestegard 1990, Mierke and Kessler 1992). The penalty function used is similar to that used for NOE distance restraints.

$$V_J = 1/2 \ K_J(J_{exp} - J_{theo})^2$$

where K_J is the force constant, J_{exp} the experimentally determined coupling constant and J_{theo} the theoretical coupling constant calculated from the Karplus equation.

This allows the coupling constants to be utilized directly in the refinement procedure without any approximations about the range of appropiate dihedral angles. Even more this approach allows the utilization of multiple coupling constants about a single dihedral angle since many Karplus equations for homo- and hetero-nuclear coupling constants are known. The resulting sum of the potential energy of different couplings describing the same dihedral angle has in most cases a reduced number (often only one) of minima. The force constants for the punishing function has to be adjusted so that the molecule can overcome the barriers between the minima during the MD simulation. As an example the energy profiles for multiple ^3J-couplings for the backbone ϕ angle of Ala^7 in cyclosporin A are plotted in Fig. 10 (Eberstadt et al 1992).

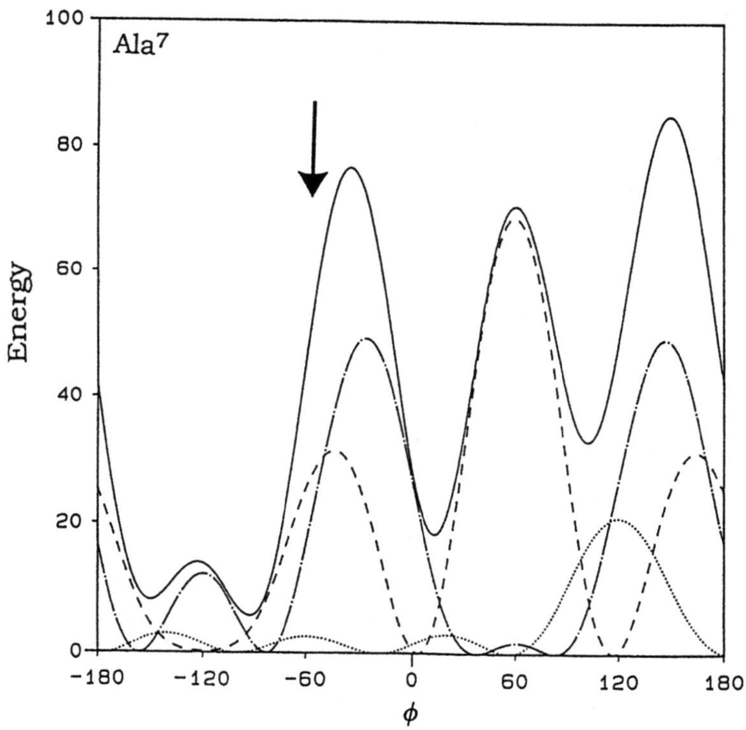

Fig. 10: *The potential energy of the coupling constant penalty term for the ϕ angle of Ala^7 in cyclosporin A. The energy profile is plotted for $^3J_{HN,H\alpha}$ (-·-·-), $^3J_{HN,C\beta}$ (·····) and $^3J_{CO(i-1),H\alpha(i)}$ (----) and the sum (——)*

In the same manner, the above mentioned angular dependence of $^1J_{C\alpha,H\alpha}$ from ϕ and ψ can be easily implemented into molecular dynamics calculations. As discussed above only limited conformational restraints are available for the ψ torsion and the use of the one-bond coupling has been shown to be a quite powerful tool in the examination of peptide conformation (Mierke et al 1992). The two dimensional energy profile is shown in Fig. 11.

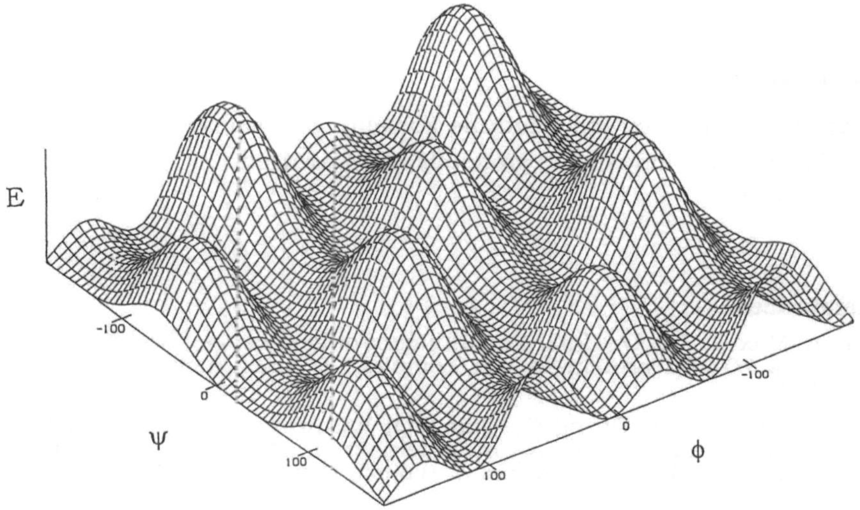

Fig. 11: *Energy profile for the one bond coupling constant between C_α and H_α of 143.2 Hz as a function of the* o *and* ψ *dihedral angles.*

Conclusions

Conformation relevant information can be obtained from NMR spectra by quantitative evaluation of NOEs and the use of multiple coupling constants for the dihedral angles. NOEs can be measured between protons which are closer in space than 5Å. Vicinal and in some cases direct couplings can be used in conformational analysis of peptides. The combined use of both parameters can improve the quality of the structure. As mentioned above, due to the flexibility of peptides and the high surface to core ratio, there is always the need to obtain as much experimental restraints as possible. It is also important to consider carefully any deviations between experimental and calculated parameters (certain NOEs or coupling constants). They implicate the first hint for conformational mobility provided the data are accurately determined and assigned.

Recently, NOEs and J-couplings have been also used in distance geometry and ensemble methods (Mierke et al 1994a, Mierke et al 1994b). Both methods screen a larger conformational space and avoid possible pitfalls by staying in local minima.

Acknowledgment:

Financial support by the Deutsche Forschungsgemeinschaft and the Fonds der Chemischen Industrie is gratefully acknowledged.

References

Bax A, Max D, Zax D (1992) Measurement of long-range ^{13}C-^{13}C J couplings in a 20-kDa protein-peptide complex. J. Am. Chem. Soc. 114: 6923-6925

Billeter H, Neri D, Otting G, Qian YQ, Wüthrich K (1992) Precise vicinal coupling constants $^3J_{HN\alpha}$ in proteins from nonlinear fits of J-modulated [^{15}N,^1H]-COSY experiments. J. Biomol. NMR 2: 257-274

Bystrov VF (1976) Spin-spin coupling and the conformational states of peptide systems. Progr. NMR Spectr. 10: 41-81

Dyson HJ, Rance M, Houghten RA, Lerner RA, Wright PE (1988) Folding of immunogenic peptide fragments of proteins in water solution. Sequence requirements for the formation of a reverse turn. J. Mol. Biol. 201: 161-200

Eberstadt M, Mierke DF, Köck M, Kessler H (1992) Peptide conformation from coupling constants: scalar couplings as restraints in MD simulations. Helv. Chim. Acta 75: 2583-2592

Egli H, von Philipsborn W (1981) Conformational dependence of one-bond C_α,H spin coupling in cyclic peptides. Helv. Chim. Acta 64: 976-988

Fesik SW, Gampe RT, Eaton HL, Gemmeker G, Olejniczak ET, Neri P, Holzman TF, Egan DA, Edalji R, Simmer R, Helfrich R, Hochlowski J, Jackson M (1991) NMR studies of [U-^{13}C]cyclosporin A bound to cyclophilin: bound conformation and portions of cyclosporin involved in binding, Biochemistry 30: 6574-6583

Griesinger C, Sorensen OW, Ernst RR (1985) Two-dimensional correlation of connected NMR transitions, J. Am. Chem. Soc. 107: 6394-6396

Jardetzky O (1980) On the nature of molecular conformations inferred from high-resolution NMR, Biochem. Biophys. Acta 621: 227-232

Kessler H (1982) Conformation and biological activity of cyclic peptides. Angew. Chem. Int. Ed. Engl. 21: 512-523

Kessler H, Loosli HR, Oschkinat H (1985a) Assignment of the ^{1}H-, ^{13}C-, and ^{15}N-NMR spectra of cyclosporin A in $CDCl_3$ and C_6D_6 by a combination of homo- and heteronuclear two-dimensional techniques. Helv. Chim. Acta 68: 661-681

Kessler H, Müller A, Oschkinat H (1985b) Differences and sums of traces within COSY-spectra (DISCO) for the extraction of coupling constants: 'decoupling' after the measurement, Magn. Reson. Chem. 23: 844-852

Kessler H, Oschkinat H (1985) Simplification of spectra for the determination of coupling constants from homonuclear correlated 2D NMR spectra, Angew. Chem. Int. Ed. 24: 690-692

Kessler H, Griesinger C, Wagner K (1987) Conformation of side chains in peptides using heteronuclear coupling constants obtained by two-dimensional NMR spectroscopy. J. Am. Chem. Soc. 109: 6927-6933

Kessler H, Griesinger C, Lautz J, Müller A, van Gunsteren WF, Berendsen HJC (1988) Conformational dynamics detected by nuclear magnetic resonance NOE values and J coupling constants. J. Am. Chem. Soc. 110: 3393-3396

Kessler H, Köck M, Wein T, Gehrke M (1990a) Reinvestigation of the conformation of cyclosporin A in chloroform. Helv. Chim. Acta 73: 1818-1832

Kessler H, Gehrke M, Lautz J, Köck M, Seebach D, Thaler A (1990b) Complexation and medium effects on the conformation of cyclosporin A studied by NMR spectroscopy and MD calculations. Biochem. Pharmacol. 40: 169-173

Kim J, Prestegard HJ (1989) Measurement of vicinal couplings from cross peaks in COSY spectra, J. Magn. Reson. 84: 9-13

Ko SY, Dalvit C, (1992) Conformation of cyclosporin A in polar solvents, Int. J. Peptide Protein Res. 40: 380-382

Köck M, Kessler H, Seebach D, Thaler A (1992) Novel backbone conformation of cyclosporin A: the complex with lithium chloride. J. Am. Chem. Soc. 114: 2676-2686

Kurz M, Schmieder P, Kessler H (1991) HETLOC, an efficient method for determinating heteronuclear long-range couplings with heteronuclei in natural abundance. Angew. Chem. Int. Ed. Engl. 30:1329-1331

Loosli HR, Kessler H, Oschkinat H, Weber HP, Petcher TJ, Widmer A (1985) The conformation of cyclosporin A in the crystal and in solution. Helv. Chim. Acta 68: 682-704

Mierke DF, Kessler H (1992) Combined use of homo- and heteronuclear coupling constants as restraints in molecular dynamics simulations. Biopolymers 32: 1277-1282

Mierke DF, Golic Grdadolnik S, Kessler H (1992) Use of one bond C_α-H_α coupling constants as restraints in MD simulations. J. Am. Chem. Soc. 114: 8283-8284

Mierke DF, Kessler H (1993) Improved molecular dynamics simulations for the determination of peptide structures. Biopolymers 33: 1003-1017

Mierke DF, Kurz M, Kessler H (1994) Peptide flexibility and calculations of an ensemble of molecules. J. Am. Chem. Soc. in press

Mierke DF, Scheek RM, Kessler H (1994) Coupling constants as restraints in ensemble distance driven dynamics, Biopolymers in press

Montelione GT, Winkler ME, Rauenbuehler P, Wagner G (1989a) Accurate measurements of long-range heteronuclear coupling constants from homonuclear 2D NMR spectra of isotope-enriched proteins. J. Magn. Reson. 82: 198-204

Montelione GT, Wagner G (1989b) Accurate measurements of homonuclear H^N-H^α coupling constants in polypeptides using heteronuclear 2D NMR experiments. J. Am. Chem. Soc. 111: 5474-5475

Müller L (1987) P.E.COSY, a simple alternative to E.COSY, J. Magn. Reson. 72: 191-196

Neuhaus DM, Wagner G, Vasak M, Kägi JHR, Wüthrich K (1985) Systematic application of high-resolution, phase-sensitive two-dimensional ^{1}H-NMR techniques for the identification of the amino-acid-proton spin systems in proteins rabbit metallothionein-2. Eur. J. Biochem. 151: 257-273

Otting G, Senn H, Wagner G, Wüthrich K (1986) Editing of 2D ^{1}H NMR spectra using X half-filters. Combined use with residue-selective ^{15}N labeling of proteins. J. Magn. Reson. 70: 500-505

Rose GD, Gierasch LM, Smith JA (1985) Turns in peptides and proteins. Adv. Protein Chem. 37:1-109

Schmieder P, Kurz M, Kessler H (1991) Determination of heteronuclear long-range couplings to heteronuclei in natural abundance by two- and three-dimensional NMR spectroscopy, J. Biomol. NMR 1: 403-420

Schmieder P, Kessler H (1991) Determination of the Φ angle in a peptide backbone by NMR spectroscopy with a combination of homonuclear and heteronuclear coupling constants, Biopolymers 32: 435-440

Titman JJ, Neuhaus D, Keeler J (1989) Measurements of long-range heteronuclear coupling constants. J. Magn. Reson. 85: 111-131

Titman JJ, Keeler J (1990) Measurement of homonuclear coupling constants from NMR correlation spectra. J. Magn. Reson. 89: 640-646

Vuister GW, Bax A (1993) Quantitative J correlation: a new approach for measuring homonuclear three-bond $J(H^NH^\alpha)$ coupling constants in ^{15}N-enriched proteins. J. Am. Chem. Soc. 115: 7772-7777

Weber C, Wider G, von Freyberg B, Treber R, Braun W, Widmer H, Wütrich K (1991) The NMR structure of cyclosporin A bound to cyclophilin in aqueous solution, Biochemistry 30: 6563-6574

Wüthrich K (1986) NMR of Proteins and Nucleic Acids; Wiley-Interscience, New York 1986

Methods used in the Assignments of the ^{1}H and ^{15}N resonances of the c-Abl *src* homology 2 (SH2) domain.

Michael Overduin, Carlos B. Rios, Bruce J. Mayer, David Baltimore, and David Cowburn
Laboratories of The Rockefeller University
1230 York Avenue
New York, New York 10021
U.S.A.

INTRODUCTION

The SH2 domain is a recognition motif of approximately 100 amino acids that serves to mediate the association of cytoplasmic proteins involved in signal transduction. SH2 domains from a number of proteins, including Crk, phospholipase C-γ, Ras GTPase-activating protein and Abl, have been shown to bind specifically protein sequences that have been phosphorylated on tyrosine residues (Anderson *et al.*, 1990; Margolis *et al.*, 1990; Mayer and Hanafusa *et al.*, 1990; Mayer *et al.*, 1991; Moran *et al.*, 1990). The transforming activity of Abl depends on the highly conserved FLVRESES motif, a central element in the binding of tyrosine phosphorylated ligands (Mayer et al, 1992).

The crystal structure of the Src SH2 complexed with low affinity phosphopeptides (Waksman *et al.*, 1992), and the solution structures of the uncomplexed Abl (Overduin *et al.*, 1992a; Overduin *et al.*, 1992b) and N-terminal p85 phosphatidylinositol 3'-kinase (Booker *et al.*, 1992) SH2 domains have been simultaneously determined. The interactions of the Src SH2 with the phosphotyrosine include a number of hydrogen bonds and amino-ring interactions to conserved basic residues. The crystal structures of the Src (Waksman *et al.*, 1993) and Lck (Eck *et al.*, 1993) SH2 domains bound to high affinity phosphopeptides have identified additional interactions that could account for the selective targeting of SH2 domains to particular tyrosine phosphorylated protein ligands.

The assignment of the ^{1}H and ^{15}N chemical shifts of the 109 amino acids of the Abl SH2 domain represents the initial stage in the determination of the three dimensional structure. This paper illustrates the use of 2 and 3 D methods with natural and ^{15}N labelled proteins. Further details of secondary structure analysis and complete structure calculation have been published (Overduin *et al.*, 1992a,b).

NATO ASI Series, Vol. H 87
NMR of Biological Macromolecules
Edited by C. I. Stassinopoulou

MATERIALS AND METHODS

Materials Sample preparation has been previously described (Overduin, 1992a,b). A sample with [^{15}N] leucine labeling was obtained using M9 medium without NH_4Cl but with approximately 100 mg/ml of each amino acid including [^{15}N] L-leucine (Cambridge Isotope Labs).

Sample preparation Samples contained up to 4 mM Abl SH2 product in phosphate buffered saline or 50 mM sodium arsenate buffer at pH 7.2 (uncorrected for isotope effects) with 50 μM sodium azide and 10% D_2O/90% H_2O. Amide exchange data was collected from a 350 μM, [^{15}N]SH2 sample that had been lyophilized once before D_2O exchange. D_2O samples were prepared by lyophilizing three times and dissolving the residue in 99.996% D_2O.

NMR Spectroscopy. NMR experiments were run on a GE OMEGA 500 spectrometer. Quadrature detection was achieved by the States-TPPI method (Marion et al., 1989a). The water signal was suppressed by using selective on-resonance irradiation during a relaxation delay of 0.6 to 1.5 s and in NOESY experiments also during the mixing period. A 60 ms SCUBA pulse train (Brown *et al.*, 1988) was employed in NOESY and DQF-COSY experiments to allow recovery of saturated H^{α}'s. Experiments were typically run at 25°C with sweep widths of 7017 and 2000 Hz for ^{1}H and ^{15}N. The internal ^{1}H chemical shift reference used was sodium 2,2-dimethyl-2-silapentane-5-sulfonate and indirect referencing was used for ^{15}N (Live *et al.*, 1984). Homonuclear experiments were run both in H_2O and D_2O using standard pulse sequences and phase cycling. A range of t_1 increments from 200 to 512, each consisting of 2048 complex points were typically acquired, with 96-256 scans per increment. In D_2O DQF-COSY spectra 4096 points were acquired. Z-filtered HOHAHA experiments (Braunschweiler and Ernst, 1983; Davis and Bax, 1985) used an MLEV-17 sequence for spin lock with mixing times (τ) of 35, 65 and 85 ms in D_2O and between 30 and 90 ms in H_2O. NOESY spectra (Macura *et al.*, 1981) were recorded in D_2O with τ's of 40, 80, 120, 180 and 240 ms and in H_2O with τ's of 40, 80, 100, 120 and 180 ms.

Heteronuclear experiments consisted of HMQC, HMQC-*J*, HSMQC, ^{15}N filtered HOHAHA and NOESY, 2D and 3D HMQC-NOESY, and 3D HMQC-HOHAHA experiments (Marion *et al.*, 1989b; Gronenborn et al, 1989; Kay *et al.*, 1990; Forman-Kay *et al.*, 1990). The τ's in HMQC-NOESY's were 120 ms, and in 3D HMQC-HOHAHA the τ was 30 ms. The 3D spectra were recorded with a 34 x 2 x 64 x 2 x 2048 complex matrix, with 32 scans per increment. Weak ^{15}N-H^{α} correlations as well as the strong ^{15}N-H' correlations were observed in standard HMQC-*J* spectra (Kay and Bax, 1990) . The amide hydrogen exchange rates were measured by following the intensity of crosspeaks in HMQC experiments initiated at 5, 11, 22, 42, 83, 123 min and 12 h after exchanging a fully protonated, lyophilized sample with 99.996% D_2O.

All data was processed with RUNMR software (Cowburn *et al.*, 1991). Signal processing typically consisted of a Gaussian apodization in ω_2 and a cosine or cosine squared apodization in combination with an inverted exponential function ($1\text{-}ce^{-t}$, where c is a constant) in ω_1. A single zerofilling and polynomial baseline correction were typically used in ω_1, while linear baseline correction was used in ω_2 where necessary.

RESULTS AND DISCUSSION

Resonance Assignment The assignment of the amino acid spin systems of the c-*abl* SH2 domain involved several stages. First sidechain spin systems were identified by tracing scalar connectivities in D_2O DQF-COSY and HOHAHA spectra according to standard methods (Wüthrich, 1986). Next the H^{α} and sidechain proton resonances were connected to their amide proton resonances in DQF-COSY, homonuclear and ^{15}N-filtered HOHAHA, and 3D HMQC HOHAHA spectra, all collected in H_2O. Although this conventional assignment strategy provided assignments for most residues, resonance overlap and weak amide proton resonance intensities precluded the full assignment of several residues. For example, in HMQC spectra only 92 of 104 expected ^{15}N-H' crosspeaks from backbone amide protons were observed (see Figure 1), all but one of which have been sequentially assigned.

Careful analysis of HMQC-*J* spectra revealed that two of these apparently single ^{15}N-H' crosspeaks contained a pair of overlapped crosspeaks. In particular, the assignments Tyr^{23} and Leu^{6} were complicated by the overlap of both their ^{15}N-H' and H'-H^{α} crosspeaks with other spin systems. The remainder of the absent ^{15}N-H' crosspeaks are derived from residues that are exposed in the tertiary structure (Overduin et al, 1992b) and their amide resonances are therefore likely diminished in intensity due to rapid exchange at neutral pH and structural mobility.

A number of approaches were used to bypass the unresolved or weak amide resonances obstructing the assignment of several residues. The slow exchange of 26 amide protons in the Abl SH2 domain (see Figure 2) was exploited by obtaining correlations free of overlap from more rapidly exchanging amide protons in HOHAHA and NOESY spectra of samples that had recently undergone D_2O exchange. The higher resolution of correlations obtained from such 2D data was particularily useful for correlating poorly ressolved H^{α} resonances with amide resonances.

Selective ^{15}N-labeling of the leucine residues was employed in order to discriminate the ten leucine spin systems (see Figure 3) whose protons typically exhibited weak couplings. Additional, very weak ^{15}N-H' correlations from two valines, an isoleucine and the C-terminal

Asp[109] reflect metabolic scrambling of the specific label. Only nine intense ^{15}N-H' crosspeaks are observed in this HMQC spectrum of the [^{15}N]-leucine labeled sample, since Leu[6]'s weak crosspeak remains overlapped with Leu[25]'s. Consequently, complete leucine assignments depended on the analysis of HMQC-*J* spectra, as described below.

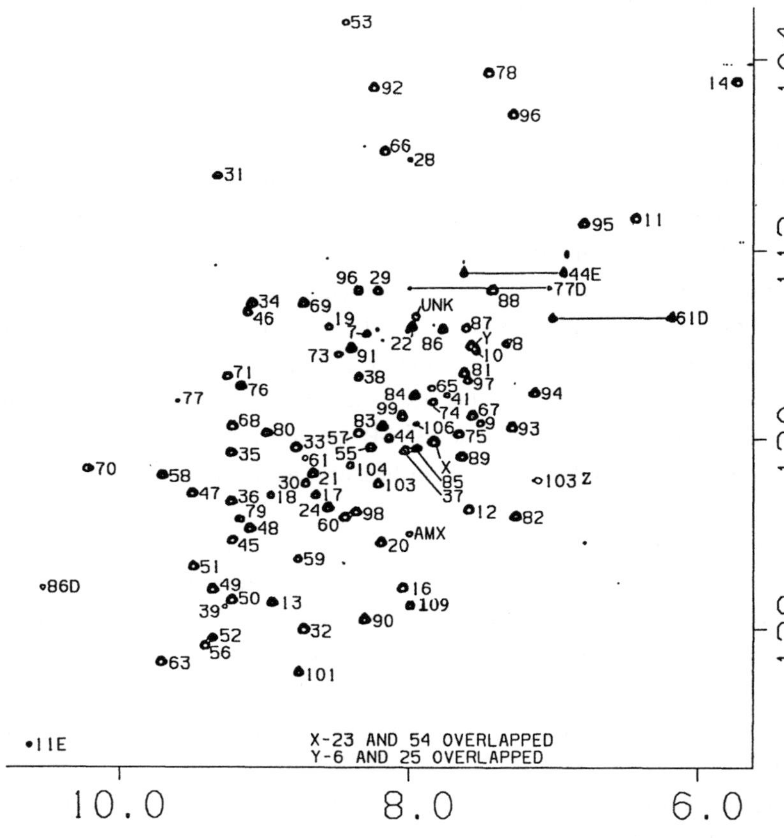

Fig. 1. HMQC Spectrum Displaying ^{15}N-H' Correlations. The ^{15}N-H' correlations are labeled by the residue number. The sidechain NH_2 and aromatic NH groups are indicated by the residue number and the letter "D" and "E" to indicate δ and ε positions, respectively. Two pairs of residues; Tyr[23] and Arg[54], and Leu[6] and Leu[25], have overlapped ^{15}N-H' correlations and are indicated by "X" and Y", respectively. The spectrum was obtained from a uniformly ^{15}N labeled 2.0 mM SH2 sample in PBS (90% H_2O, 10% D_2O) at 25°C with 2048 ω_2 points x 256 ω_1 points x 128 scans. The ω_2 (horizontal) and ω_1 (vertical) axes provide H^1 and ^{15}N chemical shifts, respectively.

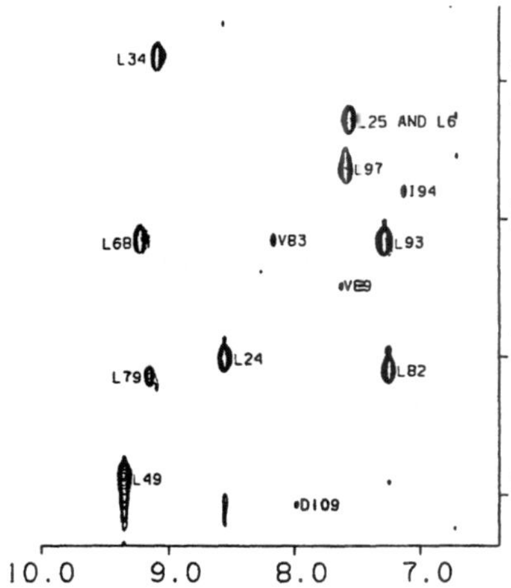

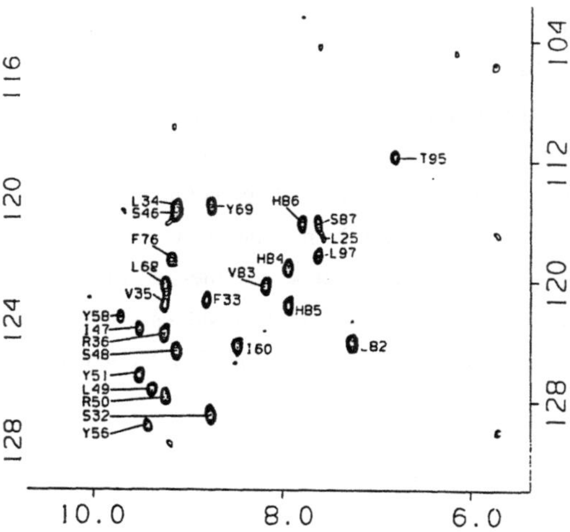

Fig. 2. (Right) HMQC spectrum displaying ^{15}N-H' correlations of slowly exchanging amide protons. A total of 26 ^{15}N-H' correlations of slowly exchanging amide protons are assigned by the amino acid number and type. The signals of Leu^{34}, Arg^{36}, Ser^{46}, Ile^{47}, Ser^{48}, Leu^{49}, Arg^{50}, Tyr^{51}, Tyr^{56}, Tyr^{58}, Tyr^{69}, Leu^{82}, and Val^{83} persisted for over 12 h. The spectrum was acquired 5 min after D_2O from a uniformly ^{15}N labeled 350 µM SH2 sample in PBS freshly exchanged with D_2O at 25°C with 2048 ω_2 points x 32 ω_1 points x 4 scans. The ω_2 and ω_1 axes provide H^1 and ^{15}N chemical shifts, respectively.

Fig. 3.(Left) HMQC Spectrum of a [^{15}N Leu] SH2 Sample. Nine leucine ^{15}N-H' correlations are labeled by their residue number. The Leu^{25} and weaker Leu^{6} correlation are overlapped. Some metabolic incorporation of ^{15}N label into other residues is evident. The spectrum was taken of a 700 µM [^{15}N Leu] Abl SH2 domain sample in PBS (90% H_2O, 10% D_2O) at 25°C with 2048 ω_2 points x 128 ω_1 points x 256 scans.

Typically HMQC-*J* spectra are collected to measure $^3J(H^{\alpha}\text{-}H')$ coupling constants from ^{15}N-H' crosspeaks (Kay and Bax, 1990). Crosspeaks correlating ^{15}N and H^{α} resonances were also discovered in HMQC-*J* spectra, providing new assignment information. A standard HMQC-*J* experiment provided 55 ^{15}N-H^{α} correlations (see Figure 4). These additional correlations were very useful in confirming the assignments of amide proton resonances, 36% of which are not resolved within 0.010 ppm. For example, Leu^{6}'s assignment posed a particular problem since its ^{15}N-H' and H'-H^{α} crosspeaks overlap those of Leu^{25} and Lys^{67}, respectively, and its ^{15}N-H'-H^{α} crosspeak was not apparent in 3D HMQC-HOHAHA spectra. The presence of the overlapped crosspeaks could be deduced from the presence of the resolved Leu^{5} ^{15}N-H^{α} crosspeak indicated in Figure 4. The assignment of the amide resonance of Tyr^{23} was also difficult since its ^{15}N-H' and H'-H^{α} crosspeaks completely overlap those of Arg^{54} and Ser^{74}, respectively and its ^{15}N-H'-H^{α}

crosspeak was not apparent in 3D HMQC-HOHAHA spectra. However, the two overlapped ^{15}N-H' crosspeaks of Tyr^{23} and Arg^{54} could be discriminated by the shape of the crosspeak in resolution enhanced HMQC-*J* spectra due to their different *J*-coupling constants. The spin systems were confirmed through their sequential assignment as described elsewhere (Overduin et al, 1992b). The chemical shifts of assigned spin systems are listed in Table 1. In total, ^{1}H and ^{15}N resonances of 99 of the total 109 amino acid residues have been assigned. For 96 of these residues, all the backbone ^{1}H resonances and over 96% of their sidechain proton resonances residues could be assigned. Remaining unassigned residues occur in loop regions and in the N- and C-terminal segments.

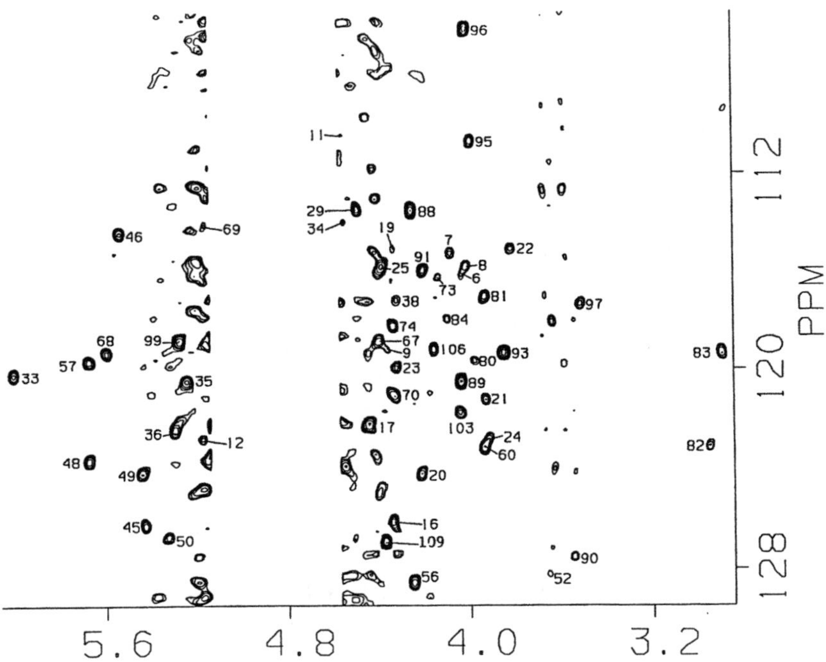

Fig. 4. Upfield region of an HMQC-*J* spectrum displaying ^{15}N-H$^{\alpha}$ and correlations. The ^{15}N-H$^{\alpha}$ crosspeaks of 52 residues are labeled by their residue number. The weak ^{15}N-H$^{\alpha}$ crosspeaks of His^{85}, Leu^{79} and Arg^{75} were also identified in the HMQC-*J* spectra but are not shown. The intense residual water resonance that obscures many of the remaining crosspeaks has been removed for clarity. The spectrum was obtained from a uniformly ^{15}N labeled 2.0 mM SH2 sample in PBS (90% H_2O, 10% D_2O) at 25°C with 2048 ω_2 points x 80 ω_1 points x 160 scans. The ω_2 and ω_1 axes provide H^1 and ^{15}N chemical shifts, respectively.

TABLE 1. Shifts of Abl SH2

residue	$^{15}N^{a}$	H'^{b}	H^{α}	H^{β}	other
leu^{6}	116.4	7.54	4.03	1.83	H^{γ} 1.46; H^{δ} 0.83, 0.65
glu^{7}	115.5	8.29	4.08	2.06, 1.95	H^{γ} 2.30, 2.30
lys^{8}	116.0	7.32	4.01	1.56, 1.14	H^{γ} 0.91, 0.91; H^{δ} 1.51, 1.51; H^{ϵ} 2.84, 2.84
his^{9}	119.4	7.50	4.36	1.98, 0.84	H^{δ} 6.73; H^{ϵ} 7.80
ser^{10}	116.3	7.53	3.32	3.53, 3.53	
trp^{11}	110.7	6.43	4.56	3.61, 2.66	H^{δ} 7.54; H^{ϵ} 10.62, 7.40; H^{η} 6.67, 6.89; H^{ζ} 6.47
tyr^{12}	123.1	7.58	5.18	3.18, 2.49	H^{δ} 6.98, 6.98; H^{ϵ} 6.76, 6.76
his^{13}	127.0	8.95	4.50	3.29, 2.68	H^{δ} 6.78; H^{ϵ} 7.42
gly^{14}	105.0	5.73	4.32, 3.49		
pro^{15}			4.90	2.17, 1.98	H^{γ} 2.18, 2.08; H^{δ} 3.85, 3.68
val^{16}	126.4	8.04	4.34	1.85	H^{γ} 1.06, 0.90
ser^{17}	122.4	8.65	4.41	3.89, 3.83	
arg^{18}	122.4	8.96	3.48	1.80, 1.58	H^{γ} 1.02, 0.50; H^{δ} 2.90, 2.82
asn^{19}	115.2	8.55	4.33	2.73, 2.66	
ala^{20}	124.4	8.20	4.22	1.53	
ala^{21}	121.5	8.67	3.94	1.51	
glu^{22}	115.3	7.97	3.82	2.29, 2.10	H^{γ} 2.93, 2.95
tyr^{23}	120.1	7.83	4.33	3.22, 3.18	H^{δ} 7.14, 7.14; H^{ϵ} 6.82, 6.82
leu^{24}	123.0	8.57	3.92	1.92, 1.35	H^{γ} 1.89; H^{δ} 1.35, 1.03
leu^{25}	116.1	7.56	4.38	1.69, 1.52	H^{γ} 1.54; H^{δ} 0.69,0.64
gly^{28}	108.24	7.99	4.11, 4.00		
ile^{29}	113.7	8.21	4.47	2.00	H^{γ} 1.34, 0.95, 3(H^{γ}) 0.88; H^{δ} 0.64
asn^{30}	121.9	8.72	4.78	3.25, 3.09	
gly^{31}	108.9	9.32	4.31, 3.95		
ser^{32}	128.1	8.73	6.13	4.04, 4.02	
phe^{33}	120.4	8.79	6.01	3.04	H^{δ} 7.58, 6.88; H^{ϵ} 6.81, 6.68; H^{η} 6.73
leu^{34}	114.3	9.09	4.55	1.33, 1.01	
val^{35}	120.6	9.23	5.25	2.61	H^{γ} 1.29, 1.16
arg^{36}	122.6	9.23	5.30	2.19, 1.30	H^{γ} 2.45, 1.50; H^{δ} 3.39
glu^{37}	120.5	8.03	4.75	1.73, 2.03	H^{γ} 2.24, 2.37
ser^{38}	117.4	8.35	4.31	3.97, 3.84	
glu^{39}	127.1	9.27	4.18	2.03, 2.03	H^{γ} 2.15, 2.27
ser^{41}	118.2	7.73	4.87	3.78, 3.59	
pro^{42}			4.37	2.35, 1.91	H^{γ} 2.10, 1.99; H^{δ} 3.65, 3.55
gly^{43} c			4.80, 3.91		
gln^{44}	120.0	8.14	4.70	2.30, 2.13	H^{γ} 2.37; H^{ϵ} 6.93, 7.62
arg^{45}	124.3	9.21	5.44	1.76,1.76	H^{γ} 1.40, H^{δ} 3.29, 3.14
ser^{46}	114.6	9.11	5.54	3.76, 3.30	
ile^{47}	122.4	9.50	4.97	1.74	H^{γ} 1.33, 0.54, 3(H^{γ}) 0.95; H^{δ} 0.94
ser^{48}	123.9	9.10	5.68	3.53, 3.32	
leu^{49}	126.4	9.36	5.43	1.92, 1.39	H^{γ} 1.55; H^{δ} 0.91, 0.79
arg^{50}	126.9	9.23	5.34	2.20, 2.20	H^{γ} 1.50, 1.50; H^{δ} 3.14, 2.71
tyr^{51}	125.4	9.49	4.70	2.89, 2.77	H^{δ} 6.96, 6.96; H^{ϵ} 6.85, 6.85
glu^{52}	128.4	9.36	3.646	1.51, 1.51	H^{γ} 2.03, 2.03
gly^{53}	102.4	8.44	4.11, 3.49		
arg^{54}	120.2	7.82	4.65	1.78, 1.64	H^{γ} 1.61, 1.38; H^{δ} 3.14
val^{55}	120.4	8.26	4.76	1.69	H^{γ} 0.81, 0.39
tyr^{56}	128.7	9.41	4.25	2.74, 2.74	H^{δ} 7.15, 7.15; H^{ϵ} 7.04, 7.04
his^{57}	119.8	8.35	5.68	2.78, 2.75	H^{δ} 7.00; H^{ϵ} 7.36
tyr^{58}	121.6	9.70	4.77	2.86, 2.60	H^{δ} 6.98, 6.98; H^{ϵ} 6.75, 6.75
arg^{59}	125.2	8.77	4.53	1.82, 1.75	H^{γ} 1.62; H^{δ} 3.28, 3.24
ile^{60}	123.4	8.45	3.94	1.73	H^{γ} 1.62, 0.69, 3(H^{γ}) 0.82; H^{δ} 0.76

asn^{61}	121.0	8.74	4.66	2.03, 1.05	H^{δ} 6.19, 7.00
thr^{62}	113.7	8.35	5.07	3.97	H^{γ} 1.32
ala^{63}	129.5	9.71	4.81	1.73	
asp$^{65\,c}$	117.9	7.84			
gly^{66}	107.9	8.16	4.35, 3.60		
lys^{67}	119.1	7.56	4.41	1.82, 1.70	H^{γ} 1.49; H^{δ} 1.46, 1.26; H^{ε} 2.86
leu^{68}	119.5	9.22	5.60	1.62, 1.37	H^{γ} 1.87; H^{δ} 0.91, 0.86
tyr^{69}	114.3	8.73	5.17	3.19, 3.13	H^{δ} 6.55, 6.55; H^{ε} 6.55, 6.55
val^{70}	121.3	10.21	4.32	1.97	H^{γ} 0.80, 0.65
ser^{71}	117.3	9.25	4.90	4.07, 4.07	
ser$^{72\,c}$			3.53	3.90, 3.81	
glu^{73}	116.4	8.49	4.13	2.12, 1.89	H^{γ} 2.23
ser^{74}	118.5	7.83	4.33	3.78, 3.37	
arg^{75}	119.9	7.65	5.10	1.46	H^{γ} 1.35,1.02; H^{δ} 2.82, 2.66
phe^{76}	117.8	9.16	4.87	3.46, 2.58	H^{δ} 7.23,7.23; H^{ε} 6.99,6.99;H^{η} 6.06
asn^{77}	118.4	9.60	4.71	3.24, 2.92 ;	H^{δ} 8.00, 7.03
thr^{78}	104.6	7.45	4.95	4.74	H^{γ} 1.27
leu^{79}	123.4	9.18	4.03	1.80, 1.48	H^{γ} 1.85; H^{δ} 1.12, 1.06
ala^{80}	119.8	8.99	3.98	1.52	
glu^{81}	117.2	7.61	3.92	2.35, 2.35	H^{γ} 2.28,2.23
leu^{82}	123.3	7.26	2.91	2.09,1.76	H^{γ} 1.81; H^{δ} 1.18, 0.72
val^{83}	119.6	8.18	2.91	1.20	H^{γ} -0.12, -0.24
his^{84}	118.2	7.96	4.09	3.14, 3.05	H^{δ} 6.99; H^{ε} 7.69
his^{85}	120.4	7.94	4.02	2.68, 2.61	H^{δ} 5.29; H^{ε} 7.62
his^{86}	115.4	7.76	5.02	3.09, 2.58	H^{δ} 6.98; H^{ε} 8.29
ser^{87}	115.3	7.60	4.69	3.93, 3.81	
thr^{88}	113.7	7.42	4.26	4.11	H^{γ} 1.05
val^{89}	120.8	7.63	4.03	1.48	H^{γ} 0.63, 0.53
ala^{90}	127.7	8.31	3.56	1.08	
asp^{91}	116.2	8.40	4.19	3.04, 2.15	
gly^{92}	105.2	8.25	4.42, 3.47		
leu^{93}	119.6	7.28	3.85	1.71, 0.89	H^{γ} 1.09; H^{δ} 0.33,-0.44
ile^{94}	118.1	7.13	3.65	1.17	H^{γ} 0.87,0.45, 3(H^{γ})0.03; H^{δ} 0.55
thr^{95}	110.9	6.79	3.99	3.37	H^{γ} 1.09
thr^{96}	106.4	7.26	4.01	3.76	H^{γ} 0.69
leu^{97}	117.5	7.59	3.53	0.91,-1.00	H^{γ} 0.98; H^{δ} 0.16, 0.16
his^{98}	123.1	8.37	4.78	2.81, 2.66	H^{δ} 6.64; H^{ε} 7.72
tyr^{99}	119.1	8.04	5.26	3.08, 2.81	H^{δ} 7.14, 7.14; H^{ε} 6.85, 6.85
pro^{100}			3.88	1.80, 1.69	H^{γ} 2.29, 2.03; H^{δ} 3.99, 3.70
ala^{101}	129.9	8.76	4.56	1.15	
pro^{102}			4.41	2.29, 1.86	H^{γ} 2.14, 2.04; H^{δ} 3.98, 3.73
lys^{103}	122.0	8.22	4.04	1.46, 1.12	H^{γ} 0.35,0.35; H^{δ} 1.56,1.43; H^{ε} 2.77,2.61
arg^{104}	121.2	8.41	4.37	1.91, 1.80	H^{γ} 1.67; H^{δ} 3.22, 3.22
ile^{106}	119.4	7.95	4.16	1.81	H^{γ} 1.32,1.21, 3(H^{γ})0.83; H^{δ} 0.82
his^{107}	119.2	8.38	4.61	3.09, 3.01	H^{δ} 6.92; H^{ε} 7.71
arg^{108}	115.8	8.18	4.36	1.81, 1.69	H^{γ} 1.57; H^{δ} 3.15
asp^{109}	127.07	7.99	4.37	2.53, 2.68	

Footnotes. (a) **Referenced** indirectly (Live et al., 1984); (b) Referenced to TSP.

CONCLUSIONS

The assignment of the ^{1}H and ^{15}N spectra of the SH2 domain of c-Abl has been accomplished. The similarity of the SH2 domain tertiary structures that have been determined (Booker *et al.*, 1992; Overduin *et al.*, 1992b, Waksman *et al.*, 1992, Eck *et al.*, 1993) imply that these assignments should be useful in the assignment of other members of the large SH2 family and in investigating their interactions with ligands.

ACKNOWLEDGMENTS Supported by grants (DB, CA 51462; DC, DK-20357, GM-47021) and fellowships (BM, CA 0887501; CR, GM 14313) from NIH. NMR resources were purchased with grants from NIH, the Keck Foundation, and NSF. We are grateful to Prof. John Kuriyan, and Drs. Nalin Pant and Yuying Gosser for discussion.

REFERENCES

Anderson , D., Koch, C. A., Grey, L., Ellis, C., Moran, M. F., and Pawson, T. (1990) *Science*, 250, 979-982.

Booker, G. W., Breeze, A. L., Downing, A. K., Panayotou, G., Gont, I., Waterfield, M. D., and Campbell, I. D. (1992) *Nature* 358, 684-687

Braunschweiler, L. and Ernst, R. R. (1983) *J. Magn. Reson.*, 53, 521-528.

Brown, S. C., Weber, P. L., and Mueller, L. (1988) *J. Magn. Reson.* 77, 166-169.

Cowburn, D., Glushka, J., DiGennaro, F., and Rios C. B. (1991) in *Computational Aspects of the Study of Biological Macromolecules by NMR*, J. C. Hoch, Ed., Plenum Press, New York.

Constantine, K. L., Ramesh V, Banyai, L., Trexler M, Patthy L, Llinas M. (1991) *Biochemistry* 30, 1663-1672

Davis, D.G., and Bax, A. (1985) *J. Am. Chem. Soc.* **107,** 2820-2821

Eck, M. J., Shoelson, S. E., and Harrison, S. C. (1993) *Nature,* **362**, 87-91.

Forman-Kay, J. D., Gronenborn, A. M., Kay, L. E., Wingfield, P. T., and Clore, G. M. (1990) *Biochemistry,* 29, 1566-1572.

Gronenborn, A. M., Bax, A., Wingfield, P. T., and Clore, G. M. (1989) *FEBS Lett.,* 243, 93-99.

Kay, L. E., and Bax, A. (1990) *J. Magn. Reson.*, 86, 110-126.

Live, D. H., Davis, D. G., Agosta, W. C., and Cowburn, D. (1984) *J. Amer. Chem. Soc.*, 106, 1939-1941.

Macura, C., Huang, Y., Suter, D., and Ernst, R.R. (1981) *J. Magn. Reson.*, 43, 259-281.

Margolis, B., Li, N., Koch, A., Mohammadi, M., Hurwitz, D.R., Zilberstein, A., Ullrich, A., Pawson, T. and Schlessinger, J. (1990) *EMBO J.,* 9,4375-4380.

Marion, D., Ikura, M., Tschudin, R., and Bax, A. (1989a) *J. Magn. Reson.*, 85, 393-399.

Marion, D., Driscoll, P. C., Kay, L. E., Wingfield, P. T., Bax, A., Gronenborn, A. M., and Clore, G. M. (1989b) *Biochemistry,* 28, 6150-6156.

Mayer, B. J. and Hanafusa, H. (1990) *Proc. Natl. Acad. Sci. USA,* 87, 2638-2642.

Mayer, B. J., Jackson, P. K., and Baltimore, D. (1991) *Proc. Natl. Acad. Sci. USA.,* 88, 627-631

Mayer, B. J., Jackson, P. K., Van Etten, R. A. and Baltimore, D. (1992) *Mol. Cell. Biol.*, 12, 609-618.

Moran, M. F., Koch, C. A., Anderson, D., Ellis, C., England, L., Martin, G. S., and Pawson, T. (1990). *Proc. Natl. Acad. Sci. USA,* 87, 8622-8626.

Overduin, M., Mayer, B.J., Rios, C. B., Baltimore, D., and Cowburn, D. (1992a) *Proc. Natl. Acad. Sci. USA,* 89, 11673-11677.

Overduin, M., Rios, C. B., Mayer, B.J., Baltimore, D., and Cowburn, D. (1992b) *Cell,* 70, 697-704.

Waksman, G., Kominos D., Robertson S. C., Pant, N., Baltimore D., Birge R. B, Cowburn D., Hanafusa H., Mayer, B. J., Overduin, M., Resh, M. D., Rios, C. B., Silverman, L., and Kuriyan, J. (1992) *Nature,* 358, 646-653.

Waksman, G, Shoelson, S., Pant, N., Cowburn D., and Kuriyan, J. (1993) *Cell,* 72, 779-790.

Wüthrich, K. (1986) *NMR of Proteins and Nucleic Acids.* Wiley, New York.

NMR and Unpaired Electrons in Biomolecules

Ivano Bertini[1], Claudio Luchinat[2] and Roberta Pierattelli[1]

[1]Department of Chemistry, University of Florence, Via G. Capponi 7, 50121 Florence, Italy.

[2]Institute of Agricultural Chemistry, University of Bologna, Viale Berti Pichat 10, 40127 Bologna, Italy.

1. The effect of unpaired electrons on the NMR parameters.

Electron spins and nuclear spins.

Electron spins scare the NMR community whereas they should not (Bertini & Luchinat, 1986). The theory is just an extension of NMR theory which includes the spin of the electrons just like other nuclear spins.

The major differences between electron spins and nuclear spins are: the large magnetic moment associated with electrons (658 times that of the proton); the capability of electrons of being into orbitals; the presence of orbital angular momenta. The orbital angular momenta allows an efficient coupling between the spin and the lattice: fluctuations in the lattice induce spin transitions through the spin orbit operator (Banci et al., 1991b). Therefore electron relaxation is by far more effective than nuclear relaxation. Besides these properties, analogies are remarkable; they stem from the possibility of describing both systems with the spin Hamiltonian formalism.

Both electrons and nuclei sense an external magnetic field, B_0, shielded by orbital magnetic moment the former, and by electrons - either paired or unpaired - the latter. In order to describe the effective magnetic field, in the electron case we use to define a g tensor; in the case of nuclei a σ screening tensor. When the electron spin quantum number S is larger than or equal to 1, due to more than one unpaired electrons, there is zero field splitting (ZFS); analogously, when the nuclear spin quantum number I is larger than or equal to 1, there is the quadrupolar splitting (Q). Both ZFS and Q are traceless tensors and have identical properties.

Nuclear relaxation in solution occurs through: (a) quadrupolar electric gradient coupling, just like electron relaxation occurs through zero field splitting modulation; (b) shielding tensor anisotropy, just like electron relaxation through g anisotropy; (c) spin rotation, that causes relaxation of both nuclei and electrons.

Dipolar nucleus-nucleus interactions which cause relaxation may (d) show analogies with the electrons belonging to two different metal ions when they are located close one to the other. For instance, if a fast relaxing electron resides on metal a in a macromolecule, a nearby slow relaxing metal b may increase its relaxation though the usual dipolar coupling

$$\Delta R_{i(a)} = E_k^2 f_i(\tau_{Sa}, \omega) \quad (1)$$

NATO ASI Series, Vol. H 87
NMR of Biological Macromolecules
Edited by C. I. Stassinopoulou

where i is either 1 or 2, E_k^2 is the average of the square of the interaction energy, $f_i(\tau_{Sa}, \omega)$ are the appropriate sum of spectral density functions (see later).

Scalar relaxation (e) for nuclei depends on the Hamiltonian

$$J_{12}\mathbf{I}_1 \cdot \mathbf{I}_2 \tag{2}$$

where I_1 and I_2 are the coupled spins and J_{12} is the coupling constant; for electrons we have the analogous Hamiltonian

$$J_{12}\mathbf{S}_1 \cdot \mathbf{S}_2 \tag{3}$$

where the two electron spins are magnetically coupled through mechanisms of exchange or superexchange types.

Therefore, if one spin, either I_2 or S_2, relaxes fast and is at thermal equilibrium with the lattice, I_1 and S_1 will exchange energy with the lattice through nuclear spin or electron spin coupling.

Electron spin - nuclear spin coupling.

Having underlined the analogies of electrons and nuclei when they are separated, we should now consider what happens when they interact. The electron algebraically adds its large magnetic field to the external magnetic field, thus causing a shift which is called hyperfine shift because hyperfine is the energy range of the electron-nucleus coupling in the energy scale of electron spectroscopy. The large electronic magnetic field may cause a fluctuating magnetic field at the nucleus and cause nuclear relaxation. This contribution to relaxation is called relaxation rate enhancement due to hyperfine coupling.

Hyperfine shift.

In order to evaluate the shift we should consider that the nucleus sees an averaged electronic magnetic moment

$$\langle \mu_{ez} \rangle = g_e \mu_B \langle S_Z \rangle \tag{4}$$

where $\langle S_Z \rangle$ is the expectation value of the $\hat{S}_Z$ operator of an S multiplet, and $g_e = 2.0023$.

Let us consider first that the electron is localized on the metal ion. The interaction energy between the latter and $\langle \mu_{ez} \rangle$ is (Figure 1):

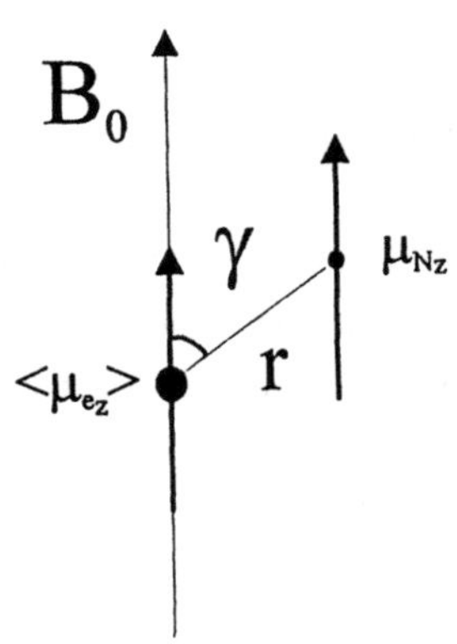

Figure 1. The nuclear magnetic moment μ_{Nz} and the average electron magnetic moment $\langle \mu_{e_z} \rangle$, both aligned with the external magnetic field B_0. The angle γ changes upon rotation of the molecule bearing the two magnetic moments.

$$E = \frac{\langle \mu_{ez} \rangle \cdot \mu_{Nz}}{r^3} \left(3\cos^2\gamma - 1 \right) \tag{5}$$

the two magnetic moments being aligned along the external magnetic field. Upon molecular tumbling the averaged energy which is given by the integral of eq. 5 is zero. However, if the magnitude of $\langle \mu_{ez} \rangle$ changes with the orientation of the molecule as a result of g being different from 2.0023 and anisotropic, the average value is different from zero. In terms of shift we have (McConnell & Robertson, 1958)

$$\left(\frac{\Delta\nu}{\nu_0} \right) = \frac{\mu_0}{4\pi} \frac{\mu_B^2 S(S+1)}{18kTr^3} \left\{ \left[2g_{zz}^2 - \left(g_{xx}^2 + g_{yy}^2 \right) \right] \left(3\cos^2\theta - 1 \right) + \right.$$

$$\left. +3(g_{xx}^2 - g_{yy}^2)\sin^2\theta \cos 2\Omega \right\} \tag{6}$$

where r, θ and Ω are the polar coordinates of the proton within the molecular axes determined by the principal directions of the g tensor.

A more complete equation is (Kurland & McGarvey, 1970)

$$\left(\frac{\Delta\nu}{\nu_0} \right) = \frac{1}{24\pi r^3} \left\{ \left[2\chi_{zz} - \left(\chi_{xx} + \chi_{yy} \right) \right] \left(3\cos^2\theta - 1 \right) + \right.$$

$$\left. +3\left(\chi_{xx} - \chi_{yy} \right) \sin^2\theta \cos 2\Omega \right\} \tag{7}$$

where χ_{kk} are the principal directions of the magnetic susceptibility tensor. Equation 7 includes second order Zeeman effects. This shift is called pseudocontact shift.

Unpaired electrons may somehow induce spin density at nuclei. Spin density is the difference between electron density with spin -1/2 and electron density with spin 1/2 in any point. Orbitals of s type have a finite non zero electron density at the nucleus. They may belong to the molecular orbital which contains the unpaired electron, and then directly place spin density at the nucleus. Alternatively, or simultaneously to the above mechanism, spin polarization from the molecular orbital containing the unpaired electron to doubly occupied molecular orbitals occurs. Spin polarization attracts the electron with the same spin as that of the unpaired electron and repels the electron with different spin. This makes the two spin orbitals of the doubly occupied orbital inequivalent, and causes spin density contribution at nuclei through the s orbitals.

Spin density ρ at the nucleus gives rise to an additional field sensed by the nucleus. The field, or shift, is bound to the spin density (or fraction of one electron)

$$-\left(\frac{\Delta B}{B_0} \right) = \left(\frac{\Delta\nu}{\nu_0} \right) = -\frac{a_c}{\hbar\gamma_I B_0} \langle S_Z \rangle \tag{8}$$

where

$$a_c = \frac{\mu_0}{3S}\hbar\gamma_I g_e \mu_B \rho \tag{9}$$

Spin density at the nucleus causes the so called contact contribution to the hyperfine coupling and the shift is called contact shift. Spin density at the nucleus is just like a limit case of the coupling. Localizing the unpaired electrons at the nucleus center is an oversimplification. We should also consider the pseudocontact shift generated by the spin density several bonds away from the metal center. This is often called ligand-centered pseudocontact shift. It is usually small for protons but not for heteronuclei which may host spin density on a p orbital.

In general, both contact and pseudocontact shifts contribute to the observed hyperfine isotropic shift, and their factorization is not obvious at all. Only in some cases we may independently know which contribution is dominant. In case of metalloproteins we have many nuclei close to the metal but separated by many chemical bonds. In this case the shift is only dipolar in nature, and if we have enough data we can determine also the magnetic susceptibility tensor (Williams et al., 1985; Emerson & La Mar, 1990).

In summary, and using a different formalism, the interaction energy between S and I is given by:

$$\mathcal{H} = \mathbf{I}\cdot\mathbf{A}\cdot\mathbf{S} \tag{10}$$

where **A** is a second rank tensor which can be divided into a traceless tensor plus scalar terms:

$$\mathcal{H} = a_c\mathbf{I}\cdot\mathbf{S} + \mathbf{I}\cdot\mathbf{A}'\cdot\mathbf{S} = a_c\mathbf{I}\cdot\mathbf{S} + a_{pc}\mathbf{I}\cdot\mathbf{S} + \mathbf{I}\cdot\mathbf{A}''\cdot\mathbf{S} \tag{11}$$

where **A'** is the dipolar coupling tensor and **A"** is the traceless part of it. Eq. 7 and eq. 8 are the two explicit equations obtained by evaluating the $a_c\mathbf{I}\cdot\mathbf{S}$ and $a_{pc}\mathbf{I}\cdot\mathbf{S}$ parts of the Hamiltonian. In eq.7 and eq.8 there are no nuclear parameters; they represent the contribution to nuclear shift by a magnetic field added to the external magnetic field.

Relaxation enhancement due to hyperfine coupling.

As far as relaxation is concerned, we should again refer to eq.11. The relaxation rates in general are given by

$$R_i = E_k^2 f_{ik}(\tau,\omega) \tag{12}$$

where i is either 1 or 2, $E_k{}^2$ is the average of the squared interaction energy whose modulation causes relaxation and $f(\tau,\omega)$ is the appropriate sum of spectral density functions. If E_k is the contact energy, then the contribution to relaxation is modulated by electron relaxation T_2 or τ_s or by chemical exchange τ_M. The function f_1 contains only the zero quantum transitions

$\omega_S - \omega_I \approx \omega_S$ and f_2 contains, as usual, also the non dispersive term τ_c (Solomon & Bloembergen, 1956; Bloembergen, 1957 ; Abragam, 1961)

$$R_1^{cont} = \frac{2}{3}\left(\frac{a_c}{\hbar}\right)^2 S(S+1)\frac{\tau_c}{1+(\omega_I-\omega_S)^2\tau_c^2} \tag{13}$$

$$R_2^{cont} = \frac{1}{3}\left(\frac{a_c}{\hbar}\right)^2 S(S+1)\left(\frac{\tau_c}{1+(\omega_I-\omega_S)^2\tau_c^2}+\tau_c\right) \tag{14}$$

$$R_{1\rho}^{cont} = \frac{1}{3}\left(\frac{a_c}{\hbar}\right)^2 S(S+1)\left(\frac{\tau_c}{1+(\omega_I-\omega_S)^2\tau_c^2}+\frac{\tau_c}{1+\omega_1^2\tau_c^2}\right) \tag{15}$$

where

$$\tau_c^{-1} = \tau_s^{-1} + \tau_M^{-1} \tag{16}$$

ω_I and ω_S are the nuclear and electron Larmor frequencies, respectively, and ω_1 is the nuclear Larmor frequency in the B_1 field of the rotating frame.

When the coupling energy is dipolar in origin, we should also refer to eq.5. The average integral of ($3\cos^2\theta$-1) is zero but not the average of its square, which is 4/5. Therefore the part of the Hamiltonian which contains the traceless tensor is the one which provides dipolar relaxation. The fluctuations which modulate the coupling energy are now the electron relaxation, chemical exchange and molecular tumbling τ_r, and the single quantum transition should be included in the f_i's (Solomon, 1955)

$$R_1^{dip} = \frac{2}{15}\left(\frac{\mu_0}{4\pi}\right)^2 \frac{\gamma_I^2 g_e^2 \mu_B^2 S(S+1)}{r^6}\left(\frac{\tau_c}{1+(\omega_I-\omega_S)^2\tau_c^2} + \frac{3\tau_c}{1+\omega_I^2\tau_c^2} + \frac{6\tau_c}{1+(\omega_I+\omega_S)^2\tau_c^2}\right) \tag{17}$$

$$R_2^{dip} = \frac{1}{15}\left(\frac{\mu_0}{4\pi}\right)^2 \frac{\gamma_I^2 g_e^2 \mu_B^2 S(S+1)}{r^6}\left(4\tau_c + \frac{\tau_c}{1+(\omega_I-\omega_S)^2\tau_c^2} + \frac{3\tau_c}{1+\omega_I^2\tau_c^2} + \frac{6\tau_c}{1+(\omega_I+\omega_S)^2\tau_c^2} + \frac{6\tau_c}{1+\omega_S^2\tau_c^2}\right) \tag{18}$$

$$R_{1\rho}^{dip} = \frac{1}{15}\left(\frac{\mu_0}{4\pi}\right)^2 \frac{\gamma_I^2 g_e^2 \mu_B^2 S(S+1)}{r^6}\left(\frac{4\tau_c}{1+\omega_I^2\tau_c^2} + \frac{\tau_c}{1+(\omega_I - \omega_S)^2\tau_c^2} + \right.$$

$$\left. + \frac{3\tau_c}{1+\omega_I^2\tau_c^2} + \frac{6\tau_c}{1+(\omega_I + \omega_S)^2\tau_c^2} + \frac{6\tau_c}{1+\omega_S^2\tau_c^2}\right) \quad (19)$$

where

$$\tau_c^{-1} = \tau_s^{-1} + \tau_M^{-1} + \tau_r^{-1} \quad (20)$$

One can recognize in the above equations the square of the electronic magnetic moment and something proportional to the square of the nuclear magnetic moment γ_I. At variance with the case of the shift, eqs.13-20 contain the nuclear parameters. In the contact case, γ_I is present in a_c (see eq. 9). Note that $R_{i\rho}$ correspond to R_2 if $\omega_1\tau_c << 1$ as is usual.

Until now, for the description of relaxation, we have considered that the Zeeman electronic levels are equally populated. The correction would be negligible. However, the induced magnetic moment, as such, gives rise to dipolar relaxation upon molecular tumbling and if the tumbling time is long because the molecule is large, the effect is dramatic. The square of the induced magnetic moment depends also on B_0^2 and therefore this relaxation mechanism puts a limit in the optimal intensity of the magnetic field (see section 4). The equations for this contribution to relaxation which is called Curie relaxation because it is bound to the induced magnetic moment are (Gueron, 1975; Vega & Fiat, 1976):

$$R_1^{Curie} = \frac{2}{5}\left(\frac{\mu_0}{4\pi}\right)^2 \frac{1}{r^6} \frac{\omega_I^2 g_e^4 \mu_B^4 S^2(S+1)^2}{(3kT)^2} \cdot \frac{3\tau_r}{1+\omega_I^2\tau_r^2} \quad (21)$$

$$R_2^{Curie} = \frac{1}{5}\left(\frac{\mu_0}{4\pi}\right)^2 \frac{1}{r^6} \frac{\omega_I^2 g_e^4 \mu_B^4 S^2(S+1)^2}{(3kT)^2} \cdot \left[4\tau_r + \frac{3\tau_r}{1+\omega_I^2\tau_r^2}\right] \quad (22)$$

$$R_{1\rho}^{Curie} = \frac{1}{5}\left(\frac{\mu_0}{4\pi}\right)^2 \frac{1}{r^6} \frac{\omega_I^2 g_e^4 \mu_B^4 S^2(S+1)^2}{(3kT)^2} \cdot \left[\frac{4\tau_r}{1+\omega_I^2\tau_r^2} + \frac{3\tau_r}{1+\omega_I^2\tau_r^2}\right] \quad (23)$$

In general the contribution is sizable in the R_2 and $R_{1\rho}$ cases.

Nuclear Overhauser Effect (NOE).

The NOE is defined as the fractional variation of the intensity of one signal when another one is saturated (Overhauser, 1953a,b). In the two spin approximation the development of η as a function of the saturation time is

$$\eta_{IJ}(\tau) = \frac{\sigma_{IJ}}{\rho_{IJ} + R_p}\left(1 - e^{-(\rho_{IJ}+R_p)\tau}\right) \tag{24}$$

where I is the observed signal, J is the saturated signal, σ_{IJ} is the cross relaxation, ρ_{IJ} is the relaxation of I due to J and R_p is the paramagnetic relaxation. R_p is the dominant term in the denominator, and therefore for a long time it was believed that no NOE could be detected in paramagnetic systems. Indeed it is small but measurable. The advantage with respect to diamagnetic systems is that the two spin approximation is often a better approximation because the nucleus relaxes through coupling with the electron and spin diffusion is reduced.

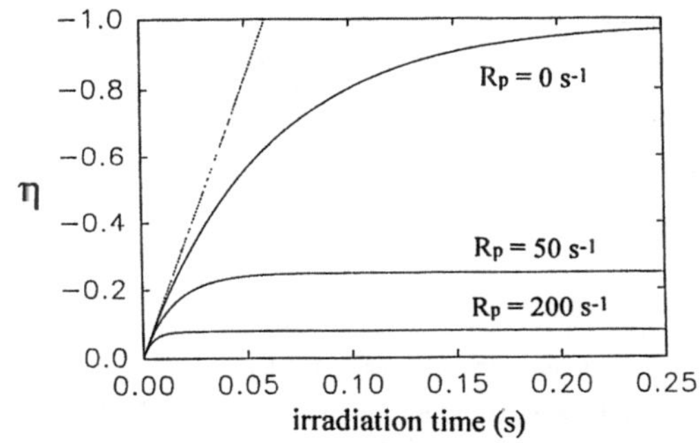

Figure 2. Build-up of NOE on a signal I as a function of the saturation time τ of signal J. Conditions: τ_r = 10 ns, for two hydrogen nuclei 1.8 Å apart. The initial slope (dotted lines) is equal to σ_{IJ}.

A typical NOE build-up curve is reported in Figure 2. At short τ values, $\eta_{IJ}(\tau) = \sigma_{IJ}\tau$ and the slope gives σ_{IJ}. For large τ, $\eta_{IJ}(\tau) = \sigma_{IJ}/R_p$. From the build-up curve we can obtain both σ_{IJ} and R_p.

The cross relaxation in case of homonuclear systems is given by:

$$\sigma_{IJ} = \left(\frac{\mu_0}{4\pi}\right)^2 \frac{2\hbar^2\gamma_I^2\gamma_J^2 J(J+1)}{15 r_{IJ}^6}\left(\frac{6\tau_c}{1+(\omega_I+\omega_J)^2\tau_c^2} + \frac{\tau_c}{1+(\omega_I-\omega_J)^2\tau_c^2}\right) \tag{25}$$

It depends on the square of the two magnetic moments, on the sixth power of the nucleus-nucleus distance, and on ω and τ_c. The latter is the reorientational correlation time. Note that R_p depends on the sixth power of the metal-nucleus distance and on the electron relaxation time as correlation time.

NOE as a function of B_0 and τ_c is reported in Figure 3. It appears that for $\omega\tau \leq 1$, the NOE is small and tends to zero. In small complexes, NOE and NOESY may easily escape detection. In this case it is appropriate to decrease the temperature or increase viscosity of the solvent in order to increase τ_c.

NOE in paramagnetic metalloproteins is still the most powerful tool to detect connectivities with the maximum signal to noise ratio. The transient NOE is measured when a 180° pulse is applied on J. This is hardly ever superior to the steady state NOE obtained by saturating J (Banci et al., 1990b), except at short irradiation time (Figure 4). Furthermore

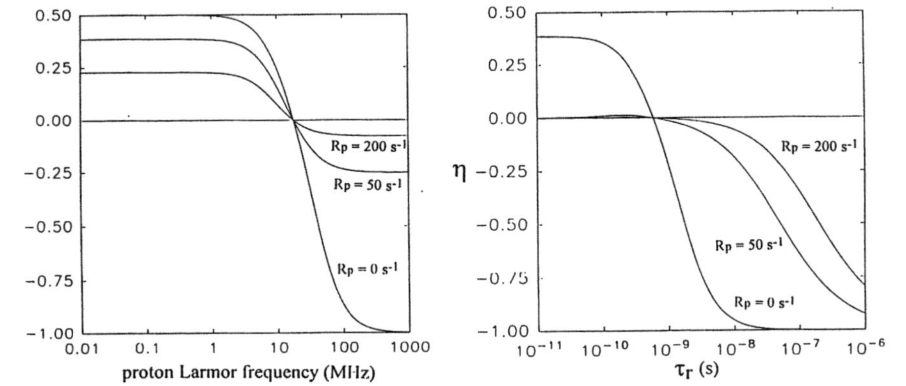

Figure 3. NOE values as a function of (A) magnetic field for τ_c = 10 ns and (B) τ_c at a field of 300 MHz., for two hydrogen nuclei at 1.8 Å distance.

selective inversion of a fast relaxing signal is hard to achieve and artifacts are a common consequence. After inversion, J transfers magnetization to I, but also relaxes with its own R_J. If J relaxes fast, then the time available to build-up NOE on I is small and the effect disappears depending on both R_J and R_I. In Figure 4 a comparison between the time dependence of transient NOE and normal NOE is shown.

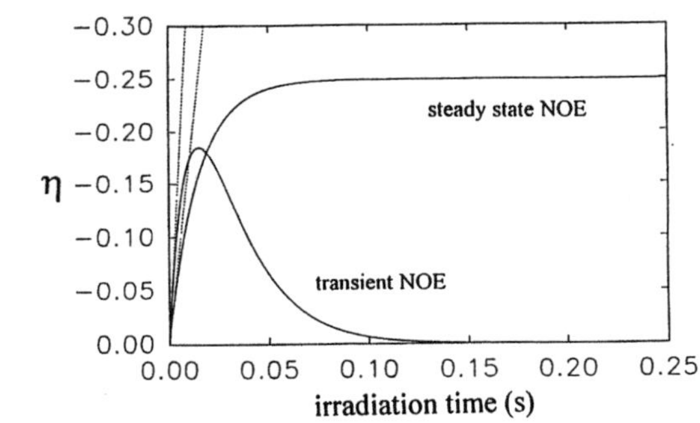

Figure 4. NOE build-up in a steady state experiment as compared to a transient NOE experiment. Conditions are the same as in Figure 2 with R_p = 50 s^{-1}. The dotted lines represent the initial build-up slopes, the faster build-up being observed for transient NOE.

The systems on which high resolution NMR is attempted are those in which the correlation time for R_p is τ_s. In this case τ_s can be determined experimentally if a large range of B_0 values are used. Often we can also learn about the B_0 dependence of τ_s. In Table 1 the values of τ_s at low B_0 are reported for a series of metal ions together with the broadening effect on a proton at 5 Å distance due to dipolar relaxation. It appears that few are the metal ions which do not cause dramatic line broadening. Among them we note iron, which is very important in biology, cobalt and nickel, which are used to substitute zinc, and lanthanides which can substitute calcium. The use of paramagnetic ions is sometimes recommended to obtain structural and dynamic information on the vicinity of the metal ion, the upper limit for the molecular size being imposed by Curie relaxation.

Magnetic coupling between electrons on different metal ions causes shortening of electron relaxation. This is general for polymetallic systems, and occurs when one metal ion in dimetallic systems is one of the above mentioned short τ_S metal ions. Therefore typically all the Fe-S proteins are suitable for investigation (Bertini et aL, 1993d).

When the metal ions have long τ_S, they have been used as relaxing agents to study for instance substrate in excess which exchanges rapidly between bound and free states.

Spin labels have the longest τ_S (10^{-5} - 10^{-7} s) and they can be used to measure the R_p of nearby (<20 Å) protons and to obtain structural information.

Metal ion	τ_s (s)	Line broadening (Hz)
Ti^{3+}	10^{-10}-10^{-11}	200-20
V^{2+}	5×10^{-10}	4000
V^{3+}	5×10^{-12}	30
Cr^{2+}	10^{-11}	200
Mn^{2+}	10^{-8}-10^{-10}	100000-1500
Fe^{3+} (HS)	10^{-10}-10^{-11}	2000-200
Fe^{3+} (LS)	10^{-11}-10^{-12}	20-5
Fe^{2+} (HS)	10^{-11}-10^{-12}	150-30
Co^{2+} (HS)	10^{-11}-10^{-12}	100-20
Co^{2+} (LS)	10^{-9}-10^{-10}	1500-200
Ni^{2+}	10^{-10}-10^{-12}	500-10
Cu^{2+}	1-3×10^{-9}	1500-3500
Ru^{3+}	10^{-11}-10^{-12}	20-5
Gd^{3+}	10^{-8}-10^{-9}	200000-30000
Dy^{3+}	8×10^{-13}	150
Ho^{3+}	8×10^{-13}	150
Tb^{3+}	8×10^{-13}	100
Tm^{3+}	8×10^{-13}	70
Yb^{3+}	1×10^{-12}	30

Table 1. Estimated room temperature electron relaxation times for some metal ions at low magnetic field (10^{-2}-10^{-3} T), together with the broadening effect on a proton at 5 Å distance.

2. Recording the spectra

1D spectra can be recorded essentially as in the case of diamagnetic systems, with some additional devices and precautions. A characteristic feature of paramagnetic systems is the large spectral width. Very short pulses may be needed, which can be obtained with special power supplies and probes. In order to sample FID's on a large spectral width, low dynamic range ADC (analog to digital converter) can be used. However, insufficient dynamic range may be a serious problem because a broad paramagnetic line has a small signal to noise ratio and may be lost if the intensity of the diamagnetic peaks is not reduced through presaturation or special pulse sequences.

In 2D spectra we should take into consideration the fast relaxing nature of nuclei. We have to reach measurable magnetization or coherence transfer before return to equilibrium occurs. In every case the experiment must be optimized to match the T_1's (NOESY) or T_2's (COSY) of the nuclei of interest. In TOCSY the spin-lock of a large spectral width may heat the sample. One advantage is that we can accumulate with short recycle time.

3. Strategies for assignment

Beside intensity, the oldest criterion for peak assignment is: the broader the signal, the closer the nucleus to the paramagnetic center. The same holds for R_1. As mentioned, this criterion should be used with caution and not for quantitative purposes. First of all, the Curie

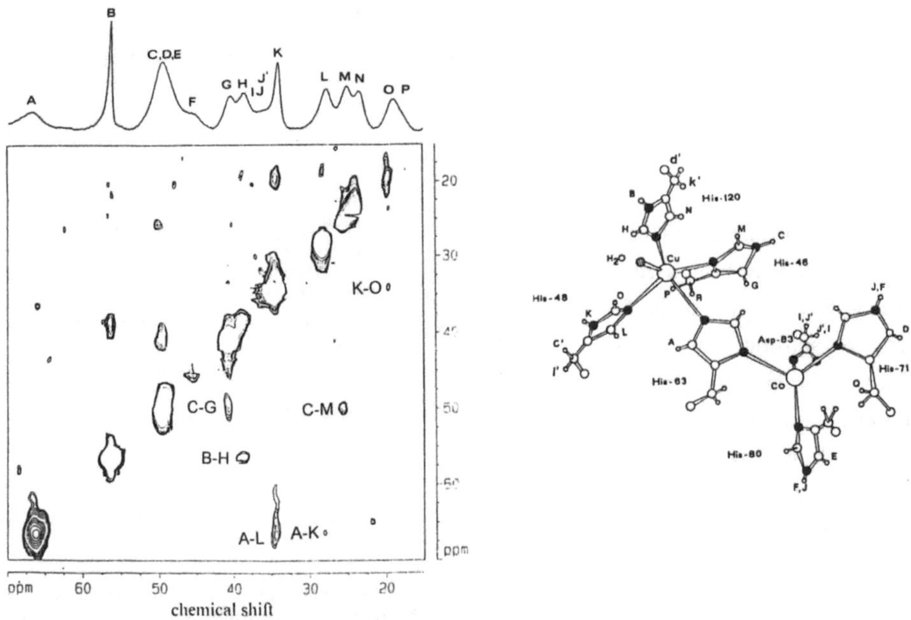

Figure 5. 600 MHz ^1H NMR NOESY spectrum of Cu_2Co_2SOD in H_2O showing cross peaks between fast relaxing hyperfine shifted signals. The 1D spectrum is reported as reference. The experiment was recorded at 300 K using 4 ms of mixing time and 185 ms of relaxation delay. A schematic drawing of the active site is also reported.

relaxation contribution to R_2 should be factored out by performing measurements at various magnetic fields. Then ligand centered effects should be considered, especially for heteronuclei.

A safer criterion for assignment is NOE. With T_1 of the order of 10 ms, NOE's provide accurate distances. In the case of copper-zinc Superoxide Dismutase (SOD, a protein of MW 32,000), zinc has been substituted by cobalt which is a suitable metal ion for NMR studies. Exchange

Table 2. Assignment of ^1H NMR hyperfine shifted signals of Cu_2Co_2SOD.

Signal	Shift (ppm)	T_1 (ms)	Assignment
A	66.2	1.5	His 63 $H\delta_2$
B	56.5	7.8	His 120 $HN\delta_1$
C	50.3	4.2	His 46 $HN\varepsilon_2$
D	49.4	3.8	His 71 $H\delta_2$
E	48.8	4.6	His 80 $H\delta_2$
F	46.7	2.1	His 71 $HN\varepsilon_2$ (His 80 $HN\varepsilon_2$)
G	40.6	3.5	His 46 $H\delta_2$
H	39.0	1.8	His 120 $H\varepsilon_1$
I	37.4	1.7	Asp 83 $H\beta_1$ (Asp 83 $H\beta_2$)
J'	35.6	1.7	Asp 83 $H\beta_2$ (Asp 83 $H\beta_1$)
J	35.4	–	His 80 $HN\varepsilon_2$ (His 71 $HN\varepsilon_2$)
K	34.5	8.0	His 48 $HN\delta_1$
L	28.4	4.3	His 48 $H\delta_2$
M	25.3	2.7	His 46 $H\varepsilon_1$
N	24.1	2.9	His 120 $H\delta_2$
O	19.6	1.9	His 48 $H\varepsilon_1$
P	18.7	1.6	His 46 $H\beta_1$
Q	-6.2	2.4	His 71 $H\beta_2$
R	-6.2	2.4	His 46 $H\beta_2$

magnetic coupling makes also copper suitable for NMR (Bertini et al., 1985). The spectrum and the NOESY map are reported in Figure 5, and the assignment in Table 2 (Banci et al., 1993d). The assignment is obtained by comparing calculated distances from NOE's to distances observed in the X-ray structure (Tainer et al., 1982).

In the case of heme proteins, local information can be obtained especially when low spin iron(III) is present (La Mar & Walker, 1979; Satterlee, 1986).

In Figure 6 the NOESY and COSY spectra of the cyanide derivative of Lignin Peroxidase (LiP) are shown (Banci et al., 1991c). Signal A shows NOESY cross peaks with 4Hα, $2H\beta_{trans}$ and $4H\beta_{cis}$. The fact that A senses two vinyl groups identifies A as $3CH_3$. Thus the relative orientation of the vinyl chain is obtained. Signal B shows NOESY with 7Hα propionates. The signals of the distal histidines are also easily assigned because the exchangeable NH gives two NOEs to two vicinal protons. Also the signals of the proximal histidine have been detected and identified (de Ropp et al., 1991). They are characterized by small T_1 values. The distance from $3CH_3$ and $2H\beta_{trans}$ are not consistent with molecular modeling. From a NOE build-up we learn that the reorientational correlation time τ_c is shorter than molecular tumbling. This is due to fast rotation around the C-CH σ bond at position 2.

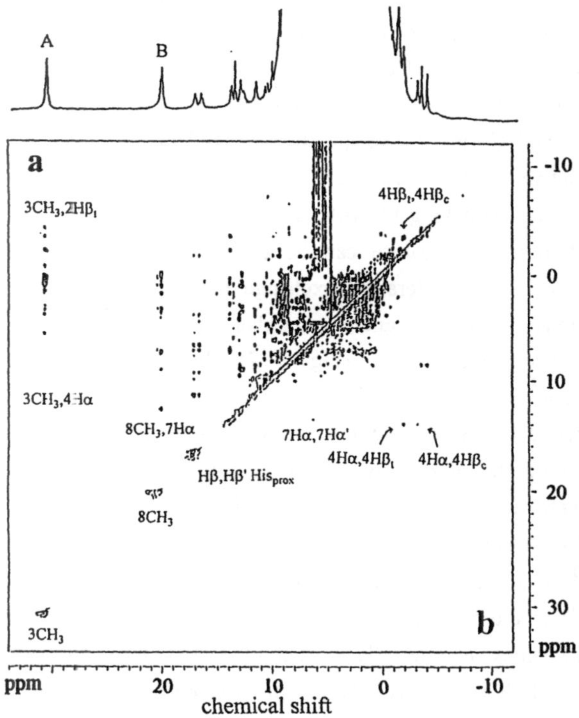

Figure 6. (a) 600 MHz NOESY spectrum of LiP-CN⁻ in D_2O solution obtained with a 15 ms mixing time; (b) magnitude COSY spectra of LiP-CN⁻ (Banci et al., 1991c).

In the case of Manganese Peroxidase (MnP) which is a protein which binds Mn(II) as substrate it has been possible to obtain information on the manganese binding site by exploiting the broadening effects of the slow relaxing S=5/2 (Banci et al., 1992e).

^1H NMR have been particularly useful in mixed valence Fe-S proteins containing iron at different oxidation states. The reduced $Fe_2S_2^+$ cluster contains one Fe^{3+} and one Fe^{2+}, the former with S=5/2 and the latter with S=2. In the spectra we observe four signals far downfield with Curie type temperature dependence, and several signals less shifted with anti Curie type temperature dependence (Dunham et al., 1971, Bertini et al., 1984). Such signals belong to protons of coordinated cysteines. This feature can be exploited by taking into consideration exchange magnetic coupling. The magnetic coupling is of antiferromagnetic type (Figure 7). In the ground state the large spin is oriented along the external magnetic field but the small spin is oriented the other way around (Figure 7A). Therefore the nuclei sensing Fe^{3+} will interact with a normal ⟨Sz⟩, whereas the nuclei sensing Fe^{2+} will interact with an inverted ⟨Sz⟩. If only the ground state were populated, nuclei sensing Fe^{2+} would be upfield (Figure 7B). In the extreme limit of infinite temperature (kT >> J) both ⟨Sz⟩ values have the values which they would have in the absence of magnetic coupling (Banci et al., 1990a). So, even the protons sensing Fe^{2+} would be downfield (Figure 7D). The actual location of the signals of anti Curie type is the result of partial occupation of excited energy levels (J ≈ kT, Figure 7C); upon increasing the temperature from case B to case C these signals tend to go far downfield.

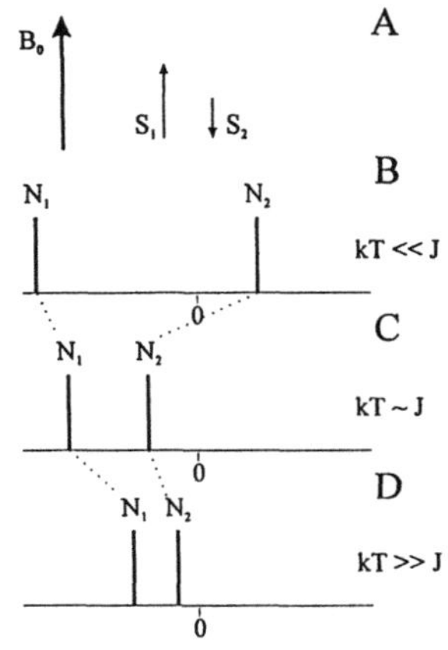

Figure 7. The S=1/2 ground state of an antiferromagnetically coupled S_1=5/2-S_2=2 system can be viewed as constituted by the larger magnetic moment oriented along the external field and the smaller magnetic moment oriented against the external magnetic field. The shift of nuclei sensing either S_1 or S_2 are shown for different temperatures.

Once we have assigned the protons to each metal domain, we note from the X-ray structure that two protons of two different βCH_2 moieties are close one to another. NOE connectivities has been observed between two βCH_2's between signals with anti Curie behavior. Therefore the iron(II) in the structure has been identified (Dugad et al., 1990; Skjeldal et al., 1991). This is an important result for NMR spectroscopy!

In the case of Fe_nS_4 clusters with n > 2 there are usually no cysteine protons close to each other to help for a spectroscopic sequence specific assignment of the protons close to the metal ion. The strategy that we have developed is based on measuring NOE's from the βCH_2's and on recognizing the signals experiencing NOE as part of TOCSY patterns. For example, the ^1H NMR spectrum of a $Fe_4S_4^{3+}$ cluster containing two Fe^{3+} ions and two $Fe^{2.5+}$ ions is shown in Figure 8. The downfield shifted signals are due to cysteinyl βCH_2 protons sensing the mixed

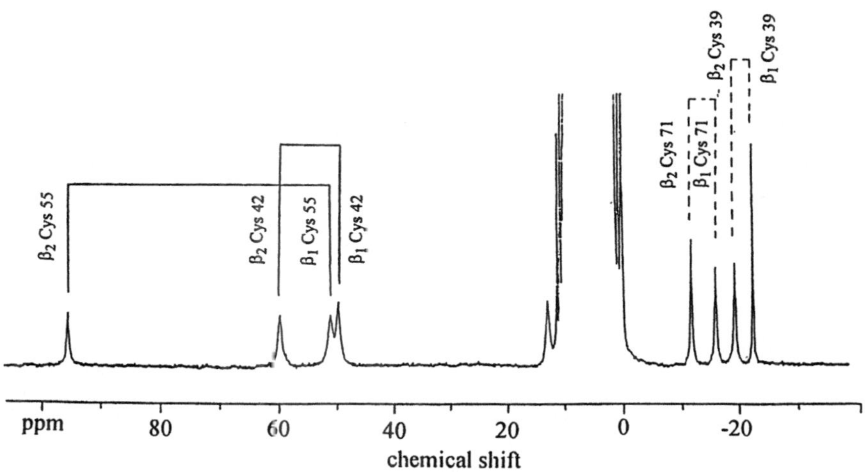

Figure 8. 600 MHz ^{1}H NMR spectrum of the oxidized form of the HiPIP II from *E. halophila* (Banci et al., 1993b).

valence pairs with a subspin of S = 9/2 and the upfield shifted signals to cysteinyl βCH_2 protons of the iron(III) domain with subspin S = 4 (Bertini et al., 1992a). NOE's from βCH_2 pairs and signals in the diamagnetic region are observed. Through TOCSY, the latter signals have been assigned to various type of residues. So, if the structure is known or a good model is available (Banci et al., 1993b), the assignment of the protons around the cluster is performed (Figure 8) (Banci et al., 1991a, 1993c; Bertini et al., 1992b; Bertini et al., 1993a,c,f). Figure 9 shows which are the cysteines bound to $Fe^{2.5+}$ and those bound to Fe^{3+}.

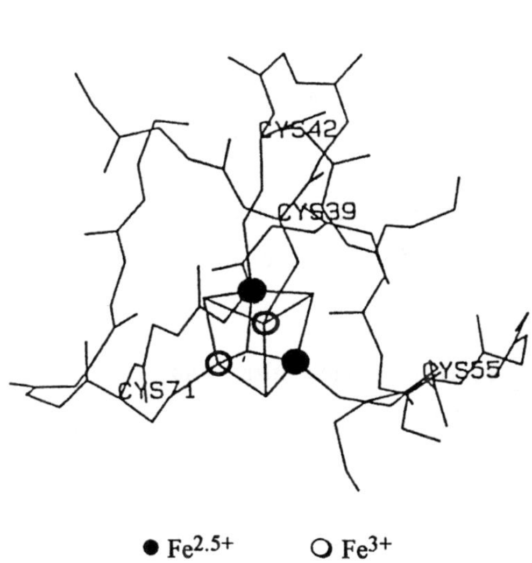

Figure 9. Location within the proton frame of the ferric and mixed valence pairs in the Fe_4S_4 cluster of oxidized HiPIP II from *E. halophila* (Banci et al., 1993b).

We have used molecular dynamics simulation (Amber IV) to generate structural models by starting from homologous structures for which the X-ray structure is available, and then we have compared interproton distances with those found by NOE's. The results are extremely good. In the case of HiPIP from *Cromatium vinosum* we have refined the existing X-ray structure with molecular dynamics (Figure 10). The refined structure is in a much better agreement with experimental NOE's (Banci et al., 1992).

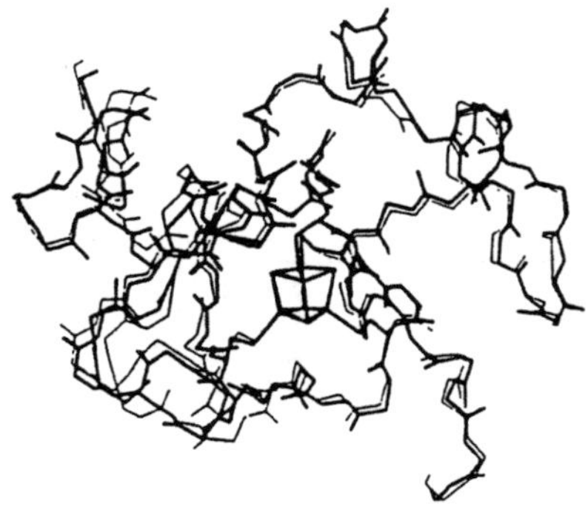

Figure 10. Comparison between X-ray (thin line) and MD-refined structure (bold line) of the HiPIP from *C.Vinosum*. NOEs indicate that the latter is in better agreement with the actual solution structure at least as far as the neighborhood of the cluster is concerned (Banci et al., 1992).

In the case of $Fe_4S_4^{2+}$ with all the iron ions at the oxidation state 2.5+ we could also follow the shift as it depends on the dihedral angle θ between the M-S-C and S-C-H planes. The following equation is found to hold:

$$\delta = A\sin^2\theta + B\cos\theta + C$$

The values of A, B and C are 11.5, -2.9 and -3.7 ppm respectively (Bertini et al., 1993b).

The $\sin^2\theta$ dependence implies an involvement of the p_z orbitals of sulfur in the spin density transfer. In the case of amines the dependence is of $\cos^2\theta$ type (Ho & Reilley, 1969).

4. Further peculiarities and perspectives

Curie relaxation is a phenomenon with important practical consequences. If we compare the spectra at 90, 200 and 600 MHz of the cobalt(II)-substituted Carbonic Anhydrase (CA) adduct with NCS^- (Figure 11), we see that at low fields there are more far shifted signals that at high field. The signals at 123, 95 and -31 ppm observed at 90 MHz

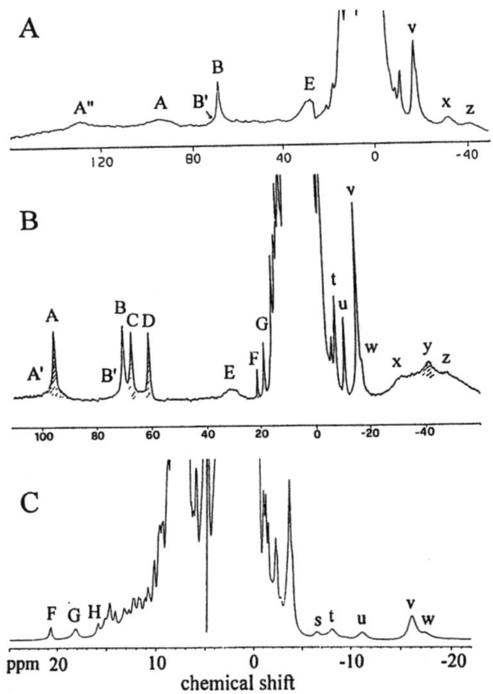

Figure 11. ^{1}H NMR spectra of the NCS^- adduct of cobalt(II)-substituted CA at (A) 90 MHz, (B) 200 MHz, and (C) 600 MHz, showing an expansion of the 25/-20 ppm region. The spectra were recorded at 300 K. Spectrum (A) was recorded in D_2O. The shaded signals in spectra (B) and (C) correspond to protons which are exchangeable in a D_2O solution. Note the progressive loss of Curie relaxation-broadened signals with increasing magnetic field.

belong to *ortho-like* protons* of histidines and those observed at 200 MHz at 94.1, 69.2, 66.4, and 60.2 ppm belong to *meta-like* protons. If we want to obtain structural information on the core of the protein, we should use different fields to investigate different spherical shells which involve signals with different relaxation properties (Bertini et al., 1993g).

Curie relaxation is important also in COSY spectroscopy. Curie relaxation and proton-proton relaxation depend on the same spectral density function. Therefore the linewidth is due to:

$$R_2 = \left(E_{Curie} + E_{dip}\right)^2 f_2(\tau,\omega)$$

and we have a cross product between the two terms. Since the sign of E_{dip} for one nuclear spin depends on the spin state of the other, the cross product can have two different signs and this causes the two components of the doublet of an AX pair to be inequivalent in terms of linewidths. If the two components of doublets have very large linewidths, then COSY cross peaks due to scalar coupling can be extremely weak, and they are often overwhelmed by cross peaks arising from cross-correlation between proton-proton dipolar interaction and Curie relaxation (Bertini et al., 1993h).

A further important progress in the NMR of paramagnetic systems could be achieved with the application of 3D spectroscopy. The advantage of using 3D experiments in paramagnetic systems, besides a generic increase in spectral resolution, is that cross peaks should be easily observable between signals buried under the diamagnetic envelope, when edited along the dimension of the well resolved far hyperfine shifted signals (Banci et al., 1993a). This partial increase of resolution should make the detection of connectivities possible which may easily be lost in 2D maps of the diamagnetic region because of signal overlapping and/or broad linewidths. A further advantage is that each section plane is cleaned from all the uncoupled signals, and peaks are present only from signals which experience two-center or three-center interactions. The disadvantage of the smaller digital resolution is much less of a problem for broad signals even when buried under the diamagnetic envelope. An encouraging example comes from the application of a 3D NOESY-NOESY sequence on MetMyoglobin-CN^- (Banci et al., 1993a). In Figure 12 a $\omega_1\omega_2$ cross section at the frequency ω_3 (25.7 ppm) of $5CH_3$ is reported. All the expected 2D cross peaks arising from $5CH_3$ are observed along the two cross-diagonal lines. 3D cross peaks are also evident. In particular the 3D cross peaks among $5CH_3$, 6Hα, 6Hα', among the $5CH_3$, 4Hα, and Hβ meso protons, and among the $5CH_3$, 6Hβ, and 6Hβ' protons are detected with good signal-to-noise ratio, despite the paramagnetic nature of the signals and the fact that the coupling is transferred during two evolution and two mixing times. The inset shows the simulated $\omega_1\omega_2$ cross section obtained using the experimental relaxation parameters of the $5CH_3$, 6Hα and 6Hα' coupled spins, with their geometric

* We define *ortho-like* protons as the ring protons three bonds away from the metal ion and *meta-like* protons as those four bonds away from the metal ion.

properties and with the experimental signal-to-noise ratio. The experimentally observed different intensities on the two sides of the diagonal are correctly predicted by the simulation. Is important to note that the simulated spectra allows us to properly chose the experimental parameters, such as the mixing times.

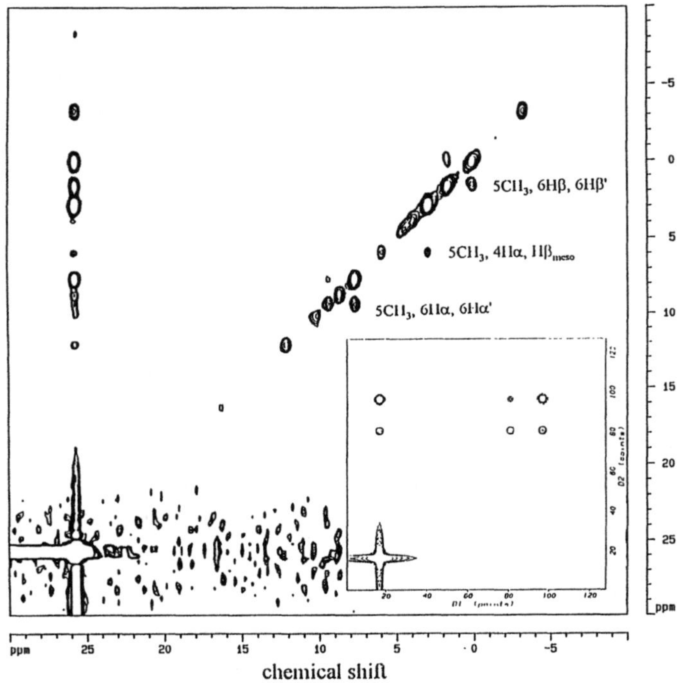

Figure 12. $\omega_1\omega_2$ cross section at the frequency ω_3 (25.7 ppm) of $5CH_3$ of a 3D NOE-NOE spectrum recorded on MetMyoglobin-CN^-. The inset shows the simulated $\omega_1\omega_2$ cross section obtained using the experimental relaxation parameters of the $5CH_3$, 6Hα and 6Hα' coupled spins, with their relative distances and with the experimental signal-to-noise ratio.

The bridge between diamagnetic and paramagnetic signals can be built with a NOE-NOESY experiment, obtained by editing a NOESY spectrum with a monodimensional NOE difference experiment (Bertini et al., 1993e). In this way, by recording the difference NOESY spectrum, it is possible to obtain information on the coupling of the signals dipolarly coupled with the fast relaxing ones, despite they are buried in the diamagnetic envelope of the protein. This experiment seems to be particularly useful to study large molecular weight proteins, since NOEs and NOESY intensities increase with the MW, and spin diffusion plays a key role in this experiment.

Hetcor experiments have been up to now much less common then homonuclear experiments in paramagnetic proteins. However, with the improvement of the isotope enrichment capabilities, it is expected that hetcor experiments will play a more important role in the future. Indeed, heteronuclear one-bond coupling constants are one or more order of magnitude larger than proton-proton coupling constants, and the corresponding scalar correlation experiments should suffer much less from the presence of broad lines.

Extensive hetcor experiments have been successfully performed on weakly paramagnetic iron-sulfur protein (Oh, B.-H. et al., 1990); ^{1}H-^{15}N cross peaks with the distal ring nitrogen of coordinated histidines have been observed (Bertini et al., unpublished results), natural abundance ^{1}H-^{13}C cross peaks have been observed for heme signals in low spin heme proteins (Yamamoto, Y. 1987; Timkovich, R. 1991; Banci et al., unpublished results), and for the β CH_2 groups of coordinated cysteines in iron sulfur proteins (Bertini et al., 1993b).

References

Abragam, A. (1961) in *The Principles of Nuclear Magnetism*, Oxford University Press, Oxford.

Banci, L., Bertini, I., & Luchinat, C. (1990a) *Struct. Bonding 72*, 113-135.

Banci, L., Bertini, I., Luchinat, C., & Piccioli, M. (1990b) *FEBS Lett. 272*, 175-180.

Banci, L., Bertini, I., Briganti, F., Luchinat, C., Scozzafava, A., & Vicens Oliver, M. (1991a) *Inorg. Chem. 30*, 4517-4524.

Banci, L., Bertini, I., & Luchinat, C. (1991b) in *Nuclear and Electron Relaxation. The Magnetic Nucleus-Unpaired Electron Coupling in Solution*, VCH, Weinheim.

Banci, L., Bertini, I., Turano, P., Tien, M., & Kirk, T.K. (1991c) *Proc. Natl. Acad. Sci. USA 88*, 6956-6960.

Banci, L., Bertini, I., Carloni, P., Luchinat, C., & Orioli, P.L. (1992a) *J. Am. Chem. Soc. 114*, 10683-10689.

Banci, L., Bermel, W, Luchinat, C., Pierattelli, R., & Tarchi, D. (1993a) *Magn. Reson. Chem.* in press.

Banci, L., Bertini, I., Capozzi, F., Carloni, P., Ciurli, S., Luchinat, C., & Piccioli, M. (1993b) *J. Am. Chem. Soc. 115*, 3431-3440.

Banci, L., Bertini, I., Ciurli, S., Ferretti, S., Luchinat, C., & Piccioli, M. (1993c) *Biochemistry 32*, 9387-9397.

Banci, L., Bertini, I., Luchinat, C., Piccioli, M., & Scozzafava, A. (1993d) *Gazz. Chim. Ital. 123*, 95-100.

Banci, L. Bertini, I., Bini, T., Tien, M. & Turano, P. (1993e) *Biochemistry 32*, 5825-5831.

Bertini, I., Lanini, G., Luchinat, C. (1984) *Inorg. Chem. 23*, 2729-2730.

Bertini, I., Lanini, G., Luchinat, C., Messori, L., Monnanni, R., & Scozzafava, A. (1985) *J. Am. Chem. Soc. 107*, 4391-4396.

Bertini, I. & Luchinat, C. (1986) in *NMR of Paramagnetic Molecules in Biological Systems*, Benjamin/Cummings, Menlo Park, CA.

Bertini, I., Capozzi, F., Ciurli, S., Luchinat, C., Messori, L., & Piccioli, M. (1992a) *J. Am. Chem. Soc. 114*, 3332-3340.

Bertini, I., Capozzi, F., Luchinat, C., Piccioli, M., & Vicens Oliver, M. (1992b) *Inorg. Chim. Acta 198-200*, 483-491.

Bertini, I., Capozzi, F., Luchinat, C., & Piccioli, M. (1993a) *Eur. J. Biochem. 212*, 69-78.

Bertini, I., Capozzi, F., Luchinat, C., Piccioli, M., & Vila, A.J. (1993b) *J. Am. Chem. Soc.*, in press.

Bertini, I., Ciurli, S., Dikiy, A., & Luchinat, C. (1993c) *J. Am. Chem. Soc.*, in press.

Bertini, I., Ciurli, S., & Luchinat, C. (1993d) *Angew. Chem.*, in press.

Bertini, I., Dikiy, A., Luchinat, C., Piccioli, M., & Tarchi, D. (1993e) *J. Magn. Reson.*, in press.

Bertini, I., Gaudemer, A., Luchinat, C., & Piccioli, M. (1993f) *Biochemistry*, in press.

Bertini, I., Jonsson, B.-H., Luchinat, C., Pierattelli, R., & Vila, A.J. (1993g) *submitted.*

Bertini, I., Luchinat, C., & Tarchi, D. (1993h) *Chem. Phys. Lett. 203*, 445-449.

Bloembergen, N. (1957) *J. Chem. Phys. 27*, 572-573.
de Ropp, J.S, La Mar, G.N., Wariishi, H., & Gold, M.H. (1991) *J. Biol. Chem. 266*, 15001-15008.
Dugad, L.B., La Mar, G.N., Banci, L., & Bertini, I. (1990) *Biochemistry 29*, 2263-2271.
Dunham, W.R., Palmer, G., Sands, R.H., & Bearden, A.J. (1971) *Biochim. Biophys. Acta 253*, 373-384.
Emerson, S.D. & La Mar, G.N. (1990) *Biochemistry 29*, 1556-1566.
Gueron, M. (1975) *J. Magn. Reson. 19*, 58.
Ho, F.F.-L. & Reilley, C.N. (1969) *Anal. Chem. 41*, 1835-1841.
Kurland, R.J. & McGarvey, B.R. (1970) *J. Magn. Reson. 2*, 286-301.
La Mar, G.N. & Walker, F.A. (1979) *The Porphyrins IV*, 61-157.
McConnell, H.M. & Robertson, R.E. (1958) *J. Chem. Phys. 29*, 1361.
Oh, B.-H., Mooberry, E.S., & Markley, J.L. (1990) *Biochemistry 29*, 4004-4011.
Overhauser, A.W. (1953a) *Phys. Rev. 89*, 689.
Overhauser, A.W. (1953b) *Phys. Rev. 92*, 411.
Satterlee, J.D. (1986) *Annu. Rep. NMR Spectrosc. 17*, 79-178.
Skjeldal, L., Westler, W.M., Oh, B.-H., Krezel, A.M., Holden, H.M., Jacobson, B.L., Rayment, I., & Markley, J.L. (1991) *Biochemistry 30*, 7363-7368.
Solomon, I. (1955) *Phys. Rev. 99*, 559.
Solomon, I. & Bloembergen, N. (1956) *J. Chem. Phys. 25*, 261-266.
Tainer, J.A., Getzoff, E.D., Beem, K.M., Richardson, J.S., & Richardson, D.C. (1982) *J. Mol. Biol. 160*, 181-217.
Timkowich, R. (1991) *Inorg. Chem. 30*, 37-42.
Vega, A.J. & Fiat, D. (1976) *Mol. Phys. 31*, 347.
Williams, G., Moore, G.R., Porteous, R., Robinson, M.N., Soffe, N., & Williams, R.J.P. (1985) *J. Mol. Biol. 183*, 409-428.
Yamamoto, Y. (1987) *FEBS lett. 222*, 115-119.

LONG TIME PROTEIN DYNAMICS

Angelo Perico
Centro di Studi Chimico-Fisici
di Macromolecole Sintetiche e Naturali, CNR
Via De Marini 6, 16149 Genova, Italy

The long time dynamics (time scale greater than picoseconds) of macromolecules in solution are governed by diffusion. In water the protein configuration diffuses in a complex conformational energy landscape with a large number of energy valleys separated by energy barriers (Young et al. 1991). Local dynamics in polypeptides, as described by NMR relaxation or fluorescence anisotropy, are very sensitive to the structure of the residues, to the inter-residue correlations and to the sequence cooperativity. This dependence of conformation and dynamics on local features is responsible for the great variety of protein biological functions observed in nature.
As the dynamics of proteins on these long time scales are not accessible to molecular dynamics (MD) simulations, reduced model descriptions should be introduced. Recently the ORZLD procedure (Perico et al 1985; Perico 1989a) was derived to produce a hierarchy of models which retain a maximum of universality compatible with the individual description of the aminoacids. The ORZLD theory is based on the optimized Rouse-Zimm approximation to the Smoluchowski diffusion equation to take into account not only large scale motions but also the long time motions of single bonds or group of bonds. This reduced description selects only part of the total degrees of freedom, as "slow variables", taking into account the remaining part as heat bath and neglecting the resulting complicated memory functions. Progressive inclusion of more degrees of freedom into the important slow variables enables to partially evaluate the effect of the neglected memory functions (Hu et al 1991). A hierarchy of models can be built choosing the

NATO ASI Series, Vol. H 87
NMR of Biological Macromolecules
Edited by C. I. Stassinopoulou

proper slow variables: virtual bonds (connecting consecutive residues), backbone bonds, backbone bonds plus lateral chain bonds.

Models in the ORZLD hierarchy can be properly employed to derive the time correlation functions (TCF) measured in relaxation experiments. The ORZLD theory was summarized in the first lecture. After definition of the slow variables, say n-1 bond vectors $\mathbf{l}_i$ of mean square lenght l_i^2, the local dynamics are given in terms of n individual friction coefficients ζ_i of each group of atoms corresponding to the chosen slow variables, of the bond vector correlation matrix $<\mathbf{l}_i \cdot \mathbf{l}_j>$ and of the inverse bead distances $<1/R_{ij}>$. These normalized matrices defined as

$$U_{ij}^{-1} = <\mathbf{l}_i \cdot \mathbf{l}_j>/l^2 = \int d(\{\mathbf{R}_i\}) \exp(-V(\{\mathbf{R}_i\})/K_BT)(\mathbf{l}_i \cdot \mathbf{l}_j)/l^2 \qquad (1)$$

and

$$<1/R_{ij}> = \int d(\mathbf{R}_i)(1/R_{ij}) \exp(-V(\{\mathbf{R}_j\})/K_BT) \qquad (2)$$

are calculated using a proper approximation to the intramolecular potential $V(\{\mathbf{R}_i\})$. Here l^2 is an average bond length defined as

$$l^2 = (n-1)^{-1} \sum_{i=1}^{n-1} <l_i^2> \qquad (3)$$

Knowing U^{-1}, $<1/R_{ij}>$ and ζ_i the second order TCF $P_2^i(t)$

$$P_2^i(t) = 3/2<\cos^2\theta^i(t)> - 1/2 \qquad (4)$$

with $\theta^i(t)$ the angle the bond vector $\mathbf{l}_i$ rotates in time t, and its correlation time

$$\tau^i = \sigma T^i = \int_0^\infty P_2^i(\sigma t) d\sigma t \qquad (5)$$

are obtained according to the procedure outlined in the first lecture. Attention is concentrated here on the TCF $P_2^i(t)$ as it

can be experimentally tested by fluorescence anisotropy on single TRP containing polypeptides or by NMR relaxation. In addition, the local persistence lenghts are introduced as the equilibrium average of the projection of the bond vector l_i on the end-to-end vector of the polymer chain
1h8

$$P_n^i/l = \sum_{j=1}^{n-1} \langle l_i \cdot l_j \rangle / l^2 = \sum_{j=1}^{n-1} U_{ij}^{-1} \qquad (6)$$

These static quantities are properties of the matrix U^{-1} which gives a description of the local flexibility of particular interest for chains of different units as polypeptides or proteins (Perico 1989b). A first model in the ORZLD hierarchy for polypeptides (Perico 1989b) is obtained choosing as slow variables the virtual bonds joining successive α-carbon atoms in the peptide chain, supposed to have the same mean square length l^2. The friction coefficient is assumed uniform

$$\zeta = 6\pi\eta_0 l/4 \qquad (7)$$

according to the estimation given for synthetic polymers in theta solutions (Perico et al 1986). This is a first rough approximation which underestimates the friction coefficient of the residue. In the following, this approximation will be removed in favour of individual friction coefficients better describing the bulkiness of the residue. The inverse bead distances are approximated by the Gaussian value

$$\langle 1/R_{ij} \rangle = (6/\pi)^{1/2} (\langle R_{ij}^2 \rangle)^{-1/2} \qquad (8)$$

with

$$\langle R_{ij}^2 \rangle = \sum_{p,q=i+1}^{j} U_{pq}^{-1} \qquad (9)$$

A first approximation to the matrix U^{-1} is obtained by the Flory method (Flory 1969) taking the 1,1 element of the equilibrium average of the product of T_i matrices:

$$U_{ij}^{-1} = (<T_i \cdots\cdots T_{j-1}>)_{11} \qquad (10)$$

The matrix T_i is the 3x3 transformation matrix from the reference frame of virtual bond i+1 to that of virtual bond i. The matrix T_i is a function of the two strongly interdependent dihedral angles (φ_i, ψ_i) describing rotations around the skeletal bonds forming the junction between the two peptide units i, i+1. For random polypeptides, couples of rotational angles of successive residues are almost independent. Assuming this statistical independence equation (10) becomes

$$U_{ij}^{-1} = (<T_i><T_{i+1}> \cdots <T_{j-1}>)_{11} \qquad (11)$$

For homopolypeptides the units are equivalent and

$$U_{ij}^{-1} = (<T>^{|i-j|})_{11} = \sum_a S_{1a} d_a^{|i-j|} S_{a1}^{-1} \qquad (12)$$

with S and (d_a) the matrix of eigenvectors and the eigenvalues of the matrix <T>. In the case of random copolypeptides

$$<T> = \sum_a W_a <T_a> \qquad (13)$$

with W_a the fraction of residue a.
The equilibrium average $<T_i>$ is obtained as

$$<T_i> = Z_i^{-1} \int d\varphi_i \int d\psi_i \exp(-V(\varphi_i, \psi_i)/K_B T) T_i(\varphi_i, \psi_i) \qquad (14)$$

with Z_i the partition function

$$Z_i = \int d\varphi_i \int \psi_i \exp(-V(\varphi_i, \psi_i)/K_B T) \qquad (15)$$

and is calculated by substituting the φ and ψ integrations with discrete sums. A first calculation was performed using the <T> matrices given by Flory for some aminoacids (Flory 1969).

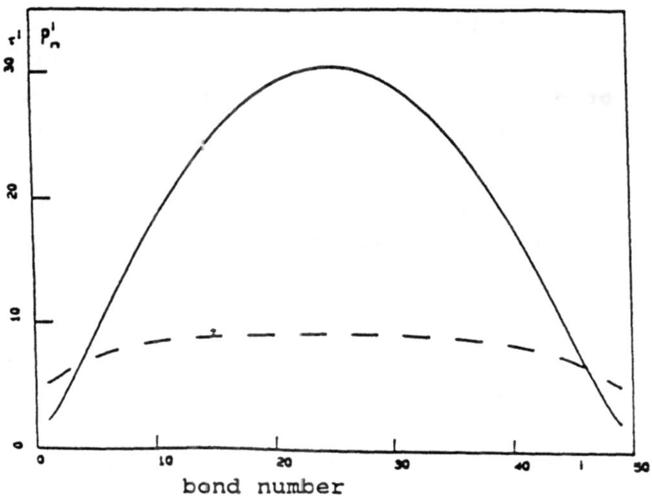

Figure 1. The bond correlation time τ^i (-) and local persistence lengths P_n^i/l (--) as a function of the bond index i for poly-L-alanine, n=50.

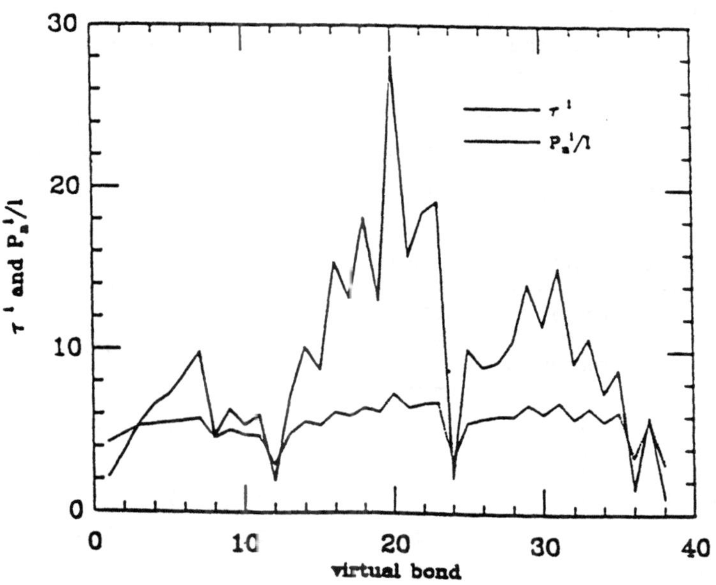

Figure 2. Dimensionless τ^i and P_n^i/l values for porcine ACTH (1-39) as a function of position in aminoacid sequence. Calculated with ζ_r=0.25 and assuming $\zeta_i=\zeta$.

In Figure 1, sequence patterns for normalized correlation times τ^i and persistence lenghts P_n^i/l are reported for a poly-L-alanine of 50 residues. The curves are bell-shaped with a maximum value in the middle of the chain and with the absence, typical of homopolymers, of any particular domain of flexibility and mobility. The dynamic pattern is more pronounced than the static one. Results for some homopolypeptides and random copolypeptides, not reported here, show that the mobility decreases from polyglycine to poly-L-alanine to poly-L-proline and increases with the content in glycyl fraction in copolypeptides glycyl-L-alanyl (Perico 1989b).

The models for protein dynamics, discussed in the following, were developed in comparison with fluorescence anisotropy experiments on single TRP containing polypeptides: [ACTH] (1-39), glucagon (1-29) and some of their fragments together with some synthetic 17 peptides (Lin Chen et al 1987; Hu et al 1990).

A first level of the ORZLD hierarchy uses only uniform friction coefficients and gross details of the polymer described as a freely rotating chain with stiffness parameter g and persistence length $l(1-g)^{-1}$ (Lin Chen et al 1987; see the previous lecture). In this simple model, local dynamics are correlated to the persistence and contour length of the fragment and to the distance of the TRP to the polymer end.

Table I. Observed and calculated correlation times.

Sample	Number of residues	TRP position	τ_{exp}(ns)	τ_{calc}	g
ACTH (1-39)	39	9	1.10±0.04	1.08	0.865
ACTH (1-24)	24	9	1.11±0.04	1.10	0.882
ACTH (1-13)	13	5	0.53±0.01	0.52	0.920
ACTH (1-10)	10	2	0.38±0.01	0.39	0.980
ACTH (5-10)	6	2	0.25±0.01	0.14	0.999
Glucagon (1-29)	29	5	0.97±0.02	0.95	0.907

The values of the stiffness parameter g, required to fit the data for the longer peptides ACTH (1-39), ACTH (1-24) and glucagon (1-29), lie in the range 0.87-0.91. This corresponds to reasonable persistence lengths in the range of 7-10 residues. In addition, Table I shows how much the relaxation is sensitive to the position along the chain resulting much faster for a TRP located near the chain ends. In fact, although glucagon is five residues longer than ACTH (1-24), the closer proximity of TRP to the glucagon end shortens the calculated anisotropy decay in accordance with the experimental findings. These results demonstrate how large effects in local dynamics should be expected depending on the position, on the contour length and on the sequence.

A second level, adopted to describe random polypeptides, uses virtual bond slow variables, full conformational energy maps of each aminoacid in the sequence, calculated by ECEPP or other force field packages, statistical independence between residues and average and individual friction coefficients for each aminoacid (Hu et al 1990). This description is realistic provided the chain has random coil configuration without specific residue interactions such as those promoting local helix formation. The basic input U^{-1} is calculated again by equation (11) and the inverse hydrodynamic distances are obtained from U^{-1} by gaussian approximation of equation (8). The average $\langle T_i \rangle$ is obtained by (14), using the energy map of the specific aminoacid, calculated by the empirical conformational energy program for peptides (ECEPP) by Sheraga and coworkers (Sippl et al 1984). This conformational energy has the following form

$$V(\varphi,\psi) = 332.0 q_i q_j / DR_{ij} + FA/R_{ij}^{12} - C/R_{ij}^{6} + \\ + q_{HB}/R_{HB}^{12} - b_{HB}/R_{HB}^{10} + U_c(1 \pm \cos n\theta) \qquad (16)$$

including electrostatic interaction, the non bonded 6-12 energy, the hydrogen bond 10-12 energy and a correction for the torsional energy. Under these conditions, all the side chain atoms are treated explicitly but they are kept fixed at such conformations as to yield a Flory parameter C_∞, describing the stiffness of the infinite homopolypeptide, of the order 9.0, which is the value generally believed to characterize several of this class of molecules. Different approaches to the friction coefficients are considered including uniform friction coefficients from equation (7): individual friction coefficients evaluated according to the Stokes law

$$\zeta_i = 6\pi\eta_o r_i \tag{17}$$

with r_i estimated from the hydrodynamic volumes of the 20 aminoacids; uniform friction coefficients given as the average

$$\zeta = n^{-1} \sum_{i=1}^{n} \zeta_i \tag{18}$$

of the individual friction coefficients of equation (17).

In the case of large individual friction coefficients or average friction coefficient, attention must be paid to reduce the hydrodynamic interaction strength

$$\zeta_r = \zeta/6\pi\eta_o l \tag{19}$$

to avoid divergencies in the hydrodynamic interaction matrix (Perico et al 1975).

Figure 2 reports the dynamic pattern τ^i for ACTH (1-39) using uniform friction coefficients from equation (7). It is noteworthy that each virtual bond relaxes individually in such a way that the dynamic pattern displays irregular pronounced maxima and minima. This is the main difference from the case of homopolypeptides calculated using the same type of approximations (see Figure 1). The local persistence length, presents a similar pattern but much less pronounced. Three dynamical domains appear clearly in ACTH, well separated by two deep minima.

In Figure 3 the same dynamical pattern for ACTH is compared to those for the four ACTH fragments of Table I to show how, in

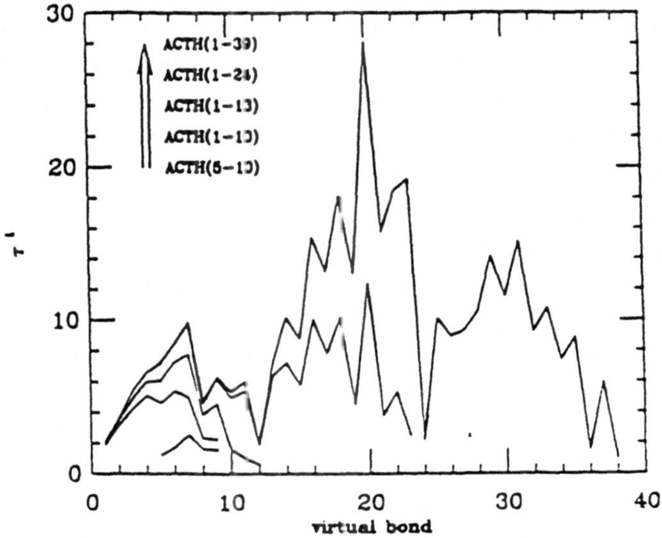

Figure 3. τ^i for various fragments of the ACTH peptides.

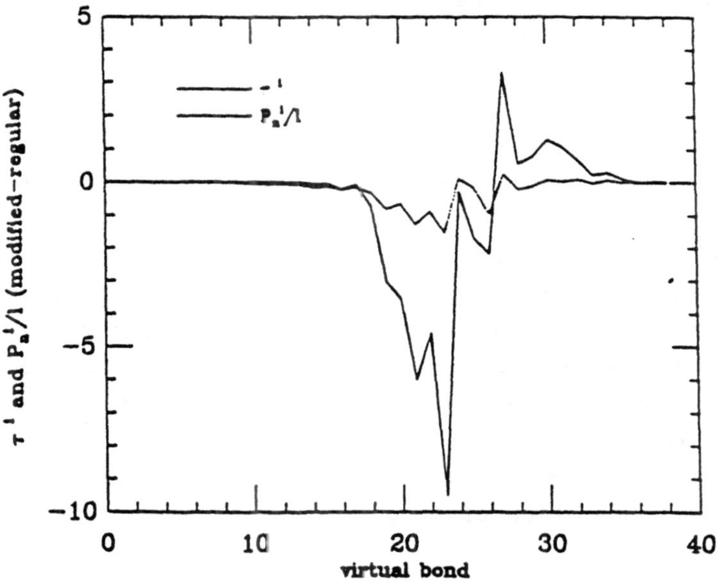

Figure 4. Swapping the amino acid sequences in position 24-25-26 from PRO-ASN-GLY to PRO-GLY-ASN in ACTH (1-39). The changes in τ^i and P_n^i/l are shown as differences between the modified and normal sequence values.

the ORZ description with statistical independence between residues, the structure of the pattern is well maintained throughout those fragments. There is only an overall speed up of the local motions and an increasing importance of end effects due to the shortening of the peptides.

The changes induced by altering the sequence of ACTH (1-39) from PRO-ASN-GLY in position 24-25-26 to PRO-GLY-ASN is reported in Figure 4 as a modified τ^{i}-regular τ^{i}. The striking feature is that this one residue shift extends its dynamical influence over more than 15 residues (around two persistence lengths). Note that the vicinity of GLY causes a greater mobility in PRO in the modified ACTH. The dynamical aspects could be of relevance in the interpretation of the biological function in natural and modified proteins.

In Figure 5 the effect of individual large friction coefficients is compared with the case of uniform low friction coefficients from equation (7). There is a remarkable slowing down of the dynamics due to more realistic consideration of large individual friction coefficients. Therefore an accurate description of the residue friction coefficients is important for a quantitative evaluation of local dynamics. Nevertheless the calculated correlation times for the TRP of all the fragments in this model still remains a factor 2 lower than the value measured by fluorescence anisotropy.

Other models assume a rigid conformation for the protein as determined by the known X-ray structure. In addition, models are considered where parts of this rigid conformation are relaxed and treated as random coil spacers. Results for BPTI protein, considered as a rigid body, are reported in Figure 6 (Perico et al in preparation). When fluctuations are allowed starting from these X-ray or X-ray-minimized configurations, a new promising generation of long time ORZ dynamic models could be generated.

The most sophisticated models for protein dynamics (Hu et al 1991) use in the derivation of U^{-1} and $\langle 1/R_{ij} \rangle$, molecular dynamics simulations to take into account specific inter-residue interactions, side chain correlations and influence of solvent motions. The statistical averages are substituted by time

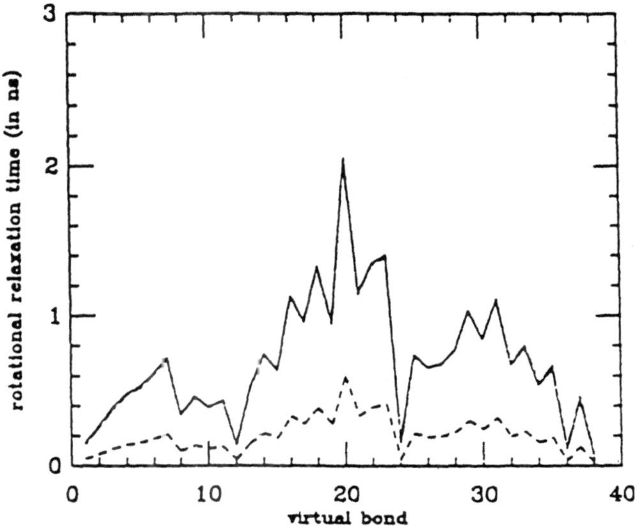

Figure 5. Correlation times τ^i for ACTH (1-39) as a function of virtual bond locations. Lower trace : $\zeta_i=6\pi\eta_o l\zeta_r$ and $\zeta_r=0.25$. Upper traces: individual ζ_i

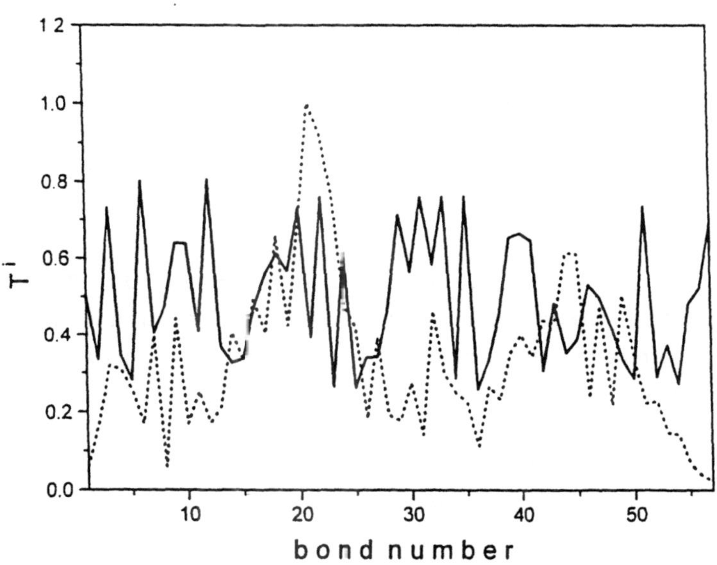

Figure 6. The correlation times T^i (ns) for BPTI. Full curve: rigid X-ray structure; broken curve: random BPTI with statistically independent residues

averages over the MD trajectory (90 psec trajectories calculated by CHARMm packages) and individual friction coefficients are calculated by the van der Waals additive method or by the accessible surface area method (Kostov et al 1993). Due to the complexity of the simulations (a total of 2945 atoms, including H_2O) used to calculate the input to the long time dynamics theory, only the small fragment of six aminoacids ACTH (5-10) was considered. Three models enable to progressively include more and more protein degrees of freedom into the space of the active variables: the simplest model has 5 virtual bonds, the second model has 15 backbone bonds and the third model has all 36 bonds including those in the side chain (see Figure 7).

The coupling of MD simulations to the ORZ long time dynamics has the following advantages:

- the correlations between neighbour residues as well as long range interactions such as hydrogen bonds are taken into account;
- the motions of the lateral chains and of the solvent H_2O molecules are explicitly included;
- the inverse distances $<1/R_{ij}>$ are evaluated exactly avoiding the gaussian approximation.

For these reasons this coupling of MD simulations to long time model theory seems to be the most promising tool to derive protein dynamics.

In Figure 8, the correlation times in nanoseconds, T_i, are reported for the three models with an increasing number of slow variables: several remarkable aspects are displayed by this dynamic pattern. The backbone bonds (1-15) relax faster than the virtual bonds, having more possibilities to move. The relaxation of the backbone bonds in the backbone bond model is faster than in the all bond model, thus showing that the presence of lateral chains slows down the motion. The sawtooth pattern for backbone bond relaxation is easily understood in terms of rigidity of the peptide bonds: each of these rigid bonds (2,5,8,11 and 14) has connections with two rather flexible bonds resulting in a faster relaxation than for a flexible bond (1,3,4,6,7,9,10,11,12,13 and 15) connected to a rigid and

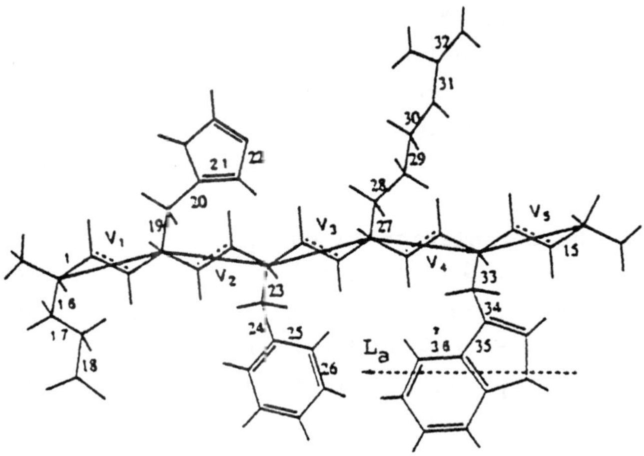

Figure 7. Molecular structure of ACTH (5-10) (GLU-HIS-PHE-ARG-TRP-GLY). V1 to V5 are virtual bonds, the bonds numbered 1 to 15 are backbone bonds, and the bonds numbered 16 to 36 are side chain bonds.

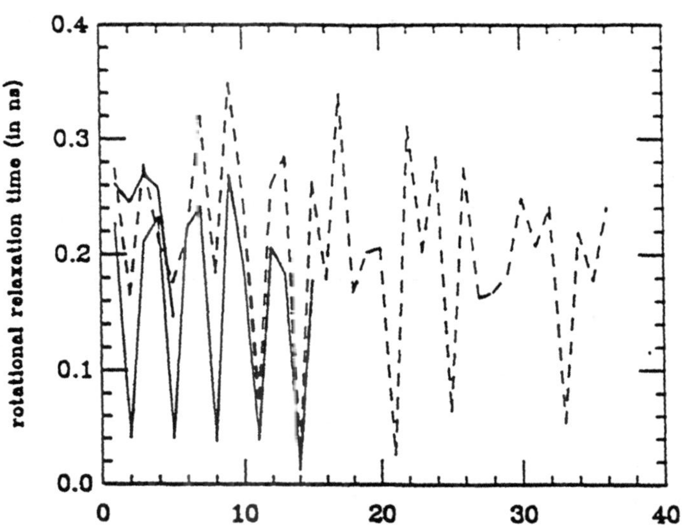

Figure 8. The T^i for all three models. The solid curve is for the virtual bond model, the lower solid curve is for the backbone bond model, and the dashed curve is for all bond model. Backbone bonds: 1-15, side chain bonds 16-36.

a flexible bond. But the main result is that in this small fragment there is no obvious time scale separation between motion of the backbone bonds and side chain bonds (16-36). Therefore, an accurate description of the local dynamics requires the inclusion in the slow variables of all the bonds in the peptide.

In Figure 9, the correlation times T_i for the same five virtual bonds in the three models are reported. In the all bond model the dynamic pattern is much slower and also displays more structure due to the explicit inclusion of side group motions that strongly couple to the relaxation of the virtual bond orientational correlations.

In Table II the correlation times calculated in different ORZ models are compared with the fluorescence anisotropy experimental correlation times for TRP. The most complete model (all bond model) gives, for the bond strictly attached to the ^1L_a dipole moment in the TRP residue,a correlation time almost coincident with that resulting from fluorescence anisotropy experiments.

Table II. Correlation times for ACTH (5-10)

Model	bond	T(ns)
virtual b. (ECEPP)[1]	4-5	.100
virtual b. (MD-ORZ)[2]	4-5	.202
backbone b. (MD-ORZ)[2]	12-13	.195
all b. (MD-ORZ)[2]	36	.240
all b. (MD-ORZ)[3]	36	.250
f. anisotropy exp.	TRP	$.250 \pm .001$

1 - uniform friction coefficient eq. (7)

2 - individual friction coefficients: van der Waals additive method

3 - individual friction coefficient ASA method, probe radius=0.

It is noteworthy that this remarkable result is obtained in a theory without any adjustable parameter. Actually one of the most questionable points is related to the choice of the individual friction coefficients. A detailed study of this aspect has been carried out (Kostov et al 1993). The effect on the long time ORZ dynamics of the hydrodynamic volume method, of the additive van der Waals method and of the accessible surface area (ASA) method, were carefully considered. It comes out that additive van der Waals method and ASA method with a zero probe radius give almost the same results and are the most promising.
In conclusion, it is worthwhile to underline that NMR relaxation techniques could be used as successfully as fluorescence anisotropy techniques to probe these local dynamics by measuring T_1, T_2 and NOE of ^{13}C in natural abundance or after specific labelling. This could be a challenging test for the ORZLD models of protein dynamics.

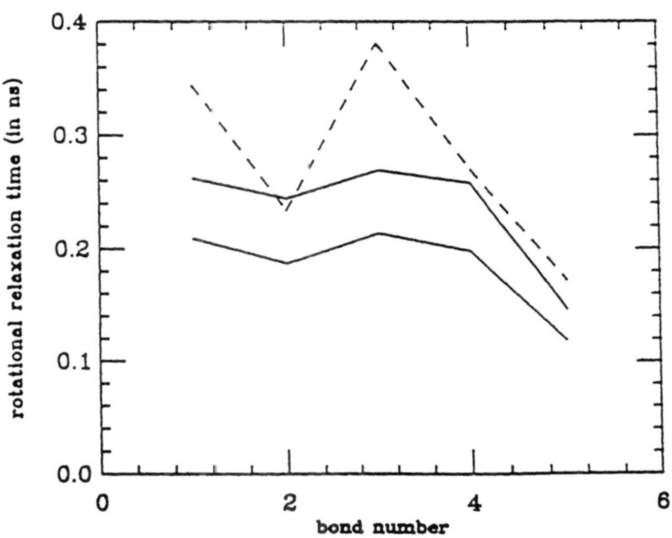

Figure 9. The T^i for virtual bond 1-5 as calculated from the three models. Upper curve: all bond model; lower curve: backbone bond model; intermedium curve: virtual bond model

REFERENCES

Flory PJ (1969) Statistical Mechanics of chain molecule,Interscience Publishers John Wiley 1969 New York

Hu Y, MacInnis JM, Cherayl BJ, Fleming GR, Freed KF, Perico A (1990) Polypeptide dynamics: experimental test of an optimized Rouse-Zimm Type model. J Chem Phys 93: 822-836

Hu Y, Fleming GR, Freed KF, Perico A (1991) Theory of long time peptide dynamics: test of various reduced descriptions and role of internal variables. Chem Phys 158: 395-408

Kostov K, Freed KF, Perico A (1993) Effect of various frictional models on long time peptide dynamics, Biopolymers 33 1423-1429

Lin Chen XQ, Petrich JW, Fleming GR, Perico A (1987) Picosecond fluorescence studies of polypeptide dynamics: fluorescence anisotropies and lifetimes. Chem Phys Lett 139: 55-61

Perico A, Piaggio P, Cuniberti C (1975) Dynamics of chain molecules. I. Solutions to the hydrodynamic equation and intrinsic viscosity. J Chem Phys 62: 4911-4918

Perico A, Guenza M (1985) Viscoelastic relaxation of segment orientation in dilute polymer solutions. J Chem Phys 83: 3103-3109

Perico A, Guenza M (1986) Viscoelastic relaxation of segment orientation in dilute polymer solutions. II. Stiffness dependence of fluorescence depolarization. J Chem Phys 84: 510-516

Perico A (1989a) Segmental relaxation in macromolecules. Accounts Chem Res 22: 336-342

Perico A (1989b) Local dynamics in biological macromolecules. Biopolymers 28: 1527-1540

Perico A, Guenza M, Mormino M in preparation

Sippl MJ, Némethy, Scheraga HA (1984) Intermolecular potentials from crystal data. 6. Determination of empirical potentials for O-H···O=C hydrogen bonds from packing configurations. J Phys Chem 88, 6231-6233

Young RD, Frauenfelder H, Johnson JB, Lamb DC, Nienhaus GU, Philipp R, Scholl R (1991) Time and temperature dependence of large-scale conformational transitions in myoglobin. Chem Phys 158: 315-327

LOCAL DYNAMICS OF MACROMOLECULES

Angelo Perico
Centro di Studi Chimico-Fisici
di Macromolecole Sintetiche e Naturali, CNR
Via De Marini 6, 16149 Genova, Italy

The viscoelastic properties of polymer materials and the biological function of biopolymers are mainly influenced by the dynamics on a long time scale (much greater than picoseconds). In solution and in the melt these dynamics are governed by diffusion.

The large scale motions in random coils are strongly molecular weight dependent and largely independent of the detailed conformational structure. Therefore they are described by simple models based on gaussian connectivity and delta function excluded volume interactions (Doi et al. 1986). On the contrary, as spatial and time scales decrease (to the nanosecond-picosecond time domain) the motions become strongly dependent on conformational potential details and, for long flexible polymers, molecular weight independent. In this case, the role of the structure of the monomers, of the lateral chains, and of the correlations inter monomers becomes important and knowledge of the sequence is essential.

Therefore in principle local dynamics require a detailed description of the motion of all the atoms in a complicate intramolecular potential. On a picosecond time scale and for polymers of small molecular weight, these dynamics can be approached by molecular dynamics (MD) and brownian dynamics (BD) simulations. But for long times and large macromolecules, MD and BD become prohibitive and a reduced theoretical description is required.

These reduced descriptions should maintain a high degree of universality to be applied to different macromolecules in different conditions. Using the projection operator technique applied to the Smoluchowski diffusion equation, a generalized Rouse-Zimm approximation (ORZ) can be generated ignoring the

NATO ASI Series, Vol. H 87
NMR of Biological Macromolecules
Edited by C. I. Stassinopoulou

memory functions (Bixon et al. 1978) and exactly solved for the experimentally important local time correlation functions (Perico et al. 1985).

The role of the barrier crossing on memory functions has been analyzed using simple one variable models for 2-4 and periodic potentials. It was shown that the ORZ approximation, which is first order in a continued fraction expansion, becomes increasingly poor as the barrier height increases (Perico et al. 1993). The higher the barrier height, the higher the order required in the continued fraction expansion. On the contrary, recent BD simulations on multidimensional models of octane, including evaluation of the memory functions, show that the ORZ approximation remains good (Chang et al. 1993). This is probably due to the fact that, in the multidimensional octane problem, the time scale of local motions is long compared to that of barrier crossing, while the one-dimensional problem treats the hardest case of comparable time scales.

A polymer of a given architecture (linear, branched, star polymer) is described as a sequence of n beads of coordinates $\mathbf{R}_i$ and friction coefficients ζ_i. The beads are connected by bond vectors $\mathbf{l}_i$. Here, for the sake of simplicity, a uniform friction coefficient, $\zeta_i = \zeta$, is assumed, and the same mean square length l^2 for all the bonds.

The resulting ORZLD dynamics (Perico 1989a) is described by the Langevin equation:

$$(\partial/\partial t)\ \mathbf{R}_i\ (t) + \sigma \sum_{j=o}^{n-1} (HA)_{ij}\ \mathbf{R}_j\ (t) = \mathbf{v}_i\ (t) \qquad (1)$$

with A the matrix of order n describing the chain connectivity

$$A = M^T \begin{matrix} 0 & 0 \\ 0 & U \end{matrix} M \qquad (2)$$

The inverse of U is the equilibrium bond correlation matrix

$$U_{ij}^{-1} = \langle \mathbf{l}_i \cdot \mathbf{l}_j \rangle / l^2 \qquad (3)$$

and M describes the bead-to-bond transformation matrix, which

is a function of the architecture of the chain.
The matrix H describes the preaveraged hydrodynamic interaction matrix.

$$H_{ij} = \delta_{ij} + \zeta_r <1/R_{ij}> (1-\delta_{ij}) \tag{4}$$

and

$$\zeta_r = \zeta/6\pi\eta_o l \; ; \quad \sigma = 3\, K_B T/l^2\, \zeta \tag{5}$$

are the hydrodynamic interaction strength and the bond rate constant. The equilibrium average $<\mathbf{l}_i \cdot \mathbf{l}_j>/l^2$ (and $1/R_{ij}>$, similarly) is calculated as

$$<\mathbf{l}_i \cdot \mathbf{l}_j>/l^2 = N \int d\, \{\mathbf{R}_i\} \exp\,(-V(\{R_i\})/k_B T)\; (\mathbf{l}_i \cdot \mathbf{l}_j)/l^2 \tag{6}$$

using a proper approximation to the exact intramolecular potential $V(\{\mathbf{R}_i\})$.
The useful concept of local persistence length is introduced as a property of the matrix U^{-1}:

$$P_n{}^i/l = \sum_{j=1}^{n-1} <\mathbf{l}_i \cdot \mathbf{l}_j>/l^2 \tag{7}$$

Equation (1) can be solved exactly, under the hypothesis of gaussian random processes, to obtain large scale or local scale dynamic properties. On a local scale, the time correlation functions (TCF) of particular importance are the first order

$$P_1^i(t) = <\cos\Theta^i(t)> \tag{8}$$

measured in dielectric relaxation, and the second order

$$P_2^i(t) = 3/2<\cos^2\Theta^i(t)> - 1/2 \tag{9}$$

measured in fluorescence anisotropy and NMR relaxation, describing the orientational relaxation of bond $\mathbf{l}_i$ with Θ^i the

angle the bond vector $\mathbf{l}_i$ rotates in time t.
The exact results (Perico et al. 1985) are:

$$P_1^i\ (t) = (1-x^2)\ [1-(2/\pi)\ \arctan x] + (2/\pi)\ x \tag{10}$$

$$P_2^i\ (t) = 1 - 3\ \{x^2 - x^3(\pi/2)[1-(2/\pi)\ \arctan x]\} \tag{11}$$

$$x = [1-(M_1{}^i(t))^2]^{.5}\ /M_1{}^i(t) \tag{12}$$

The fundamental TCF for the bond vector $\mathbf{l}_i$ TCF

$$M_1^i = <\mathbf{l}_i(t)\cdot\mathbf{l}_i(0)>/l_i^2> \tag{13}$$

is in turn calculated as a function of the eigenvalues and eigenvectors of HA, λ_a, $\{Q_{ia}\}$:

$$M_1{}^i(t) = \sum_{a=1}^{n-1} (Q_{ia} - Q_{i-1,a})^2\ \mu_a{}^{-1}\ \exp(-\lambda_a\sigma t), \tag{14}$$

and $l^2\mu_a^{-1}$, is the mean square length of mode a.
Equations (10)-(12) show that in the ORZ approximation $P_1^i(t)$ and $P_2^i(t)$ are universal functions of $M_1^i(t)$. In Figure 1 the universal plot of $P_1^i(t)$ and $P_2^i(t)$ against $M_1^i(i)$ is presented: this plot is independent of the position of the relaxing segment i and of the specific ORZ model. Note that P_2 and P_1 have a very different behaviour in all the time range. In addition, P_1 differs from M_1 and P_2 displays strong differences from M_1^2 and also from M_1^3 (the second order TCF for a sphere or a rod). This general conclusion is in striking contradiction with the simple assumption, often reported in literature, according to which P_1 may be approximated by M_1 and P_2 by M_1^2, and with the models based on tetrahedral lattices, according to which $P_1 \equiv P_2$.
The theory explicitly takes into account the effect of the position of segment i and the conformational details which enter via M_1^i.

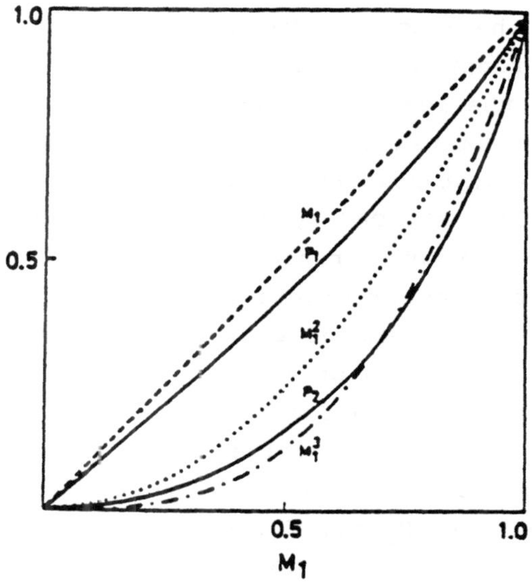

Figure 1. The first and second-order TCFs, $P_1(t)$ and $P_2(t)$ (-) and the segment vector autocorrelation function $M_1(t)$ (--) toghether with M_1^2 (t) (...) and $M_1^3(t)$ (-.-) versus M_1.

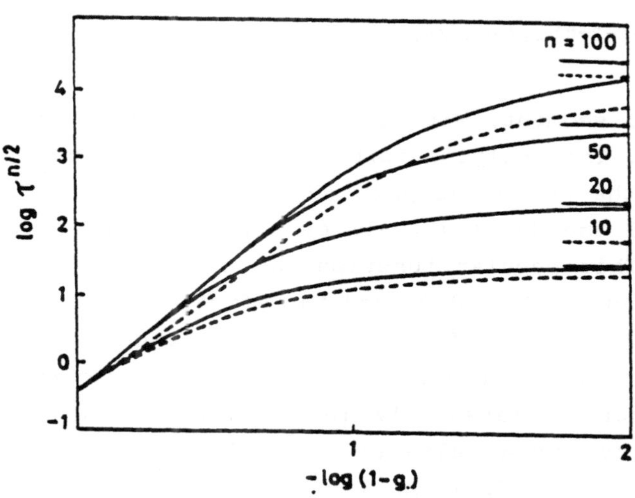

Figure 2. The correlation time $\tau^{n/2}$ for the TCF $P_2^{n/2}(t)$ vs the numerical persistence length P/l = $(1-g)^{-1}$ for several n values: rod limits (arrows); ζ_r=0 (-); ζ_r=0.25 (--).

In conclusion, the long time ORZLD dynamics of any polymer can be given as a function of the equilibrium averages $<\mathbf{l}_i \cdot \mathbf{l}_j>$, $<1/R_{ij}>$ and of the bead friction coefficients ζ_i. In the case $<1/R_{ij}>$ is approximated by its gaussian expression in terms of the mean square distances $<R_{ij}^2>$, all the conformational information enters into the ORZ dynamics via the equilibrium average of the bond vector correlation matrix, U^{-1}. In addition, the local friction coefficients are required: in the simplest model, they are approximated uniformly by fitting the viscoelastic behaviour (Perico et al. 1986), while in more refined models they are estimated from the local hydrodynamic volume, or by the van der Waals additive method or by the accessible surface area method (Hu et al. 1991; Kostov et al. 1993).

The fundamental TCF, $M_1^i(t)$, for the simplest model in the ORZ hierarchy, that is the gaussian bead-spring model in the free draining limit, takes the analytic forms (Perico 1988):

$$M_1{}^{n/2}\ (t) = \exp\ (-2\sigma t)\ I_o\ (2\sigma t) \tag{15}$$

for a central segment, and

$$M_1^1(t) = \exp(-2\sigma t) I_1(2\sigma t)/\sigma t \tag{16}$$

for the end segment, with $I_m(x)$ the modified Bessel function of order m. The analytic form in the case $i = n/2$ is coincident with the essential part derived for the conformational TCF by Hall and Helfand (Hall et al. 1982) in a Kramers type approach. This popular function was assumed to describe also the behaviour of P_1 and P_2 and therefore adopted as a basis of many empirically modified functions in the data analysis of NMR relaxation and fluorescence anisotropy experiments. But this function describes only $M_1^{n/2}$ taking into account only the connectivity for a relaxing central segment in a Gaussian chain in the free draining limit. However, the measured TCFs P_1^i, P_2^i differ strongly from M_1^i, and in addition M_1^i depends on the segment position and on the conformational details of the chain through the dependence of U^{-1} on $V(\{\mathbf{R}_i\})$.

Several approximations to the intramolecular potential are considered below.

For a bead-spring model it results

$$U^{-1}_{ij} = \delta_{ij}, \tag{17}$$

while for a freely rotating chain (FRC) with θ the valence angle and a stiffness parameter

$$g = -\cos\theta, \tag{18}$$

it results

$$U^{-1}_{ij} = g^{|i-j|} \tag{19}$$

In the case of RIS models, U^{-1} can be always written as a linear combination of FRC type components (Perico et al. 1987)

$$U^{-1}_{ij} = \sum_{h=1}^{\nu} \alpha_h g_h^{|i-j|} \tag{20}$$

with ν the product of space dimensionality multiplied by the number of rotational states, g_h the eigenvalues and α_h related to the eigenvectors of the matrix which combines the rotational matrix from the two reference frames of successive bonds with the statistical-weight matrix of the allowed rotational states.

The correlation time for the second order TCF P_2^i defined as:

$$\sigma T^i = \tau^i = \int_0^\infty P_2{}^i\,(t)\,d\sigma t \tag{21}$$

when plotted against the sequence i gives a comprehensive local dynamical pattern of a polymer.

In Figure 2, $\tau^{n/2}$ for a semiflexible chain, described as an FRC model (Perico et al. 1986), is reported against the normalized persistence length $(1-g)^{-1}$: for $g\to 0$, the ORZ model gives the flexible gaussian behaviour, and in the $g\to 1$ limit the rod behaviour is obtained within a fairly good approximation.

In Figure 3, the correlation time $\tau^{n/2}$ vs n is reported (Perico 1988) for RIS models of isotactic polystyrene (iPS), polyethylene (PE), isotactic polypropylene (iPP) and polydimethylsiloxane (PDMS). This figure shows how large the cooperativity and sensitivity to local conformational details are even for flexible polymers.

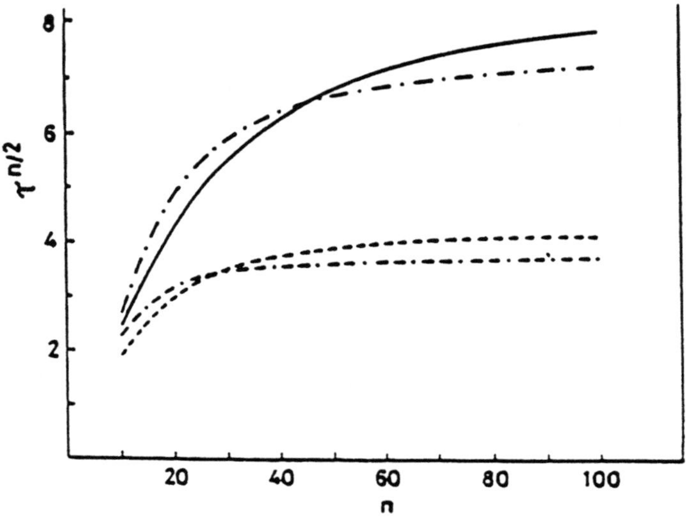

Figure 3. The correlation time $\tau^{n/2}$ for the TCF $P_2^{n/2}(t)$: iPS (-), PE (-·-), iPP (--), and PDMS (--·).

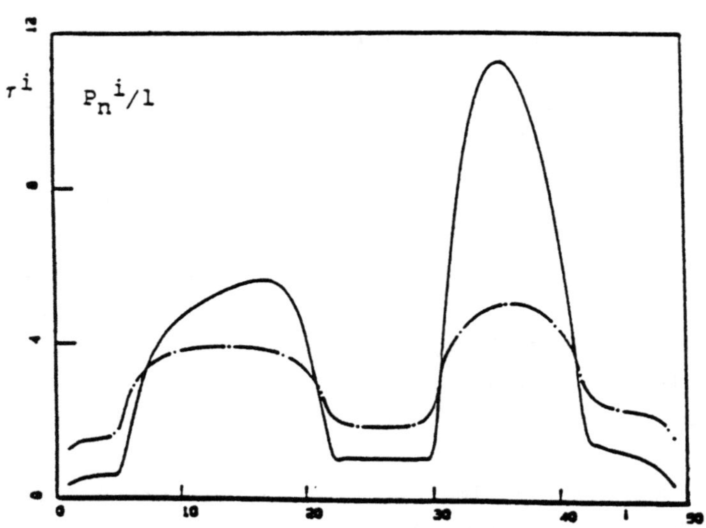

Figure 4. The bond correlation time τ^i (-) and local persistence length P_n^i/l (-·-) for a multiple domain chain in the ORZ freely rotating chain model; n=50.

For a gaussian or a semiflexible chain, as well as for any homopolymer, the function τ^i is a bell shaped curve with the maximum in the middle of the chain and strong end effects. On the contrary, if the chain is characterized by domains of different stiffness or by different units (aminoacids) the dynamic pattern displays more structure. A typical pattern is reported in Figure 4 for a model chain with five blocks of different stiffness in the FRC approximation (Perico 1989b). Note that the dynamic pattern follows that of the local persistence length but the structure is much more pronounced. In the following lecture, much more detailed models of the ORZLD hierarchy are presented for polypeptides.
Finally, the ORZLD theory is applied to give a structural interpretation to T_1 and NOE NMR experiments on two highly ordered synthetic polymers: poly[(R,S)-3,7-dimethyl-1-octene] (DMO) and poly[(R,S)-3-methyl-1-octene] (MO) (Perico et al. 1990). The experimental results describe the relaxation of ^{13}C in natural abundance of the 9 (in the case of DMO: 10) carbons of the monomeric unit (Figure 5) 1-8,3'(and 7'), obtained at three (two for MO) frequencies.
Each ^{13}C of the type j (j = 1,...8, 3',7') undergoes a dipolar relaxation which turns out to be a function of the spectral density

$$J_j(\omega) = 2R_e\int_o^\infty G_j(t)\ \exp(i\omega t) \tag{22}$$

which is the Fourier transform of the TCF $G_j(t)$ describing the dipolar relaxation of ^{13}C in position j. In the case of ^{13}C in position j=1,2, along the backbone, the relaxing dipoles are assumed rigidly attached to the virtual bond, and the relaxation is simply described by the $P_2^{n/2}(t)$ for long polymers (experimentally the correlation time for the two carbons are similar). In the case of a carbon on the lateral chain in position j it is assumed (Lipari et al. 1982)

$$G_j(t) \approx P_2^{n/2}(t)g_j(t) \tag{23}$$

with $g_j(t)$ describing the restricted motion of the lateral chain relative to the virtual backbone bond:

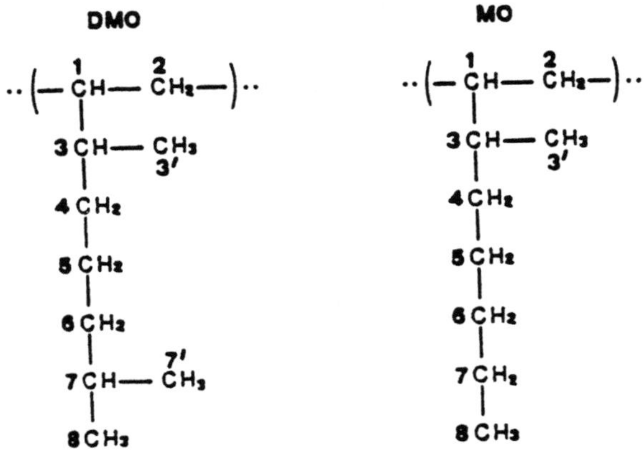

Figure 5. Structure of poly[(R,S)-3,7-dimethyl-1-octene] (DMO) and poly[(R,S)-3-methyl-1-octene] (MO) and numbering of atoms.

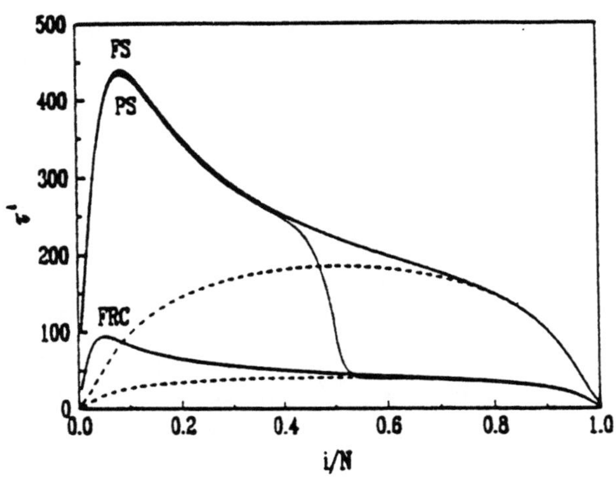

Figure 6. Bond correlation time τ^{i} as a function of i/N for an arm length N=200, f=12, and correlation at the center $\alpha=-(f-1)^{-1}$. FS: fully stretched arm model, g=0.87. PS: partially stretched arm model. FRC: freely rotating chain star model, g=0.77. Linear freely rotating chain of length N=200 (dashed curves): upper curve, g=0.87; lower curve, g=0.77.

$$g_j(t) = S_j^2 + (1 - S_j^2)\exp(-t/\tau_j) \quad (24)$$

S_j is the order parameter taking the value 0 for completely unrestricted motion and 1 in the case of a lateral chain rigidly stuck to the backbone. Fitting to all the T_1 and NOE data gives a persistence length of 39 and 43 virtual bonds for DMO and MO respectively. While these persistence lengths P seem somewhat high, due to an underestimation of the friction coefficient of the beads (here assumed uniform: $\zeta = 6\pi\eta_0 l/4$), the reduction of P in DMO could be ascribed to the perturbation of the backbone conformational order due to the greater bulkiness of its lateral chain. As to the order parameter, it correctly decreases exponentially with increasing j in the lateral chain. The use of RIS polymer models in the ORZLD theory could give more detailed descriptions of these NMR relaxation experiments.

Finally, in Figure 6, the local dynamic pattern τ^i of an arm for a partially stretched model of a regular star polymer (with number of arms f = 12) is reported (Guenza et al. 1991; Guenza et al. 1992). It could be of interest to check this type of pattern by NMR relaxation experiments on single ^{13}C labeled star polymers in dilute theta conditions.

REFERENCES

Bixon M, Zwanzig R (1978) Optimized Rouse-Zimm theory for stiff polymers. J Chem Phys 68: 1896-1902

Chang XY, Freed KF (1993) Test of theory for long time dynamics of floppy molecules in solution using brownian dynamics simulation of octane. J Chem Phys submitted.

Doi M, Edwards FS (1986) The theory of polymer dynamics, Clarendom Press 1986 Oxford.

Guenza M, Mormino M, Perico A (1991) A local approach to the dynamics of stars polymers. Macromolecules 24: 6168-6174

Guenza M, Perico A (1992) A reduced description of the local dynamics of star polymers. Macromolecules 25: 5942-5949

Hall CK, Helfand E (1982) Conformational state relaxation in polymers: time correlation functions. J Chem Phys 77: 3275-3282

Hu Y, Fleming GR, Freed KF, Perico A (1991) Theory of long time peptide dynamics: test of various reduced descriptions and role of internal variables. Chem Phys 158: 395-408

Lipari G, Szabo A (1982) Model-free approach to the interpretation of Nuclear Magnetic Resonance relaxation in macromolecules. 1. Theory and range of validity. J Am Chem Soc 104: 4546-4559

Kostov K, Freed KF, Perico A (1993) Effect of various frictional models on long time peptide dynamics. Biopolymers 33: 1423-1429

Perico A, Guenza M (1985) Viscoelastic relaxation of segment orientation in dilute polymer solutions. J Chem Phys 83: 3103-3109

Perico A, Guenza M (1986) Viscoelastic relaxation of segment orientaion in dilute polymer solutions. II. Stiffness dependence of fluorescence depolarization. J Chem Phys 84: 510-516

Perico A, Ganazzoli F, Allegra G (1987) A hierarchy of models for the dynamics of polymer chains in dilute solution. J Chem Phys 87: 3677-3686

Perico A (1988) Viscoelastic relaxation of segment orientation in dilute polymer solutions. III. Results in the optimized Rouse-Zimm approximations to the rotational isomeric state hierarchy of models. J Chem Phys 88: 3966-4000

Perico A (1989a) Segmental relaxation in macromolecules. Accounts Chem Res 22: 336-342

Perico A (1989b) Local dynamics in biological macromolecules. Biopolymers 28: 1527-1540

Perico A, Altomare A, Catalano D, Colombani M, Veracini CA (1990) Relaxation of segment orientation in dilute polymer solution. Interpretation of ^{13}C T_1 and NOE experiments on dilute poly[(R,S)-3,7-dimethyl-1-octene] and poly[(R,S)-3-methil-1-octene]. Macromolecules 23: 4912-4917

Perico A, Pratolongo R, Freed KF, Pastor RW, Szabo A (1993) Positional time correlation function for one dimensional system with barrier crossing: Memory function correction to the optimized Rouse-Zimm approximation. J Chem Phys 98: 564-573

STRUCTURAL AND CONFORMATIONAL STUDIES OF GLYCOPROTEINS AND OLIGOSACCHARIDE RECOGNITION DETERMINANTS.

Elizabeth F. Hounsell
University College London
Glycoprotein Structure/Function Group
Department of Biochemistry
Darwin Building, Gower Street
London, WC1ET 8BT UK

INTRODUCTION

Oligosaccharide chains often constitute a large proportion of the molecular mass of glycoproteins. Their relative hydrophilicity, flexibility and heterogeneity can make crystallisation difficult or impossible for X-ray studies of the native molecule and hence NMR spectroscopy is the method of choice for analysis. NMR is also useful for analysis of oligosaccharide sequences found on glycoproteins, glycolipids, proteoglycans, polysaccharides, antibiotics and as free oligosaccharides which are recognition determinants and potential drugs. Their multiple chiral centres and polyhydroxyl functionality make oligosaccharides important natural information molecules and artificial templates on which to stereochemically design ligands for different protein or DNA binding motifs.

The extent of their natural information content can be gauged from Fig. 1. which shows the common monosaccharides in mammalian glycoconjugates. Each monosaccharide has a different orientation of functional groups (H,OH, NH_2, NHAc or CO_2H) above or below the plane of the glycosidic ring. The configuration at C1, giving α or ß anomericity, adds an additional level of complexity as does the ability to have D- or L- configuration and pyranose or furanose ring forms. One or more of the OH or NH groups are sometimes naturally found substituted by SO_3^- or PO_4^-, and additional anionic groups can be added synthetically, to interact specifically with basic amino acid protein motifs, for example. NMR is the method of

NATO ASI Series, Vol. H 87
NMR of Biological Macromolecules
Edited by C. I. Stassinopoulou

choice for determining these structural parameters in particular by coupling constant analysis. Additional information on how the monosaccharides are linked together between one or more of the hydroxyl groups around the ring can be obtained by comparisons of libraries of chemical shift data and by NOE measurements. The latter analysis, performed by rotating frame (ROE) experiments for determinants of tri- to octasaccharides, can also provide information on through space distances in addition to those across the glycosidic bond which can be interpreted in terms of the dihedral φψ angles (Fig. 1).

Hexose	D-Galactose α/β	(Gal)	+		+
	D-Mannose α/β	(Man)		+	
	D-Glucos α/β	(Glc)		+	
Deoxy-hexose	L-fucose α/β	(Fuc)			
	L-Xylose β	(Xyl)		+	
Acetamido	N-acetyl-D-glucosamine α/β	(GlcNAc)	+		
-sugars	N-acetyl-D-galactosamine α/β	(GalNAc)	+		
Uronic acids	D-Glucuronic acid β	(GlcA)	+		
	D-Galacturonic acid β	(GalA)	+		
	L-Iduronic acid α	(IdoA)	+		
Sialic acids	N-acetylneuraminic acid α	(NANA)	+	+	+
	N-glycolylneuraminic acid α	(NGNA)	+	+	+

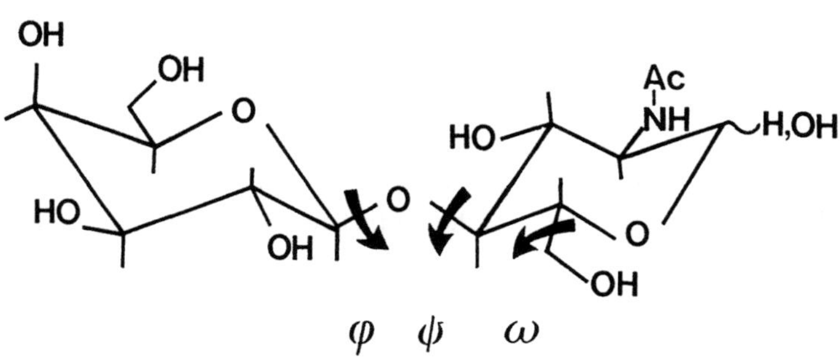

Fig. 1. A summary of monosaccharides and their substituents commonly making up mammalian glycoconjugates (glycoproteins, glycolipids and proteoglycans), and a diagram of the disaccharide Galβ1-4GlcNAc showing the dihedral angles across the glycosidic bond.

For the types of oligosaccharide recognition determinants being considered herein and for glycoproteins there is usually not enough material for natural abundance ^{13}C studies. I will therefore restrict my discussion to the most useful experiments, i.e. ^{1}H 1D spectra, homonuclear ^{1}H-^{1}H COSY, relayed COSY, TQFCOSY, TOCSY and ROESY experiments and ^{1}H detected ^{13}C heteronuclear HMQC. These experiments are illustrated with respect to the oligosaccharide chains of the receptor for epidermal growth factor (EGF), mucin glycoprotein oligosaccharides and sulphated oligosaccharide sequences involved in recognition by antibodies (Hounsell et al., 1986; Hounsell, 1987) and the envelope glycoprotein, gp 120, of the human immunodeficiency virus HIV_1.

RESULTS AND DISCUSSION

Primary Sequence Analysis

As discussed in two recent reviews (Hounsell 1993; Smith et al, 1994), NMR forms a part of the structural analysis of glycoproteins which also includes protease digestion, analysis of peptides by liquid secondary ion (LSI) mass spectrometry (MS), analysis of glycopeptides by amino acid N-terminal sequencing, release of oligosaccharide chains from protein, their chromatographic profiling and gas-liquid chromatographic (GC-MS) monosaccharide and linkage analysis. By the end of such a scheme something of the oligosaccharide structures at each glycosylation site in the glycoprotein is known. NMR analysis of the released oligosaccharides can provide a finger-print of the exact structure by comparison with the chemical shift data base which can now be carried out by computerised methods (Hounsell and Wright, 1990; van Kuik et al, 1993). More detailed chemical shift and coupling constant analysis can usually only be achieved by access to a large scale source of the oligosaccharides, the structures of which have been determined by several analytical methods including NMR. This highlights one of the differences in analysis of oligosaccharides compared to other macromolecules in that the same structure may be found on many different proteins. In addition part sequences are present on other forms of glycoconjugates e.g. on the O- and N-linked oligosaccharide cores of glycoproteins (Fig.2.), attached to lipid and within polysaccharides etc. The computerised techniques for assisted spectral interpretation are also able to carry out this part sequence analysis and are capable of giving some structural interpretation on oligosaccharide chains still linked in intact glycoproteins and in high molecular weight repeated polysaccharide sequences. In the former case the relative flexibility of oligosaccharides compared to globular proteins can allow for NMR solution state analysis, particularly when only one or two N-linked chains are present.

```
N-linked core   Galβ1-4GlcNAcβ1-2Manα1\                  ±Fucα
                                       6                  |1,6
                          ±GlcNAcβ1-4ManβGlcNAcβ1-4GlcNAcβ1 N-Asn
                                       /3
                Galβ1-4GlcNAcβ1-2Manα1

O-linked core          Galβ1-4GlcNAcβ1\
                                       6
                                        GalNAcα1-O-Ser/Thr
                                       /3
                                   Galβ1

Glycolipid                     Galβ1-4Glcβ1-Ceramide
```

Fig. 2. Core sequences to which additional oligosaccharides can be attached which may have structures in common to the different cores and to polysaccharides or free oligosaccharides.

Where enough material is available, 2D-correlated spectroscopy methods are highly desirable for chemical shift assignment. Figs 3 and 4 illustrate COSY and relayed COSY experiments for the oligosaccharide alditol Galβ1-4GlcNAcβ1-6GalNAcol (Feeney et al, 1986). This is an oligosaccharide released from a mucin glycoprotein (Hounsell et al, 1985) and represented the first complete characterisation of the GlcNAcβ1-6GalNAc core linked through the hydroxyl groups of Ser or Thr aminoacids of proteins (O-linked chains). Seven such core regions have now been identified (Table 1) in studies which are not only important in determining structure/function relationships in pathologically relevant mucin molecules, but also in membrane and serum

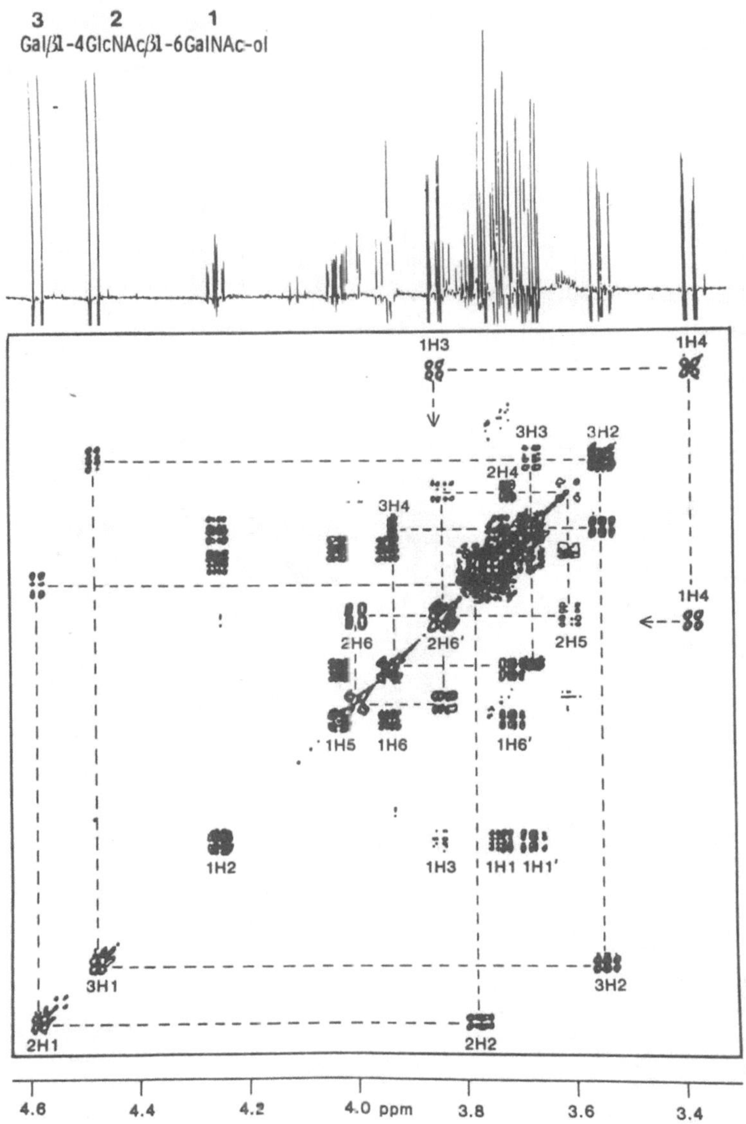

Fig.3. The 1D ^1H and ^1H-^1H DQFCOSY of the trisaccharide alditol Galβ1-4GlcNAcβ1-6GalNAcol omitting the CH_3 acetamido signal at 2-2.1ppm.

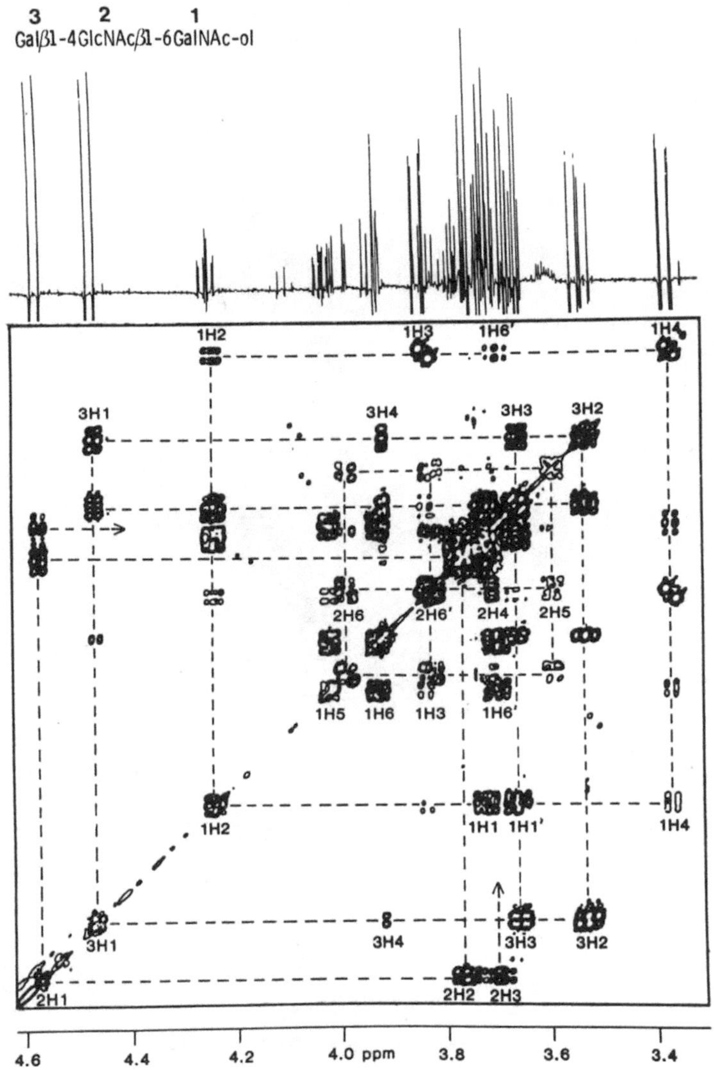

Fig.4. The 1D ^{1}H and ^{1}H-^{1}H RELAYED COSY of the trisaccharide alditol Galβ1-4GlcNAcβ1-6GalNAcol omitting the CH_3 acetmido signal at 2-2.1ppm.

glycoproteins where poly-O-glycosylation is also found and has profound effects on protein structure, antigenicity and conformation (reviewed in Hounsell et al 1985; Hounsell, 1987; Hounsell and Davies, 1993).

GalNAcα1-	0
Galβ1-3GalNAcα1-	I
Galβ1\ 3 GalNAcα1- 6 GlcNAcβ1/	II
GlcNAcβ1-3GalNAcα1-	III
GlcNAcβ1\ 3 GalNAcα1- 6 GlcNAcβ1/	IV
GalNAcα1-3GalNAcα1-	V
GlcNAcβ1-6GalNAcα1-	VI
GalNAcα1-6GalNAcα1-	VII

Table 1. - The O-linked chain core region structures linked to Ser/Thr hydroxyl groups.

Oligosaccharide sequences of interest in medical research are usually of greater complexity than that shown in Figs 3 and 4. Fig. 5 shows the 1D ^{1}H and ^{1}H - ^{1}H COSY of the heptasaccharide.

```
GalNAcα1-3Galβ1-4GlcNAcβ1-3Galβ1-4Glcα/β
          |1,2    |1,4
          Fucα    Fucα
└────────────────────────┘
```

This oligosaccharide was isolated from the faeces of breast fed human babies where the reducing sugars based on lactose (Galβ1-4Glc) are found in abundance. The relevance of this oligosaccharide to our studies is the bracketed non-reducing pentasaccharide (in the absence of lactose) which was identified by immunochemical methods on the membrane bound receptor for EGF of a cell line called A431 (Gooi et al., 1985a, 1985b). Here it is expected to be linked via the GlcNAc residue to the Man-containing core of N-linked chains.

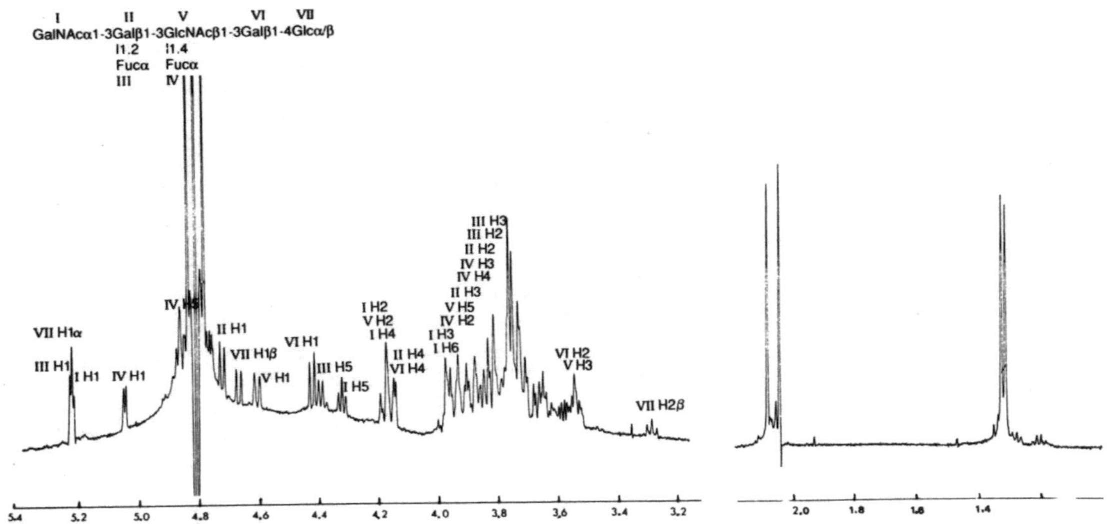

Fig.5. The 1D ^1H experiment of the ALe^b heptasaccharide with the acetone CH_3 resonance set at 2.225ppm.

Studies of the site-specific glycosylation patterns of the receptor for EGF from A431 cells and a transgenic product secreted from CHO cells (Smith et al submitted) have shown that eleven N-linked oligosaccharide chains can be present on the extracellular portion of this receptor. In A431 cells, which are a blood group A expressing fibroblast line, the ALe^b pentasaccharide underlined above is obviously able to be biosynthesised on the glycoprotein chains. The ability to isolate the free heptasaccharide in large quantities allowed for comprehensive chemical shift assignment by COSY and also by TOCSY (Fig.6) experiments.

As a third example of 2D correlated spectroscopy in an even higher molecular weight oligosaccharide/glycoconjugate, Fig. 7 shows the ^{1}H 1D spectra of two preparations of bovine lung heparin. The assignments were made by ^{1}H-^{1}H COSY and ^{1}H detected heteronuclear multiple quantum correlation (HMQC) spectroscopy. These two preparations were shown to have different activities as inhibitors of HIV infectivity (Coomb et al, submitted) and thus it was important to characterise their structure. Our present studies and those of others (Linhardt et al, 1986; Petitou et al, 1988; Horne and Gettins, 1991; Gettins and Horne, 1992; Linhardt et al, 1992; Yamada et al, 1993; Desai et al, 1993) have concentrated on chemical shift analysis of isolated purified heparin oligosaccharides (tri- to octasaccharide). These studies illustrate the newly found complexity and biosynthetic control in a proteoglycan sequences. Each of the different sequences characterised has potential for biological activity e.g. anti-viral, anti-coagulent, anti-atherogenic, anti-angiogenic and cell growth factor control. In Fig. 7 the major signals for H1-H3 shown are consistent with 2-O-sulphation (and not 6- or 3-). The H2, H3 signals for $GlcNSO_3$ and $GlcNH_2$ have different chemical shifts. Those for H1 overlapped, but could be distinguished by the HMQC experiment. In both spectra upfield of the acetone standard at 2.225 ppm are signals for minor GlcNAc components. Additional variability arises from O-sulphation of $GlcNH_2/SO_3/Ac$ and the incomplete conversion of GlcA to IdoA.

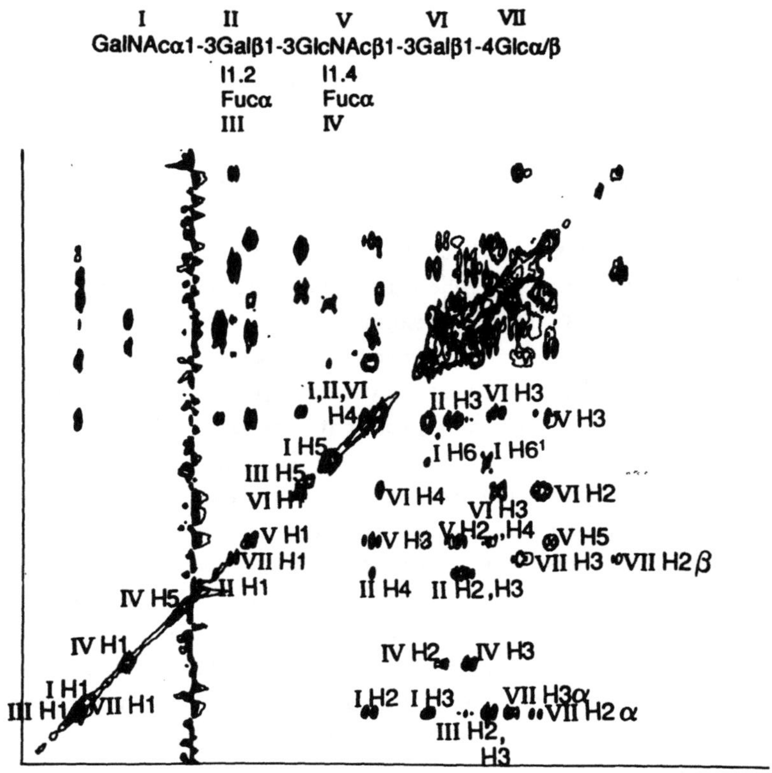

Fig.6. The ^{1}H-^{1}H TOCSY of the ALe^{b} heptasaccharide omitting the CH_3 acetamido signals at 2-2.1ppm and CH_3 fucose C6 signals at 1.2-1.3ppm.

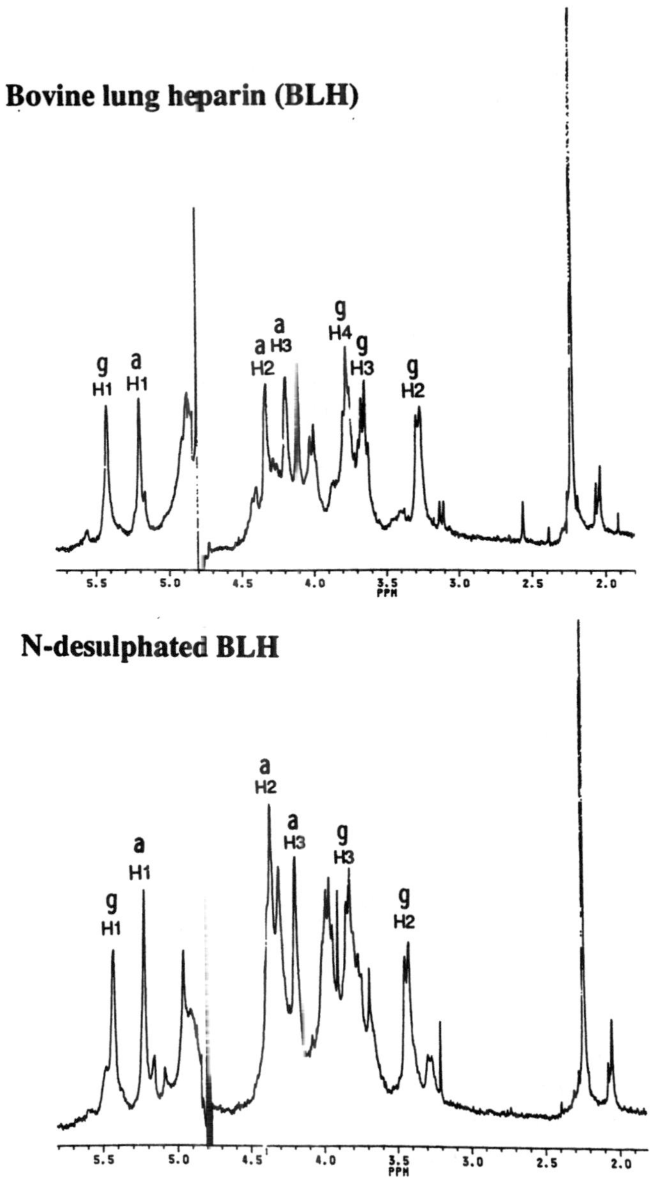

Fig.7. The 1D ^{1}H spectra of high molecular weight heparin polymers made up of repeating $GlcNSO_3/GlcNH_2\alpha1$-4(residue g) and IdoA α1-4(residue a) with variable sulphation

Conformational Analysis

In order to exploit the molecular recognition of oligosaccharides in terms of the design of pathological inhibitors, it is necessary to have detailed information about their solution conformation and molecular dynamics. This can be difficult to achieve for the more flexible backbone regions of oligosaccharide chains, but is possible for the stereochemically hindered globular-type molecular recognition determinants such as ALe^b (see above) or for regions which are constrained by adjacent protein, for example. Fig 8 illustrates a possible low energy solution conformation for the ALe^b sequence which has been determined jointly by molecular mechanics simulations and NOE studies.

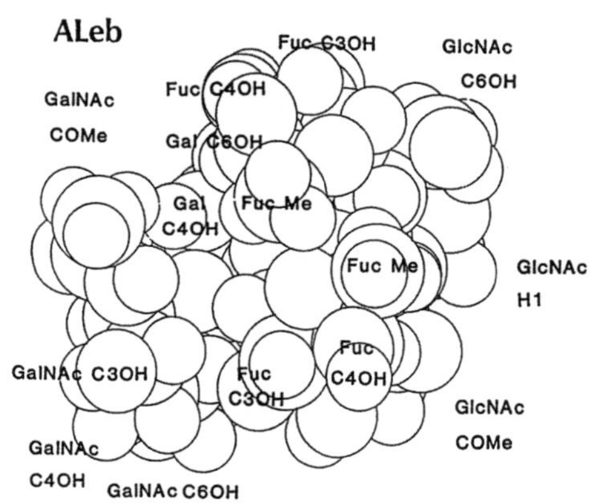

Fig. 8. A possible solution conformation for the ALe^b heptasaccharide showing GalNAc to the left and the two fucose residues in the forefront. To the right is the N-acetylglucosamine which in N-linked glycoproteins would be linked to the rest of the mannose containing core.

A general strategy for this type of analysis is to first carry out Ramachandran plots of φ ψ angles across the glycosidic bond starting from different geometries. The resulting energy contour diagram gives several minima which can be used as starting points for making the complete molecular model and subsequent energy minimisation followed by molecular dynamics. This approach is also discussed for mucin O-linked core region structures in Renouf and Hounsell, 1993. The through-space distances measured in the resulting models are then correlated with qualitative ROE's obtained by 2 x 2D HOHAHA-ROESY experiments or by analysing slices of 3D HOHAHA-ROESY experiments (C.J. Bauer and E.F. Hounsell, unpublished results). The reasons for carrying out this study was to compliment structural analysis work to describe the highly glycosylated EGF receptor. In other areas it is desirable to look at oligosaccharide conformation, for example, to design analogues of ligands for carbohydrate-binding proteins of particular importance in inflammation and lymphocyte trafficking. We are also using this approach to design sulphated oligosaccharides which will interact with basic amino acid side groups of the HIV glycoprotein gp120. Similar studies have been carried out very successfully to predict low molecular weight heparin-type oligosaccharide sequences with optimal activity as activators of anti-thrombin III (van Boeckel et al 1991).

METHODS

Oligosaccharides are evaporated three times from D_2O using a centrivap apparatus and then dissolved at 0.1mM - 10mM in D_2O containing 0.1-1 μl acetone as internal standard and transferred to 5mm NMR tubes. 500-MHz ^{1}H-NMR spectra were obtained using a Bruker AM500 spectrometer, a pulse interval of 8.19 s, and flip-angle of 65°. In the highest resolution 1D experiments, a spectral width of 2kHz was used with 32 K data points, giving a digital resolution of 0.12 Hz/pt. Up to 500 transients were averaged and the free induction decays were multiplied by a Gaussian resolution-enhancement function. Chemical shifts were measured in ppm from the signal for internal acetone and given with reference to DSS (4,4-dimethyl-4-silapentane-1-sulphonate) taken as 2.225 ppm from acetone at 295 K. Spectral simulations were carried out using the PANIC programme which is part of the standard Bruker software.

As described by Feeney et al (1986) the 2D experiments were all variants of the original two-pulse correlation experiment (COSY) which allows pairs of spins that have a resolved scalar coupling to be identified. Two types of signal are important in the experiments involving pairs of coupled spins (cross-peaks) and transitions within a single multiplet (diagonal peaks). Only the first convey useful correlation information. Three variants of the basic experiment were employed. (a) A "double-quantum filter" was used together with phase-cycling and data processing to give a phase-sensitive spectrum. The combination of these two techniques has the advantage that it gives pure absorption-phase line-shapes for all the peaks in the 2D spectrum. (b) A simplified correlation spectrum can be obtained by means of a pulse sequence which eliminates all splittings due to proton-proton coupling from one of the dimensions of the 2D spectrum (F_1-decoupling). This makes it easier to extract information from crowded spectral regions. (c) Introduction of an additional coherence transfer step gives the relayed correlation experiment. Extra information is available from this RELAYED-COSY experiment, as cross-peaks can occur between pairs of spins which are not coupled to each other but which are both coupled to a third spin.

Total correlation spectroscopy (TOCSY), rotating frame nuclear Overhauser enhancement spectroscopy (ROESY) and heteronuclear multiple quantum correlation (HMQC) experiments were carried out at 295K on a Varian Unity 600 MHz spectrometer. TOCSY experiments were acquired using on MLEV-17 mixing sequence of duration 100ms and field strength approximately 7KHz. ROESY experiments used a 100-300ms mixing time provided by a train of small flip angle pulses. HMQC spectra were acquired with Garp decoupling during acquisition and were optimised for a coupling constant of 165 Hz. Suppression of the water signal was achieved by selective presaturation in all of the NMR experiments.

Molecular structures were modelled using Insight II and Discover (Biosym Technologies Ltd, Basingstoke, UK). Monosaccharide building blocks were constructed for use with the AMBER forcefield, with carbohydrate specific parameters introduced. Charges for substituents were obtained from equivalent groups within the AMBER library with modifications to maintain neutrality carried out locally without affecting the rest of the molecule. The structures were minimized and used to construct di- and oligosaccharide plots (IUPAC definitions: $\varphi = O_5\text{-}C_1\text{-}O_x\text{-}C_x$, $\psi = C_1\text{-}O_x\text{-}C_x\text{-}C_{x-1}$) were generated for each disaccharide using the 'rotors' command (Biosym) to vary the glycosidic angles independently. The dihedrals were maintained using a force of 500kcal/rad^2 and the τ angle set to 116.9° by fixing atoms C_1, O_x and C_x. At each grid point the disaccharide was minimized to a maximum derivative of 0.1kcal/Å. Successive dihedral changes were made with reference to the starting structure to avoid propagation of irreversible structural changes. A contour plot was constructed showing the energy levels at 1 kcal intervals up to 10kcal above the minimum energy value.

Conformations defining local energy minima were further minimized, with angular constraints removed, to a maximum derivative of 0.001 kcal/Å. The resulting structures were used to construct larger oligosaccharides by adding monosaccharide units at dihedral angles derived from the disaccharide minimizations. All angular combinations corresponding to local energy minimum conformations were minimized for each structure investigated and the resulting lowest energy conformations used as starting points for the construction of larger oligosaccharides. Molecular dynamics simulated annealing was carried out on the lowest energy structures of the oligosaccharides investigated with, for example, 10 ps simulations at 500K followed by simulated annealing for 1 ps at 10 degree intervals down to 10K.

ACKNOWLEDGEMENTS

This work could not have been carried out without the collaboration of Chris Bauer, Tom Frenkiel and James Feeney. I also wish to thank David Renouf for our computer graphics studies and Almitra Malhotra for preparing the typescript

REFERENCES

Coombe DR, Harrop HA, Hounsell EF, Bauer CJ, Feeney J, Mahmood N, Mulloy B, Parish CR and Rider CC (submitted) The anti-HIV-1 activity of chemically modified heparins: correlation between GP-120 V3 loop binding and inhibition of cellular HIV-1 infection in vitro. Biochemistry

Desai UR, Wang HM, Kelly TR and Linhardt RJ (1992) Structure elucidation of a novel acidic tetrasaccharide and hexasaccharide derived from a chemically modified heparin. Carbohydr. Res. 241: 249-259

Feeney J, Frenkiel TA and Hounsell EF (1986) Complete ^{1}H-N.M.R. assignments for two core-region oligosaccharides of human meconium glycoproteins, using 1D and 2D methods at 500 MHz. Carbohydr. Res. 152: (1986) 63-72

Gettins P and Horne AP (1992) One- and two-dimensional ^{13}C-n.m.r. characterization of two series of oligosaccharides derived from porcine intestinal mucosal heparin by degradation with heparinase. Carbohydr. Res. 223: 81-98

Gooi HC, Picard JK, Hounsell EF, Gregoriou M, Rees AR and Feizi T (1985a) Monoclonal antibody (EGR/G49) reactive with the epidermal growth factor receptor of A431 cells recognizes the blood group ALe^{b} and ALe^{y} structures. Mol. Immunol. 22: 689-693

Gooi HC, Hounsell EF, Lax I, Kris RM, Libermann TA, Schlessinger J, Sato JD, Kawamato T, Mendelsohn J and Feizi T (1985b) The carbohydrate sepcificities of the monoclonal antibodies 29.1, 455 and 3C1B12 to the epidermal growth factor receptor of A431 cells. Biosci. Rep. 5: 83-94

Horne A and Gettins P (1991) ^{1}H-N.m.r. spectral assignments for two series of heparin-derived oligosaccharides. Carbohydr. Res. 225: 43-57

Hounsell EF, Lawson AM, Feeney J, Gooi HC, Pickering NJ, Stoll MS, Lui SC and Feizi T (1985) Structural analysis of the O-glycosidically linked core-region oligosaccharides of human meconium glycoproteins which express oncofoetal antigens. Eur. J. Biochem. 148: 367-377

Hounsell EF, Feeney J, Scudder P, Tang PW and Feizi T (1986) ^{1}H-NMR studies at 500 MHz of a neutral disaccharide and sulphated di-, tetra-, hexa- and larger oligosaccharides obtained by endo-ß-galactosidase treatment of keratan sulphate. Eur. J. Biochem. 157: 375-384

Hounsell EF (1987) Structural and conformational characterization of carbohydrate differentiation antigens. Chem. Soc. Rev. 16: 161-185

Hounsell EF, Jones NJ, Gooi HC and Feizi T (1988) 500-MHz ^{1}H-N.M.R. and conformational studies of fucosyl-oligosaccharides recognised by monoclonal antibodies with specificities related to Le^{a}, Le^{b}, and SSEA-1. Carbohydr. Res. 178: 67-78

Hounsell EF and Wright DJ (1990) Computer-assisted interpretation of ^{1}H-n.m.r. spectra in the analysis fo the structure of oligosaccharides. Carbohydr. Res. 205: 19-29

Hounsell EF and Davies MJ (1993) Role of protein glycosylation in immune regulation. Ann. Rheum. Dis. 52: S22-S29

Hounsell EF (1993) A general strategy for glycoprotein oligosaccharide analysis. In EF Hounsell ed. Methods in Molecular Biology Vol 14 Glycoprotein Analysis in Biomedicine, Humana Press Ch1

Linhardt RJ, Rice KG, Merchant ZM, Kim YS and Lohse DL (1986) Structure and activity of a unique heparin-derived hexasaccharide. J. Biol. Chem. 261: 1448-14454

Linhardt RJ, Wang HM, Loganathan D, and Bae JH (1992) Search for the heparin antithorombin III-binding site precursor. J. Biol. Chem. 267: 2380-2387

Petitou M, Lormeau JC, Perly B, Berthault P, Bossennec V, Sie P and Choay J (1988) Is there a unique sequence in heparin for interaction with heparin cofactor II? Structural and biological studies of heparin-derived oligosaccharides. J. Biol. Chem. 263: 8865-8690

Smith KD, Davies MJ and Hounsell EF (1994) Structural profiling of oligosaccharides. In J. Walker ed. Methods in Molecular Biology Vol 20 Protocols for Proteins and Peptides Humana Press Ch18

Smith KD, Davies MJ, Carruthers RA, Cashmore GC, Purkiss P, Lawson AM and Hounsell EF (Submitted) Analysis of the glycosylation patterns of the extracellular domain of the receptor for epidermal growth factor receptor expressed in Chinese hamster ovary fibroblasts. Anal. Biochem.

Renouf DV and Hounsell EF (1992) Conformational studies of the backbone (poly-N-acetyllactosamine) and the core region sequences of O-linked carbohydrate chains. Int. J. Biol. Macromol 15: 37-42

Yamada S, Yoshida K, Sugiura M, Sugahara K, Khoo K, Morris HR and Dell A (1992) Structural studies on the bacterial lyase-resistant tetrasaccharides derived from the antithrombin III-binding site of procine intesstinal heparin. J. Biol. Chem. 268: 4780-4787 van Boeckel CA, Grootenhuis PD, and Haasnoot CA (1991) Specificity in the recognition process between charged carbohydrates and proteins. Trends. Pharmacol. Sci. 12: 241-3

van Kuik JA, Hard K and Vliegenthart JFG (1992) A ^{1}H NMR database computer program for the analysis of the primary structure of complex carbohydrates. Carbohyd. Res. 235: 63-68

^{13}C NUCLEAR MAGNETIC RELAXATION OF CARBOHYDRATE MOLECULES IN SOLUTION.

Photis Dais
Department of Chemistry
University of Crete
71409 Iraklion, Crete
Greece

INTRODUCTION

Nuclear magnetic resonance spectroscopy is a very powerful tool for analyzing the conformation and molecular architecture of carbohydrate molecules. Both 1D and 2D methodologies have provided valuable information about small and large molecules, ranging from the anomeric configuration of a monosaccharide to the more complex problem of the sequence of monosaccharide residues that constitute an oligo-, or polysaccharides.
Another important application is concerned with the property of motional freedom to which many physical, and biological functions of carbohydrates in solution are related. The study of the motional dynamics of these systems provides, not only complementary and sometimes unique information about structure and conformation, but also a deeper insight into their physical behavior and biological properties in living organisms.
Nuclear magnetic resonance has long been employed to monitor molecular motions and has been extensively applied to dynamic problems of small and large molecules (Lyerla et al 1974). In particular, carbon-13 relaxation experiments can probe motion at several carbon sites simultaneously in the molecular frame. Each carbon nucleus is associated with a set of relaxation parameters, such as spin-lattice relaxation time (T_1), spin-spin relaxation time (T_2), and nuclear Overhauser enhancement (NOE) that reflects the average interaction of that nucleus with the environment. These relaxation parameters, are related to the spectral density function $J_i(\omega_i)$, of local magnetic fields generated by the atomic and electronic environment of the nucleus. These local fields are modulated by the tumbling motion of the molecule in solution and can promote relaxation, if they have a frequency component at or near the Larmor frequency, ω_0, of the nucleus providing, thus, the link between nuclear magnetic relaxation and molecular motion.
Assuming a purely ^{13}C-^{1}H dipolar mechanism, T_1, T_2, and NOE parameters, can be expressed in terms of spectral density function as follows:

$$\frac{1}{NT_1} = \frac{1}{20}\Omega\,[J_0(\omega_H-\omega_C) + 3J_1(\omega_C) + 6J_2(\omega_H+\omega_C)] \qquad (1)$$

NATO ASI Series, Vol. H 87
NMR of Biological Macromolecules
Edited by C. I. Stassinopoulou

$$\frac{1}{NT_2} = \frac{1}{40}\Omega\,[4J(0) + J_o(\omega_H\text{-}\omega_C) + 6J_1(\omega_H) + 3J_1(\omega_C) + 6J_2(\omega_H+\omega_C)] \tag{2}$$

$$NOE = \frac{\gamma_H}{\gamma_C}\,\frac{6J_2(\omega_H+\omega_C) - J_o(\omega_H\text{-}\omega_C)}{J_o(\omega_H\text{-}\omega_C) + 3J_1(\omega_C) + J_2(\omega_H+\omega_C)} \tag{3}$$

and, $\Omega = (\frac{\mu_o\gamma_H\gamma_C\hbar}{4\pi r_{CH}^{\,6}})^2$

Here, ω_H, ω_C, and γ_H, γ_C are the Larmor frequencies and gyromagnetic ratios of the ^{13}C and ^1H nuclei respectively, $\hbar = h/2\pi$ is the Planck's constant, N is the number of directly bonded protons to the carbon atom, and r_{CH} is the C-H internuclear distance.
For isotropic motion, the spectral density function is given by

$$J_i(\omega_i) = \frac{2\tau_c}{1 + \omega^2\tau_c^{\,2}} \tag{4}$$

τ_c is the molecular correlation time describing molecular motion at a microscopic scale.
For small and medium sized molecules in non-viscous media, the product $\omega\tau_c$ is much less than unity ($\omega\tau_c << 1$), and Eqs. 1-3 become

$$\frac{1}{T_1} = \frac{1}{T_2}\,\Omega\,\tau_c \tag{5}$$

$$NOE = \frac{\gamma_H}{2\gamma_C} = 1.988 \tag{6}$$

In this region, the so-called extreme narrowing region, high frequency motions dominate the spectral density function, and the relaxation parameters are frequency-independent.
The major objectives of this chapter is to introduce some qualitative and quantitative aspects in the relaxation behavior of carbohydrate molecules dealing with monosaccharides and oligosaccharides, to demonstrate the advantages and limitations of this approach, and to critically review several results that have been obtained in the field.

QUALITATIVE MOTIONAL DESCRIPTION.

A. MONOSACCHARIDE MOLECULES

Monosaccharide molecules in non-viscous media tumble in the extreme narrowing region.

Therefore, NOE values, assuming an asymptotic value of ca 2.0, have no diagnostic utility in describing their motional characteristics. The same is true for spin-spin relaxation times because, usually $T_2=T_1$, in this region, (Eq. 5). Moreover, T_2 values are affected preponderantly by low frequency motions, which are absent from these low molecular weight compounds. Nevertheless, spin-lattice relaxation times are affected by such intra-and inter-molecular factors as molecular weight, differences in C-H bond lengths and other specific structural features, hydrogen bonding and solvent effects, and the type of the overall and internal motion.

Relaxation theory does not take explicit account of the molecular weight-dependence of the relaxation time. This factor enters indirectly into the hydrodynamic equation relating the diffusion constant, D, with the rotational friction coefficient in terms of the volume α^3.

$$D = \frac{kT}{\xi}, \qquad (\xi = 8\pi\alpha^3\eta) \tag{7}$$

η is the viscosity of the medium and α is the radius of the molecule assumed to be spherical. In the general case, D, is a tensor and the individual components of the diagonalized diffusion tensor are given by $D_i = kT/\xi_i$. Furthermore, D_i is related to τ_c by the simple equation

$$\tau_{c,i} = \frac{1}{6D_i} \tag{8}$$

From equations 7 and 8, it follows that because large molecules tumble at a slower rate than small molecules of similar shape, the former are characterized by smaller T_1 values.

The effect of the molecular weight on the T_1 values for several carbohydrate molecules has been studied by Serianni and co-workers (Serianni et al 1982; Wu et al 1983; Serianni et al 1984). For example, C_5 glycosides tumble more slowly in solution than C_4 glycosides, and hence they have smaller T_1 values (Serianni et al 1982; Wu et al 1983)., as shown by comparison of methyl α-D-erythrofuranoside (1) and methyl α-D-ribofuranoside (2). Another example (Dais et al 1983) is the pair of orthoesters 3,4,6-tri-O-acetyl-1,2-O-(1-methoxyethylidene)-β-mannopyranose (3) and 3,4,6-tri-O-acetyl-1,2-O-(1-benzyloxyethylidene)-β-D-mannopyranose (4). In each molecule, all of the ring carbon atoms bearing a single proton have essentially the same ^{13}C-T_1 values, indicating an approximately isotropic motion.

However, the presence of the heavier, exo benzyl group in 4 results in smaller T_1 values than in 3, which has an exo methyl group. The effective correlation times calculated from the average of the T_1 values of the ring carbons were 18.2 ps and 26.0 ps for the methoxy and benzyloxy derivatives respectively. In a recent publication (Dais et al 1989) invoked molecular weight dependence of the T_1 values in conjuction with solvent effects in order to explain the dynamic behavior of a pair of anomeric glycosides, namely methyl-α- (5a) and β- (5b) D-glucopyranoside in a group of solvents that represent an extended range of such properties as viscosity and dielectric constants. The effective correlation times for an approximately isotropic motion, τ_R, ranged from 40.3 ps in methanol to 100.7 ps in acetic acid for 5a, and from 61.6 ps to 180.1 ps for 5b in the same solvents. Neither solvent viscosity, nor dielectric constant bore any direct relationship to the correlation times found from the overall motion, and attempts to correlate relaxation data with parameters, other than dielectric constant, that reflect solvent polarity, e.g. Kosover Z-values, Winstein Y-values, etc., were unsuccessful. Based on the maximum allowed error of +13% in the τ_R values derived from the propagation of the experimental error in the measured T_1 values, the rate of the overall motion for either 5a or 5b in these solvents followed the order: methanol appears ~ N,N-dimethylformamide ~ D_2O < pyridine < dimethylsulfoxide. The above sequence appears to reflect both the solvent viscosity and the molecular weight of the solvated entities. On this basis, and assuming that each hydroxyl group is hydrogen-bonded to two molecules of the solvent (Harvey et al 1978), the molecular weight of the solvated species are: in methanol (256), N,N-dimethylformamide (364), water (144), pyridine (496), and dimethylsulfoxide (312).

Another intramolecular factor reflected in the measured relaxation times is the C-H bond length, which appears in Eqs. 1 and 2. Nonuniformity in the bond lengths for the various C-H vectors in an isotropic tumbler may result in reproducible differences among their respective T_1 values. A 1% difference in bond length used in a relaxation calculation would overestimate the T_1 value by 6%. Conversely, a correlation time deduced from the experimental value of T_1 would be overestimated by 6%. This effect should be taken into consideration in analyzing apparent differences in relaxation times, either by estimating bond lengths from other sources, e.g. crystal data, infrared or microwave spectra, or performing additional relaxation experiments. Deuterium spin-lattice relaxation times of selectively deuterated compounds is a consistent method of evaluating differences in bond lengths, since ^2H T_1 values are predominantly quadrupolar and essentially unaffected by bond lengths (Lyerla et al 1974).

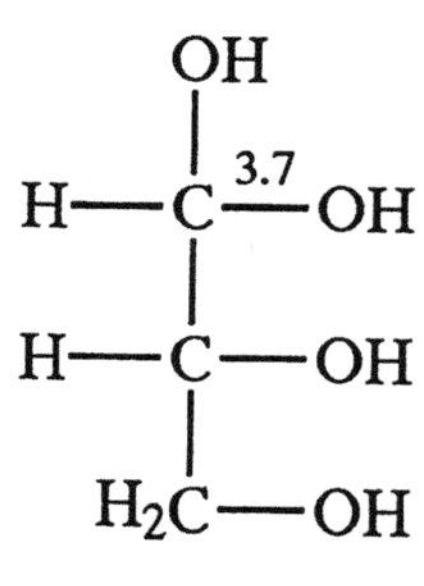

6a

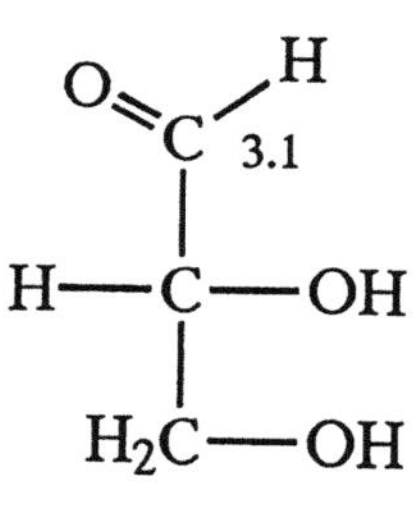

6b

The effect of C-H bond length on T_1 is evidenced (Serianni et al 1982) in a comparison of T_1 values for C-1 of the hydrate (6a) and aldehyde (6b) forms of enriched DL-[1-^{13}C] glycerose. These different T_1 values between these forms cannot be explained by molecular weight effects alone, whereas a shorter C-1-H-1 bond in the aldehyde form expected when a carbon changes from sp^3 to sp^2 hybridization, would cause C-1 to relax faster and produce the small difference in relaxation times observed.

Since T_1 depends on the inverse of the sixth power of the r_{CH} distance (Eqs. 1, and 2), a directly attached proton makes the most significant contribution to the relaxation of a given carbon relative to more distant protons in the molecule. The contribution of the latter is usually as little as 2% and may be ignored for qualitative descriptions. However, for quantitative treatment, or in particular cases where the carbon of interest is surrounded by several non-bonded protons, Eq. 9 should be used for the analysis of the relaxation data. This equation is a modification of Eq. 5, assuming that τ_c is the correlation time for each ^{13}C-^{1}H intramolecular interaction

$$\frac{1}{T_1} = \sum_{i=1}^{n} \left(\frac{\gamma_H \gamma_C \hbar}{r^3_{CH,i}}\right)^2 \tau_c \qquad (9)$$

Such an example is offered by sucrose (7) (Allerhand et al 1985). The substantially longer T_1 value (6.0 s) for the quaternary C-2' carbon relative to the T_1 values of the remaining ring carbons (average 0.5 s) reflects the lack of efficient dipolar relaxation by a directly bonded proton. However, a full NOE was found for this carbon, indicating that the surrounding non-bonded protons are effective contributors to its dipolar relaxation.

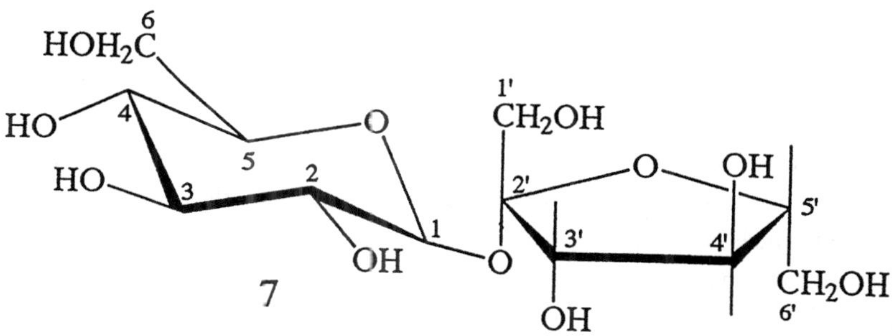

7

Specific structural features of monosaccharides are manifested in several ways, such as changes in the type of the overall motion, free or hindered internal flexibility, intra-molecular vs. inter-molecular hydrogen bonding. All pentofuranoses and pentofuranosides studied (Serianni et al 1982; Wu et al 1983; Serianni et al 1984; Dais et al 1987) appear to tumble isotropically in solution as evidenced by the similarity of the T_1 values of the ring carbons. This suggests that exocyclic hydroxymethyl groups stabilize the furanosyl ring conformation, or if any puckering motion exists, it must be very slow to affect relaxation data. For the tetrofuranosyl rings, however, lacking an exocyclic hydroxymethyl moiety, NT_1 values for the C-4 carbons (relaxed via two directly bonded protons) are larger than the average T_1 for the remaining ring carbons. This observation (Serianni et al 1984) is commensurate with internal flexibility at the C-4 site for the tetrofuranosyl ring.

The motional behavior of the majority of hexopyranoses and derivatives has been described as isotropic despite the fact that sizeable differences, well outside the experimental error, have been observed in the T_1 values of the ring carbons in several instances. Anisotropic motion has been observed for methyl-β-D-galactopyranoside (Czarniecki et al 1977) (8), methyl-β-D-glycopyranoside (5b) (Czarniecki et al 1977; Dais et al 1989), and both anomers of the 6-deoxy sugar L-fucose (α-L, 9a and β-L, 9b) (Czarniecki et al 1977). Other examples are exo (10a) and endo (10b) methyl-2,3-0-benzylidene-α-L-rhamnopyranoside, which have been described (Neszmelyi et al 1977) as isotropic tumblers, although there are large differences, as much as 26% (experimental error ± 6%) among the carbons of the pyranosyl ring, especially for the exo derivative.

The anisotropy usually observed in the overall motion of β-hexopyranosides relative to α-hexopyranosides, which reorient nearly isotropically, has been attributed (Czarniecki et al 1977) to an effect of the β-anomeric substituent on the inertial axes, sufficient to cause molecular diffusion about a preferential, but undefined, axis of rotation. Although the basic argument seems correct, it is unlikely that anisotropic motion of, e.g. compound 8, can be described in terms of a preferential axis of rotation. Moreover, such solvent effects as hydrogen bonding may alter the overall shape of the solvated species, resulting not only in a further shift of the principal axes of the diffusion tensor, but also in a different type of overall motion, as observed for molecule 5b, which rotates isotropically in protonated solvents, e.g. methanol, acetic acid, whereas a slight anisotropy is detected in D_2O, N,N-dimethylformamide, pyridine, and dimethylsulfoxide (Dais et al 1989). Anisotropic motion has also been observed for both anomers of compound 9.

Internal rotation occurring in pentofuranoses, hexopyranoses and their derivatives is manifested mainly in the NT_1 values of the exocyclic hydroxymethyl carbon (moderated by two protons), which is usually larger than those of the ring carbons. This indicates that the exocyclic carbon, apart from the overall motion, receives relaxation contribution from internal motion about the C-4-C-5, or C-5- C-6 bond. The degree of motional freedom of the hydroxymethyl group depends mainly on structural details, as well as on the existence and strength of intramolecular hydrogen bonding. Thus, the shorter NT_1 value observed (Czarniecki et al 1977) for the C-6 carbon in compound 5a and 5b relative to that in compound 8 has been attributed to a more restricted internal rotation in the former derivatives due to some intramolecular interaction (possibly hydrogen bonding) not present in galactose. Although the hydroxymethyl groups of the a- and β-anomers of methyl-D-glucopyranosides exhibit the same motional freedom in D_2O (Czarniecki et al 1977), changes of solvent are accompanied by a greater degree of motional freedom in the α- than in the β-anomer. As there is no obvious basis for assuming large differences in the rotational conformation and solvation pattern at C-6 for these monomers, the differing degrees of motional freedom consistent with the relaxation data can be attributed to variations in their principal diffusion axes, and effects on the T_1 data mentioned previously.

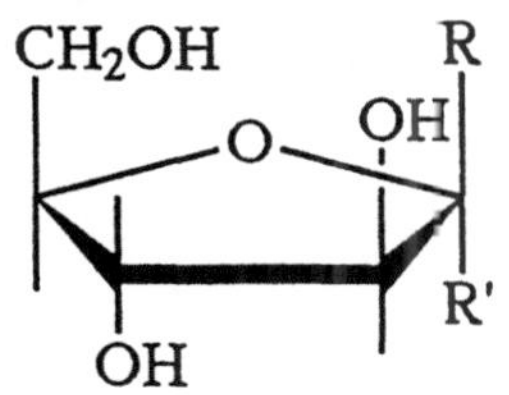

11a R = OH, R' = CH_2OH

11b R = CH_2OH, R' = 0H

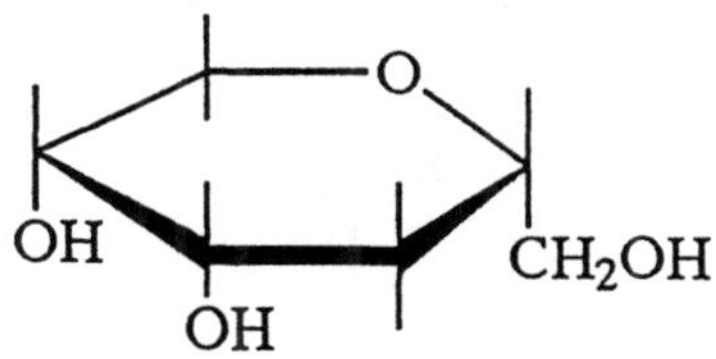

12

Another interesting example of hindered internal rotation is related to the relative stability of the β-furanose form of D-fructose (11a) in dimethylsulfoxide solutions (Dais et al 1987; Dais et al 1985), in marked contrast to the fact that the β-pyranose form (12) is by far the major tautomer in water. From the ^{13}C-T_1 relaxation data of the three major tautomers of D-fructose, namely α-(11b) and β-(11a) furanose, and β-pyranose (12), it can be concluded that these tautomers tumble isotropically in both D_2O and $(CD_3)_2SO$ solutions. T_1 values for the secondary carbons of 11a, 11b and 12 respectively are all within 1.64±0.08 s, 1.73±0.07 s, 1.55±0.05 s in D_2O, and 0.53±0.03 s, 0.54± 0.04 s, and 0.52±0.03 s in $(CD_3)_2SO$. In the absence of freedom of rotation about the exocyclic C-C bonds, the T_1 value of the primary carbons would be expected to be one-half of the secondary carbon. However, these values for all isomers in D_2O are greater than the relaxation times of the overall motion.

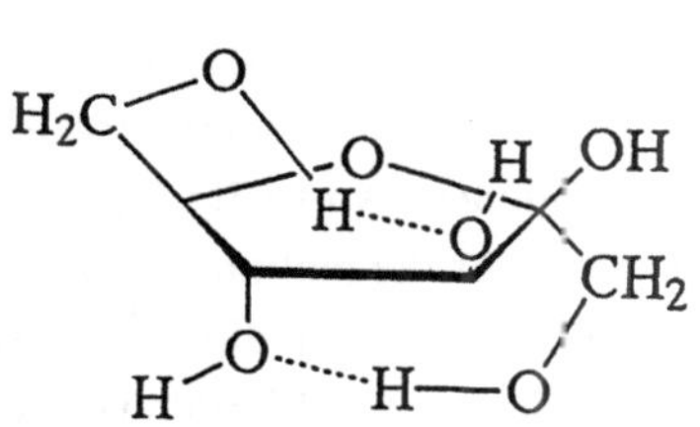

Fig. 1. Proposed intramolecular hydrogen bonding network of the β- furanose in dimethylsulfoxide solution.

This is indicative of relatively fast internal motions about the C-1-C-2 and C-5-C-6 bonds.

Also, the corresponding values for 11b and 12 in $(CD_3)_2SO$ are greater than those of the ring carbons, again indicative of fast internal rotations, as in D_2O. By contrast, the NT_1 values for the primary carbons of 11a are only slightly greater than those of the overall motion, suggesting that there is a restricted motion about the exocyclic bonds. These restricted motions have been attributed (Dais et al 1987; Dais et al 1985) to the presence of an intramolecular hydrogen-bonding network which, in turn, should be a major enhancement in stability for the β-furanose form in $(CD_3)_2SO$ solutions (Fig. 1).

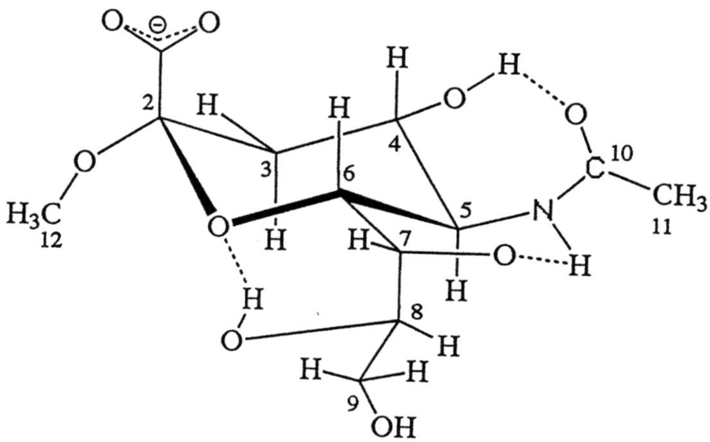

Fig. 2. Conformational model of anion 13 showing the proposed hydrogen-bond network and preferred conformation of the glycerol side chain.

A final example of restricted internal motion due to intramolecular hydrogen bonding is offered by N-acetyl derivatives of neuraminic acid (Czarniecki et al 1976; Czarniecki et al 1977) (13). Although methyl groups experience considerable free rotation about their C_3 symmetry axis, independent of the overall molecular motion, the motions of the C-7 and C-8 carbons of the side chain are isotropic with respect to the ring as evidenced by the near equality of these T_1 values with those of the ring carbons (C-3, C-4, C-5, C-6). The latter, being closely similar, indicate isotropic overall motion. This evidence of hindered motion has been explained in terms of an intra-molecular hydrogen-bonding network (Fig. 2), which locks the ring very strongly into its favorable conformation and links the motion of C-7 and C-8 to the ring through highly favorable six-membered ring interactions. In addition, further stabilization of this conformation is possible through formation of a third intramolecular hydrogen bond, between the acetamido carbonyl oxygen and the C-4 hydroxyl group in a seven-membered ring (Fig. 2).

B. OLIGOSACCHARIDE MOLECULES

Since the pioneering work of Allerhand and Doddrell (Allerhand et al 1971) in 1971 employing relaxation measurements as an effective methodology for the assignment of resonances in proton decoupled carbon-13 spectra of stachyose, raffinose, and sucrose, several publications appeared in the literature adopting the same strategy. The main qualitative conclusions from these studies can be summarized as follows:

(1) The similarity of the T_1 values for all the ring carbons in a disaccharide molecule may indicate isotropic tumbling without a preferred axis of rotation.

(2) Anisotropy in the molecular reorientation of a disaccharide molecule is expected when a C-H bond vector of the molecule lies on, or near a preferred axis of rotation. This results in a shorter T_1 value for carbons whose C-H bonds lie along the anisotropic axis than for those with other orientations.

(3) The NT_1 values of an exocyclic hydroxymethyl carbon, and/or a methyl carbon (moderated by two and three protons respectively) are in general higher than those of the ring carbons. This indicates internal reorientation about the exocyclic bond, the rate of rotation being dependent of the molecular geometry.

(4) For tri-, tetra-, and oligosaccharides a gradation in the T_1 values of the ring carbons, from the central monomer unit to the end monomer units, is observed. This reflects an increase in internal flexibility on going from the central monomer residue towards the end of the carbohydrate chain. The internal flexibility is more pronounced when one end of the oligosaccharide is anchored either by hydrogen bonding, by some other electrostatic interaction, or by attachment through chemical bond to a heavy anchor.

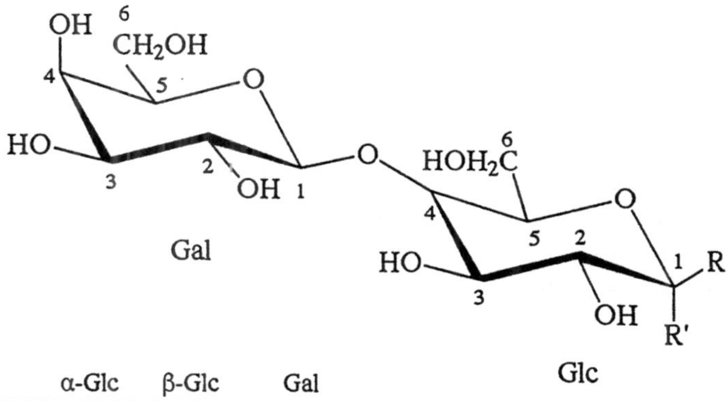

C	α-Glc	β-Glc	Gal
1	0.34	0.41	0.37
2	0.43	0.39	0.39
3	0.40	0.42	0.40
4	0.35	0.39	0.30
5	0.41	0.35	0.38
6	0.18	0.18	0.27

16a R = H, R' = OH

16b R = OH, R' = H

(5) Although no systematic studies have been carried out to examine the effect of the types of glycosidic bonds that link the various sugar monomer units in the oligosaccharides, the nature of these bonds is expected to modify the mobility along the carbohydrate chain.

Restricted internal motion has been detected in the cell membrane trisaccharides O-(N-acetyl-α-neuraminic acid)-(2→3)-O-β-D-galactopyranosyl-(1→4)-D-glucopyranose (14) (Czarniecki et al 1977) and O-(N-acetyl-α-neuraminic acid)-(2→6)-O-β-D-galactopyranosyl-(1→4)-D-glycopyranose (15) (Jaques et al 1980) distinguished by the α-(2→3) and α-(2→6) glycosidic bonding between the neuraminic acid group and the D-galactopyranosyl residue. The T_1 value for the C-6 of the galactopyranosyl residue decreases from 0.14 s for the α-(2→3) isomer to 0.08 s for the α-(2→6) linked isomer, indicating a more restricted motion of the glycosidically linked carbon in the latter.

A nuclear magnetic relaxation study (Czarniecki et al 1977) of lactose (16), which is a basic constituent disaccharide unit of all gangliosides, demonstrates that this molecule reorients anisotropically. The preferred axis of molecular reorientation appears to lie along the axis of the molecule and therefore is reflected in the shorter T_1 value at C-4 of the galactose residue. The apparent differences between the relaxation times observed for the α-(16a) and β-(16b) C-1 resonances may also be reflected in their differing C-H orientations relative to the anisotropic axis. Similar anisotropic motion was observed for the methyl-β-lactoside (Czarniecki et al 1977), methyl-β-cellobioside Berry et al 1977).

The presence of anisotropic motion along the molecular length is suggested for trisaccharide 14 as the anomeric carbons of the α- and β- isomers show different T_1 values. A smaller T_1 for the α-C-1 (equatorial, 0.11 s) relative to that of β-C-1 (axial, 0.24 s) indicates a close correspondence between the equatorial C-H vector and the preferred axis of rotation.

As noted previously, internal mobility along a carbohydrate chain is more pronounced if one end of the chain-like molecule is fixed by chemical or physicochemical bonding to a heavy anchor. A number of biologically important molecules are made up of a complex aglycone to which one or several sugar units may be radially attached. A representative example is k-strophatoside (17) (Neszmelyi et al 1977). Relaxation time measurements allow assignment of the carbon atoms belonging either to the steroid or to the oligosaccharide unit. Furthermore, the average T_1 values for the three sugar component reflect their sequence with respect to the steroid anchor, namely T_1 (terminal glucose) > T_1 (central glucose) > T_1 (inner sugar, cymarose). This trend in the T_1 values reflects an increase in flexibility along the oligosaccharide chain, from the sugar residue attached to the steroid towards the nonreducing end unit.

Segmental mobility along a carbohydrate chain is also reflected in T_1 data for 14 and 15. An increase in T_1 values observed (Czarniecki et al 1977; Jaques et al 1980), proceeding from the neuraminic acid residue through those of galactose and glucose indicates increasing mobility along the same direction. However, unlike the segmental motion in stachyose

(Allerhand et al 1971), which radiates from the central residues, the neuraminic acid component in 14 and 15 is the least mobile and appears to anchor one end of the molecule in solution.

2. QUANTITATIVE MOTIONAL DESCRIPTION

A. MONOSACCHARIDE MOLECULES.

To obtain a full understanding of the dynamic behavior of a monosaccharide molecule in solution, the experimental relaxation data should be treated quantitatively in terms of a motional model that describes the various degrees of freedom. On the other hand, the relaxation data may not uniquely define a dynamic model, and several models may be able to reproduce the experimental relaxation parameters.

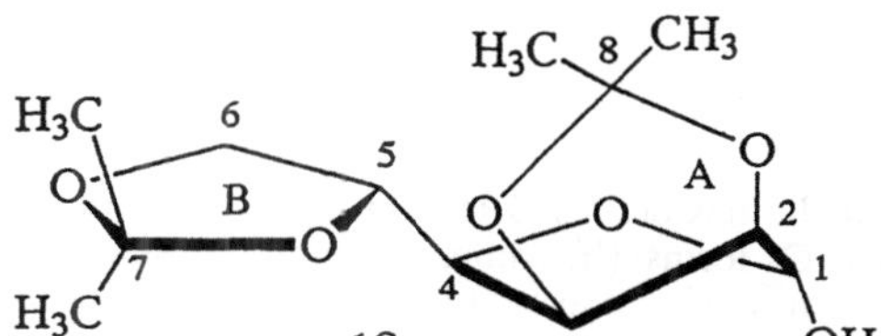

18

The problem of discriminating between models can be solved, either by inspection of the molecular shape and existing geometrical constraints or by conducting variable temperature relaxation measurements. As to the first approach, molecular shape is important for describing overall isotropic and anisotropic molecular motion, whereas geometrical constraints may restrict internal rotation of a flexible segment to a narrow angular range (e.g. rotation in-a-cone-, or restricted internal diffusion), in which case, a model based on free internal motion may not be compatible with the experimental data. The second approach is more rigorous in testing the suitability of the motional model adopted and in addition, provides activation energies, which are useful kinetic parameters of molecular reorientation in solution.

One representative example of dynamic modelling of monosaccharides will be given in this section. Other applications could be found elsewhere (Dais et al 1986); Dais et al 1986). Variable temperature ^{13}C NMR spin-lattice relaxation time measurements have been used to probe the motional behavior of 2,3:5,6-di-O-isopropylidene-α-D-mannofuranose (18) in dimethylsulfoxide solution (Dais et al 1983). This carbohydrate molecule offers structural features well suited to the study of a variety of internal motions, i.e. ring oscillation, ring puckering interconversion, and methyl internal rotation, all of which are superimposed on

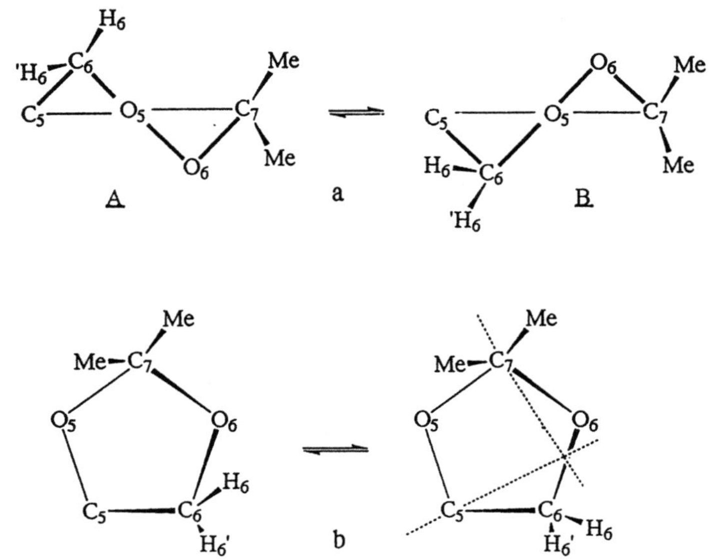

Fig. 3: Dynamic model for the flexible 5,6-O-isopropylidene ring in 18. (a) Interconversion between two antisymmetric C_2 forms; (b) Same, as viewed perpendicular to the C-5-O-5-C-7 plane. The two effective internal jump axes are indicated by dotted lines.

an isotropic overall reorientation. Isotropic overall motion is reflected in the T_1 values of the protonated ring carbons of the mannose ring, which are equal within experimental error at each temperature. An Arrhenius type plot of the correlation time of the overall motion (calculated with the use of Eq. 5) vs. reciprocal temperature yields an activation energy of 20.4 Kj/mole.

Among the various models examined to evaluate internal rotations a two-state jump model (London 1978) was found satisfactory for interpreting the oscillatory and puckering motions of the flexible 5,6-O-isopropylidene ring. In this model the C-H vector jumps between two equivalent, stable states A and B characterized by correlation times τ_A and τ_B respectively (Fig. 3). The internal motion is superimposed on an isotropic motion associated with correlation time, τ_R. The spectral density of this model is given explicitly elsewhere (Dais et al 1983; London 1978).

The diffusional model (Eq. 11), describing free rotation, was adopted for the motion of the geminal methyl groups of the rigid 2,3-O-isopropylidene ring. This model is described by the following equation (Woessner 1962)

$$\frac{1}{T_1(CH_3)} = N\left(\frac{\gamma_H \gamma_C \hbar}{r^3_{CH}}\right)^2 \left(\frac{A}{6D_R} + \frac{B}{6D_R D_i} + \frac{C}{6D_R + 4D_i}\right) \qquad (10)$$

where

$A = 1/4(3\cos^2\Delta - 1)^2$

$B = 3\sin^2\Delta\cos^2\Delta = (3/4)\sin^2(2\Delta)$

$C = (3/4)\sin^4\Delta$

A, B, and C are geometrical constants and are functions of the angle, Δ, formed between the C-H vectors and the internal rotation axis. D_R and D_i are the diffusion coefficients for the overall and internal motion. Eq. 10 assumes a diffusional process for the methyl group. Both models were applied from 15° to 80° C. The activation energies associated with these two types of motions were 26.3 Kj/mole for the 5,6-O-isopropylidene ring, and 15.9 and 14.2 Kj/mole for the exo and endo methyl groups of the 2,3-O-isopropylidene ring respectively. The activation energy associated with the temperature-dependence of the rate of overall molecular tumbling was found to agree with the hydrodynamic prediction of 19.1 Kj/mole associated with the temperature dependence of the ratio (η/T), where η is the shear viscosity of the solution.

B. OLIGOSACCHARIDE MOLECULES

Information concerning the overall motion of oligosaccharide molecules in solution, and also the relative spatial rigidity of the individual sugar monomers should be available from the spin-lattice relaxation times of the individual ring carbon nuclei. However, the dynamic modelling of these molecules based on experimental relaxation data has important limitations, which are documented here. Partitioning of the various modes of reorientation, even for the simplest member of this class, i.e. a disaccharide molecule, is not an easy task. For instance, separation of rotatory diffusion from internal oscillations around the glycosidic bonds is not feasible, because no ring carbon atom in the disaccharide moiety relaxes exclusively via the overall molecular motion. This problem becomes more serious if internal motion of exocylic substituents, e.g. a hydroxymethyl group, is considered in the process of dynamic modelling. An approximate solution to this problem could be the simulation of the rotational correlation time associated with the overall motion through hydrodynamic, or light scattering measurements. Nevertheless, these measurements afford a single, effective correlation time, which cannot explicitly describe an anisotropic motion. Moreover, the continuum model involved in the hydrodynamic theory may not be realistic for such small molecules, and give erroneously long correlation times. Although corrections to the hydrodynamic theory exist, and correlation times close to the true values can be obtained for spherical, spheroidal and ellipsoidal molecules, a quantitative determination of the magnitude and rate of oscillatory motions around glycosidic bonds requires the elaboration of a suitable dynamic model. No such model is available, however, to describe flexibility in terms of torsional angles Φ and Ψ.

Most treatments assume a rigid disaccharide molecule tumbling isotropically, or anisotropically about a preferential axis of rotation. For instance, methyl β-lactoside 19 is regarded (Berry et al 1977) as an anisotopic tumbler. Carbon atoms that bear an axially-oriented proton have closely similar T_1 values, whereas C-4', which uniquely bears an equatorially oriented proton, is characterized by a smaller T_1 value. This suggests that the tumbling motion is anisotropic, with a principal axis of rotation directed parallel to the C-4'-H-4' vector. Assuming that this axis is perpendicular to the axial C-H vectors and lies in the general direction joining C-4', C-1', C-4, and C-1, the Woessner equations give $D\perp = 6.7x10^8\ s^{-1}$, and $D// = 10.4x10^8\ s^{-1}$ ($\rho = D// / D\perp = 1.55$) which suggests a small, but definite anisotropic overall motion.

Another example is sucrose (7), the dynamics of which have been studied extensively (Allerhand et al 1971; Allerhand et al 1985; Bock et al 1982; McCain et al 1986; Kovacs et al 1989) in various solvents, as a function of temperature, concentration and magnetic field. In two studies (Allerhand et al 1985; Bock et al 1982), this molecule was described as a rigid isotropic tumbler adopting a solution conformation similar to that in the solid, except, perhaps of the loss of one intramolecular hydrogen bond. Very careful ^{13}C spin-lattice relaxation measurements undertaken by McCain and Markley (McCain et al 1986) and others (Allerhand et al 1985) for sucrose in aqueous solution revealed that this molecule tumbles anisotropically. The relaxation data have been presented in terms of normalized relaxation rates, $<1/T_1>$, defined as the relaxation rate for an individual ring carbon divided by the average $1/T_1$ value for all the ring carbons as measured during in a given run. This ensures that these values are independent of the experimental conditions and systematic errors in the relaxation measurements. On the average, two sets of ring carbons gave results well outside the experimental error. Ring carbons C-2, C-3, C-4, and C-5 of the glucose moiety were characterized by shorter $<1/T_1>$ values ($<1/T_1> = 0.97\ s^{-1}$) than those of carbons C-1 of glucose, and C-3', C-4', C-5' of the fructose residue ($<1/T_1> = 1.03\ s^{-1}$). This pattern is in agreement with the different angles formed by the two sets of C-H vectors, with a preferential axis of rotation assumed to be parallel to the crystallographic b-axis of sucrose in the crystal. The former set of vectors lie nearly parallel to each other and almost perpendicular to the long axis (b-axis) of the molecule (angle $87^o \pm 2^o$), whereas the latter set of the four vectors form a cone that has its axis parallel to the long axis so that the four bonds all make angles of 27 ± 10° to the b-axis. This pattern of relaxation times is evidence that the conformation of sucrose in water is similar to that in the crystal.

The rigidity of disaccharide molecules has been questioned in two papers (McCain et al 1986; Kovacs et al 1989) dealing with the dynamics of sucrose (7) and methyl-3-O-α-L-rhamnopyranosyl-α-D-glucopyranoside in solution. In these instances, the experimental T_1 and NOE data have been used to determine the frequency dependence and amplitude of the rotational spectral density function, J(ω). The frequency dependence of J(ω) shows the same behavior as the theoretical spectral density function for a rigid molecule, but it has a lower amplitude, ca. 89% of the theoretical rigid-molecule value. This behavior is attributed to a rapid initial decay of the correlation function caused by internal motions (e.g. vibrational and torsional oscillations), which rapidly accomplish part of the averaging of an internuclear vector over all its possible orientations. This initial decay is followed by a slower exponential decay caused by overall molecular rotation, with the usual molecular rotational correlation times.

REFERENCES

Allerhand A, Doddrell D and Komorowski D (1971) Natural abundance carbon-13 partially relaxaed Fourier transform nuclear magnetic resonance spectra of complex molecules. J. Chem. Phys. 55: 189-198

Allerhand A and Doddrell D (1971) Strategies in the application of partially relaxed Fourier transform nuclear magnetic resonance spectroscopy in assignments of carbon-13 resonances of complex molecules. Stachyose. J. Am. Chem. Soc. 93: 2777-2779

Berry JM, Hall LD and Wong KF (1977) Concerning the tumbling motion of disaccharides in aqueous solution. Carbohydr. Res. 56: C16-C20

Bock K and Lemieux RU (1982) The conformational properties of sucrose in aqueous solution: Intramolecular hydrogen bonding. Carbohydr. Res. 100: 63-74

Czarniecki MF and Thornton ER (1976) ^{13}C spin-lattice relaxation in neuraminic acids. Evidence for an unsual itramolecular hydrogen bonding network. J. Am. Chem. Soc. 98: 1023-1025

Czarniecki MF and Thornton ER (1977) Carbon-13 nuclear magnetic spin-lattice relaxation in the N-acetylneuraminic acids. Probes for internal dynamics and conformational analysis. J. Am. Chem. Soc. 99: 8273-8278

Czarniecki MF and Thornton ER (1977) Carbon-13 nuclear magnetic resonance of ganglioside sugars. Spin-lattice relaxation probes for structure and microdynamics of cell surface carbohydrates. J. Am Chem. Soc. 99: 8279-8282

Dais P, Shing TK and Perlin AS (1983) Proton spin-lattice relaxation rates and nuclear Overhauser enhancement, in relation to the stereochemistry of β-D-mannopyranose 1,2-orthoacetates. Carbohydr. Res. 122: 305-313

Dais P and Perlin AS (1983) Motional behavior of 2,3: 5,6-di-O-isopropylidene-α-D-mannopyranoside in solution. A ^{13}C spin-lattice relaxation study. Can. J. Chem. 61: 1542-1548

Dais P and Perlin AS (1985) Stabilization of the β-furanose form, and kinetics of the tautomerization on D-fructose in dimethylsulfoxide. Carbohydr. Res. 136: 215-223

Dais P and Fainos G (1986) Motional behavior of "asperlin" in solution. A ^{13}C spin-lattice relaxation study. Can. J. Chem. 69: 560-565

Dais P and Perlin AS (1986) Chemical shifts of the methyl groups in di-O-isopropylidene furanoses, and their relationship to molecular conformation and site of ring fusion. Spin-lattice relaxation measurements and motional characteristics. Carbohydr. Res. 146: 177-191

Dais P and Perlin AS (1987) Intramolecular hydrogen-bonding and solvation contributions to the relative stability of the β-furanose form of D-fructose in dimethylsulfoxide. Carbohydr. Res. 169: 159-169

Dais P and Perlin AS (1989) A ^{13}C spin-lattice relaxation study of solvent effects on the rotational dynamic of methyl glucosides. Carbohydr. Res. 194: 288-195

Harvey JM and Symons MCR (1978) The hydration of monosaccharides. An NMR study. J. Solution Chem. 7: 571-586

Jaques LW, Glant S and Weltner Jr. W (1980) Spin-lattice relaxation times for two isomers of N-acetylneuraminyl lactose. Carbohydr. Res. 80: 207-211

Kovacs H, Bangley S and Kowalewski J (1989) Motional properties of two disaccharides in solution as studied by carbon-13 relaxation and NOE outside the extreme narrowing region. J. Magn. Res. 85: 530-541

London RE (1978) On the interpretation of ^{13}C spin-lattice relaxation sesulting from ring puckering in proline. J. Am. Chem. Soc. 100: 2678-2685

Lyerla Jr. JR and Levy GC (1974) Carbon-13 nuclear spin relaxation. In: Topics in carbon-13 NMR spectroscopy. Wiley & Sons New York. Vol. 1: 79-148

McCain DC and Markley JL (1986) The solution conformation of sucrose: Concentration and temperature dependence. Carbohydrate Res. 152: 73-80

McCain DC and Marklay JL (1986) Rotational spectral density functions for aqueous fructose: Experimental determination using ^{13}C NMR. J. Am. Chem. Soc. 108: 4259-4264

Neszmelyi A, Liptak A and Nanasi P (1977) ^{13}C NMR relaxation times and chemical shifts of the exo and endo isomers of dioxolane-type benzylidene acetals of carbohydrates: Determination of the absolute configuration. Carbohydr. Res. 58: C7-C9

Neszmelyi A, Tori K and Lukacs G (1977) Use of the carbon-13 spin-lattice relaxation times for sugar sequence determination in steroidal oligosaccharides. Chem. Commun. 613-614

Serianni AS and Barker R (1982) ^{13}C spin-lattice relaxation times of [1-^{13}C]- enriched carbohydrates. J. Magn. Reson. 49: 335-340

Serianni AS and Barker R (1984) [^{13}C]- enriched tetroses and tetrofuranosides: an evaluation of the relationship between NMR parameters and furanosyl ring conformation. J. Org. Chem. 49: 3292-3300

Woessner DE (1962) Nuclear spin-relaxation in ellipsoids undergoing rotational Brownian motion. J. Chem. Phys. 37: 647-654

Wu GD, Serianni AS and Barker (1983) Stereoselective exchange of methylene protons in methyl tetrofuranosides: Hydroxymethyl group conformations in methyl pentofuranosides. J. Org. Chem. 48: 1750-1757

DYNAMIC MODELLING OF POLYSACCHARIDE MOLECULES IN SOLUTION

Photis Dais
Department of Chemistry
University of Crete
71409 Iraklion, Crete
Greece

INTRODUCTION

Geometric differences within a carbohydrate chain owing to the presence of different monosaccharide residues and/or different type of glycosidic linkages result in a wide range of conformational characteristics. For instance, linear homopolysaccharides of a given hexose, differing only in the glycosidic linkage position or anomeric configuration, show remarkable different solution properties, e.g. amylose versus cellulose, due to well known differences in their equilibrium stereochemistry (Brant et al 1990). The question arising from these experimental findings is whether these equilibrium conformational differences reflect time-dependent conformational changes in solution. The nuclear magnetic relaxation technique can provide a convenient probe of the rates of macromolecular conformational motion in the nanosecond frequency range associated with conformational rearrangements resulting from internal rotations about the glycosidic bonds of the carbohydrate chain (Lyerla et al 1974). In particular, carbon-13 relaxation experiments can probe motions at several carbon sites simultaneously in the molecular frame. Each carbon nucleus is associated with a set of relaxation parameters that reflects the average interactions of this nucleus with its environment. ^{13}C relaxation parameters, such as spin-lattice relaxation time (T_1), spin-spin relaxation time (T_2), and Nuclear Overhauser Enhancement (NOE) are related to the spectral density function, $J_i(\omega_i)$, which describes the distribution of motional frequencies and the amplitude of the local magnetic fields generated by the atomic and electronic environment of the ^{13}C nucleus, and modulated by the motion of the molecule in solution. These fluctuating local magnetic fields can promote relaxation, if they have a frequency component at or near the Larmor frequency, ω_0, of the nucleus providing, thus, the link between nuclear magnetic relaxation and molecular motion.
If the macromolecule is rigid, as for stiff carbohydrate chains, such as cellulose, contribution to the spectral density function arises necessarily from the overall rotational diffusion associated with correlation times τ_c of the order of 10^{-6} - 10^{-8} s. For relatively flexible carbohydrate chains, such as linear segments of dextran, internal motion about the flexible α-(1--- 6) bond contribute most effectively to the spectral density in the nanosecond frequency range. Unfortunately, the spectral density function, which provides information

NATO ASI Series, Vol. H 87
NMR of Biological Macromolecules
Edited by C. I. Stassinopoulou

about the frequency distribution of such local changes in conformation, is known at a very limited number of frequencies (in a given magnetic field, and for ^{13}C-^{1}H dipolar interaction, the spectral density function is known at frequencies ω_C, ω_H-ω_C, and $\omega_H+\omega_C$, where ω_C and ω_H are the ^{13}C and ^{1}H Larmor frequencies respectively), and it is not practicable to obtain $J_i(\omega_i)$ over a sufficiently wide range of frequencies. Therefore, it is more convenient in NMR relaxation to develop a theoretical time-correlation function (TCF), G(t), and obtain the spectral density function by Fourier transformation.

$$J(\omega) = \int_{-\infty}^{+\infty} G(t)\, e^{i\omega t}\, dt \qquad (1)$$

The TCF describes the time-dependence of local conformational changes in solution in the nanosecond frequency range, known as segmental motion, and reflects the conformational constraints in a carbohydrate chain. Obtaining G(t) is the crucial point for a quantitative interpretation of the relaxation data. The spectral density function and TCF are Fourier transform pairs, interrelating motional frequencies (spectral density, frequency domain) and motional rates (TCF, time domain).

For the simple case of isotropic rotational diffusion, the TCF is a single exponential function of time decaying with a time constant, τ_c, the molecular correlation time

$$G(t) = \sim \exp(-t/\tau_c) \qquad (2)$$

Substituting Eq. 2 into Eq. 1 and integrating yields the familiar expression for the spectral density

$$J(\omega) = \frac{2\tau_c}{1+\omega^2\tau_c^{\,2}} \qquad (3)$$

A polysaccharide chain can experience a variety of motions, such as local conformational changes, rotational diffusion of the whole chain, or tumbling of large segments of the chain at much lower frequencies that the backbone local rearrangements, and motions of pendent groups. In view of the wide variety of motions with different rates and mechanisms, it is not surprising that Eq. 3 is an unrealistic model for describing the actual spectral density function for a polysaccharide chain.

Interest in the motional behavior of polysaccharide molecules in solution has grown in recent years. Measurements of ^{13}C relaxation parameters have been carried out in several instances (Perlin et al 1982; Seymour et al 1981; Gorin et al 1981) for linear polysaccharides to probe variations in segmental motion of terminal groups, and for branched polysaccharides, to investigate differences in mobility between side-chain residues and those incorporated in the backbone. Experiments such as these have facilitated the

assignment of ^{13}C-resonance signals that were difficult to identify on the basis of chemical shifts, and have provided a qualitative description of mobility as reflected in the ^{13}C relaxation parameters. Although qualitative analysis of ^{13}C relaxation data appears to be useful in describing gross features of the mobility of carbohydrate chains, it is less valuable than quantitative information inherent in the measured ^{13}C relaxation parameters.
This chapter is restricted to studies of polysaccharides in dilute (or reasonably so) diamagnetic isotropic solutions where dipole-dipole coupling is averaged to zero. Furthermore, the chapter concentrates on the principles underlying the experimental and theoretical approaches to relaxation of polysaccharides, rather than providing a comprehensive survey of applications. For a survey of earlier work, the reader is referred to the literature (Perlin et al 1982; Seymour et al 1980; Gorin et al 1981).

QUALITATIVE MOTIONAL DESCRIPTION

As mentioned earlier, the time-dependence of local conformations in a carbohydrate chain can be described by a TCF that reflects the geometric constraints of the carbohydrate chain. These constraints, arising primarily from nonbonded interactions between sugar residues, restrict motion about the chemical bond of the glycosidic linkages, which modulate the frequency and the strength of the local magnetic fields. Constraints about the glycosidic bonds, usually expressed in terms of torsional angles φ and ψ (and ω for (1→6)-glycosidic bonds) may introduce severe restrictions on the domain of the conformational space accessible to the backbone units. Apart from solvent effects and intramolecular hydrogen bonding, the space available to the allowed conformations depends on the type of linkage and the nature of the monosaccharide residue in the chain (Brant et al 1990). Thus, amylose an α-(1→4)-linked glucan is expected to have a relatively smaller domain of (φ,ψ) conformational space than linear dextran, an α-(1→6)-linked glucan. In the latter polysaccharide the occurrence of three chemical bonds between the pyranose rings of adjacent sugar residues allows for a considerably wider range of orientations of these residues relative to each other (Brant et al 1990), and thus, an increase in chain mobility. Support to this expectation has been offered by comparing the variable temperature ^{13}C relaxation behavior of linear dextran and amylose solutions in dimethylsulfoxide (ca. 15% w/v) (Matsuo 1984). Dextran displayed distinctly larger T_1 values than amylose above ca. 40° C. However, both biopolymers in $(CD_3)_2SO$, and perhaps in D_2O showed similar T_1 values in the temperature range 28°-40° C. This behavior suggests (Kadkhodael et al 1991) that in dilute solutions at room temperature viscous damping effects predominate over contributions to T_1-sensitive conformational differences in the region of the glycosidic linkage. At elevated temperatures the relaxation behavior of amylose and dextran is consistent with a transition to dominance of the conformational dynamics by the geometrical characteristics of the glycosidic bonds.

An interesting procedure to gain a deeper understanding of the effect of segmental mobility in a polysaccharide chain on its relaxation behavior is to examine a homologous series of oligomeric compounds that asymptotically approach the polymer structure. In this respect, T_1, T_2, and NOE measurements have been reported for glucose oligomers corresponding in structure to pullulan (Benesi et al 1985), amylose and dextran (Kadkhodael et al 1991), and also to cellulose triacetate (CTA) (Buchanan et al 1990) in an attempt to mimic segmental motion occurring in the high molecular weight analogues. Comparison of the interior carbon resonances of the oligomers with those of the corresponding polymers gives an indication of the critical length of oligomers required to match segmental motion occurring in the polymer chain. Along these lines a thorough study has been performed (Kadkhodael et al 1991) on maltose , α-(1→4)-linked, and isomaltose, α-(1→6)-linked oligomers. The average T_1 values over the five endocyclic ring carbons of the reducing, $<T_1>_{red}$, nonreducing, $<T_1>_{non}$, and interior, $<T_1>_{int}$, residues of maltose (M_i) and isomaltose (IM_i) oligomers in D_2O were plotted as a function of the number of the monomer units (degree of polymerization, DP) at two different concentrations. Several conclusions can be derived from these relaxation measurements. (1) The average T_1 values follow the order $<T_1>_{red} > <T_1>_{non} > <T_1>_{int}$ for both maltose and isomaltose oligomers. Although it is not clear why the reducing terminal residues have longer $<T_1>$ values than the nonreducing terminal residues, the average T_1 value for the interior residues is smaller that for either terminus. This indicates an increase in internal flexibility on going from the central monomers units towards the free ends of the carbohydrate chain, as expected from earlier studies with carbohydrate oligomers (Seymour et al 1980). (2) The $<T_1>$ values measured at 20 g/ dL for all three classes of residues appear to attain asymptotic behavior with respect to increasing chain length. For $<T_1>_{red}$ and $<T_1>_{non}$ the estimated convergence is completed at about DP = 12, whereas for the $<T_1>_{int}$ the convergence is attained at DP = 8. At this DP level, local motions in the oligomeric M_i and IM_i is expected to match those occurring in the high molecular weight analogues amylose and linear dextran respectively. The polymer matched DP of $<T_1>_{int}$ for CTA (Buchanan et al 1990) is 7, whereas that for the pullulan (Benesi et al 1985) oligomers is 15. Although fewer measurements have been carried out at the lower concentration of 3 g/ dL (Kodkhodael et al 1990), it appears that the leveling off of the $<T_1>$s at this concentration is less rapid than at 20 g/ dL. (3) $<T_1>$ values in both series increase with decreasing concentration, although the sensitivity of these values to concentration is greater for IM_i than for M_i series. As a consequence the $<T_1>$s of the maltose and isomaltose series cannot be distinguished at concentration 3 g/ dL. It appears that at room temperature in dilute solutions, viscous damping effects dominate contributions from structural features in the glycosidic linkage associated with conformational changes in the ns time scale. This behavior is similar to that mentioned earlier for amylose and linear dextran under different experimental conditions. (4) The

<T_1>s for M_i measured at 20 g/ dL are distinctly larger than those of IM_i. This observation appears to be in opposite direction from that expected from the less conformationally constrained isomaltose oligomers. A plausible explanation lies on the fact that M_i series, in the DP range $i < 10$, assumes a more compact conformation in solution than the more random IM_i oligomers with a greater chain extension. Correspondingly, in this oligomeric range of DP, it is likely that the segmental mobility of IM_i oligomers to experience more strongly interchain interactions and viscosity effects that the more compact M_i compound of the same DP. (5) This explanation would account as well for the last observations that the relaxation times of the isomaltose oligomers are more sensitive to concentration than those of the maltose oligomers.

The relative conformational freedom within a linear homopolysaccharide of hexose residues linked by two different types of linkages has been examined (Benessi et al 1985) with oligomeric and polymeric forms of pullulan, a basic fragment of which (1) contains two α-(1→4)- and one α-(1→6)-linked D-glucopyranosyl residues. It is to be expected that the segmental motion of the backbone is not uniform, with regions of increased segmental motion centered around the a-(1→4)-linked residues. On this basis, the C-1 and C-4 carbons of the latter being directly attached to the mobile α-(1→6)-linked residue, should have the larger T_1 values, as it is really observed (Benesi et al 1985).

- [- α- Glcp- (1→6)- α- Glcp - (1→4)- α -Glcp - (1→4) -]$_x$ - 1

-{ - α -Glpc^l - (1→6) -]$_n$ - α - Glpc^b - (1→6) - }$_x$ - 2

m
↑
1
α - Glpc^t

- { - α - Glpc^l - (1→4) -]$_n$ - α - Glpc^b - (1→4) - }$_x$ - 3

6
↑
1
α - Glpc^t

Other applications of ^{13}C NMR relaxation measurements have been concerned with determining the sequence of the monosaccharide residues in the complex biopolymer structures (Seymour et al 1980; Dill et al 1979; Gorin et al 1979), and distinguishing between the sugar units in the side-chain rather than the main chain, or identifying a brach-point position (Seymour et al 1980; Dill et al 1979; Vignon et al 1983). This approach is based primarily on differences in segmental mobility observed along the carbohydrate chain of linear polysaccharides due to the nature of the monosaccharide units, the type of linkage, and, for branched polysaccharides, on differences in segmental motion between the main and side-chains. A systematic study (Seymour et al 1980) has been performed on α-glucans, including dextrans (2) and a synthetically prepared branched amylose (3). In these structures, t is a terminal group, b a branch-point residue, l a linear chain extending residue (the chain length is considered constant), and m is the carbon atom number at the position of branching. ^{13}C chemical shift assignment were facilitated by comparing T_1 values of the various carbon resonances of the monosaccharide units in the main and side-chains, and those of the monosaccharide residues at the branch-point. The relative magnitudes of T_1 values associated with the various carbon positions in the branched polysaccharides were reasonable indices of the relative degree of segmental freedom. These values increase in the order T_1 (b) < T_1 (l) < T_1 (t) reflecting increasing segmental mobility.

QUANTITATIVE MOTIONAL DESCRIPTION

In modelling the dynamics of a long carbohydrate chain three general types of motion are considered: (a) the overall rotatory diffusion, (b) segmental motion, and (c) the internal motion of pendant groups (e.g. hydroxymethyl group) exocyclic to the polymer backbone. Each of these motions is considered as an independent source of the motional modulation of dipole-dipole interactions, so that the composite TCF is a product of the TCFs associated with each motion.

The overall molecular tumbling is the molecular weight dependent motion, whereas segmental motion does not depend on the chain length. In addition, these two motions compete with one another in the relaxation of the ring carbons. For relatively low molecular weight polysaccharides, the overall motion is the dominant relaxation source. With increasing molecular weight this motion becomes progressively slower until the segmental contribution prevails and the relaxation parameters become molecular weight-independent. Therefore, for sufficiently high molecular weight polymers, the overall rotatory diffusion is a negligible contributor to the backbone carbons, and, as mentioned earlier, it can be safely ignored. For example, the overall motion of an amylose sample in dimethylsulfoxide ($\overline{M}_w$ = 3.3×10^5) corresponds to a long correlation time $\tau_R = 7.0\times10^{-6}$ s, as estimated from hydrodynamic measurements (Dais 1987) through the following equation

$$\tau_R = \frac{2\overline{M}_w[\eta]\eta_o}{3RT} \qquad (4)$$

here $\overline{M}_w$ is the molecular weight of the polymer, $[\eta]$ is the intrinsic viscosity of the polymer solution, and η_o is the solvent viscosity. For anisotropic motion of a stiff polymer chain, which can be approximated as a rod or a cylinder-like molecule the correlation times for rotational diffusion about the major ($\tau_{//}$) and the short ($\tau_{\perp}$) axis of the cylinder are given by the following relations

$$\tau_{//} = \frac{\pi\eta_o L^3}{18kT} \{\ln(L/b)-1.57+7[1/\ln(L/b)-0.28]^2\}^{-1} \quad (5)$$

$$\tau_{\perp} = \frac{8\pi\eta_o b^2 L}{6kT} \quad (6)$$

where L and b are the axial length and the transverse radius of the cylinder respectively.
The second independent motion requires a suitable dynamic model that reflects the geometric constraints of chain flexibility characteristic of polysaccharides. However, no model is yet available that takes polysaccharide structural details into account, and describes possible modes of reorientation in the carbohydrate chain.
Several attempts have been made to interpret relaxation data of polysaccharides by employing a variety of dynamic models suitable for local chain motions of synthetic polymers.
One motional model that has been applied to polysaccharides (Schaefer 1973) is the log (χ^2), which assumes that isotropic reorientation is characterized by a skewed distribution of correlation times, with tails toward longer correlation times. The distribution function of this model contains two parameters: The width, p, of the distribution, and the average correlation time, $\bar{\tau}$, characterizing the centre of the distribution. The larger the width parameter, p, the narrower the distribution. For very large values of p, the distribution is so narrow as to be indistinguishable experimentally from a single correlation time.
This model has been used to understand the conformational behavior of the gel-forming (1→3)-β-glucans, the aggregation among the helical segments, and its consequences for the gelation mechanism (Saito et al 1978). In agreement with expectation, $\bar{\tau}$ showed the existence of distinct differences in mobility between the glucans with finite network (DP_n=49 and 131) and the gels with an infinite network (DP_n=540). The $\bar{\tau}$ value for DP_n=540 is much larger than that of the polymer with DP_n=49 and 131. Nevertheless, an increase in the width parameter was observed with increasing molecular weight, which appears to be unreasonable for the glucans with a finite network. Moreover, the fitting parameters of the model were unable to reproduce the experimental line-widths. The same fitting problem was encountered as well in a study (Torchia et al 1977) concerned with the segmental mobility of glucosaminoglucans chains in solution and in intact tissues. In both studies a truncated log (χ^2) distribution (by excluding very long correlation times > 500-1000 ns) was invoked to reproduce the experimental data and improve the problem with the

width parameter. Although the value of the average correlation times are not exceedingly altered by this treatment, the problem of the width parameter has been improved (Saito et al 1978; Torchia et al 1977). However, truncation of the log (χ^2) distribution may be an artifact of the fitting procedure, since it would exclude slow motions that are present.

The applicability of a second model developed for synthetic polymers by Jones and Stockmayer (JS) (Jones et al 1977) to the relaxation of polysaccharides was tested for linear dextrans and amylose in $(CD_3)_2SO$ and D_2O (Matsuo 1984). However, this model appears to be limited, as it has been shown (Dais 1985) by NMR multifield relaxation experiments on amylose in $(CD_3)_2SO$ at 80° C. The ineffectiveness of the JS model for describing the dynamics of carbohydrate molecules has been rationalized (Dais 1987; Dais 1985) on the basis of the specific nature of a carbohydrate chain, as compared with a hydrocarbon chain for which the model applies. Instead it has been suggested (Dais 1987; Dais 1985), that local conformational changes in carbohydrate chains can be expressed in terms of oscillatory, or other types of motions within a disaccharide unit. These motions, the amplitude of which depends on geometric constraints within the simple kinetic unit, introduce an angular dependence of the ^{13}C-^{1}H dipole-dipole interactions, a fact that may explain the observed differences between the relaxation parameters of the C-1 carbon and those of the remaining ring carbons. An examination of the nature of the internal and overall modes of reorientation in amylose has been made (Dais 1987) by employing a variety of dynamic models.

On the basis of the multifield T_1, T_2, and NOE data, two models were found to be consistent with the nature of the internal motion: The internal conic diffusion model (Lipari et al 1980; Lipari et al 1981) and the internal 2-state jump model (London et al 1981). Both internal motions are superimposed on an anisotropic overall motion of helical segments within the amylosic chain. In the first model (Lipari et al 1980; Lipari et al 1981), depicted schematically in Fig. 1a, internal motion is described as a wobbling in-a-cone, so that the C-H vector moves freely at a given rate, τ_w, inside the conical boundary defined by an angle Θ, but has zero probability of being found outside the boundary. β is the angle that the director of the cone makes with the z axis, τ_z and $\tau_x = \tau_y$ are the correlation times for rotation about the z and x (or y) axes of the cylinder (helical segment).

In the second model (London et al 1981) (Fig. 1b) the relaxation vector jumps internally between stable states A and B, with lifetimes τ_A and τ_B, respectively. The jump range is defined by angle 2Θ (i.e. the jump between $-\Theta$ and $+\Theta$ with respect to the bisector). The correlation times τ_z, τ_x, and τ_y have been defined previously. β_1 describes the angle between the z axis and the internal jump axis, β_2 describes the angle between the jump axis and the C-H vector. The orientation of the bisector relative to the z axis is described by the angle α.

Experimental relaxation data of amylose were nicely reproduced by using those two models (Dais 1987). However, the wobbling in-a-cone model is favored for two reasons: First, it requires fewer adjustable parameters than the bistable-jump model to fit the data, and second, the jump model requires β_2 = 48.9° for the C-1 carbon, which is very different from that (37°) observed in the crystal structure of amylose-dimethylsulfoxide complex (Winter et al 1974).

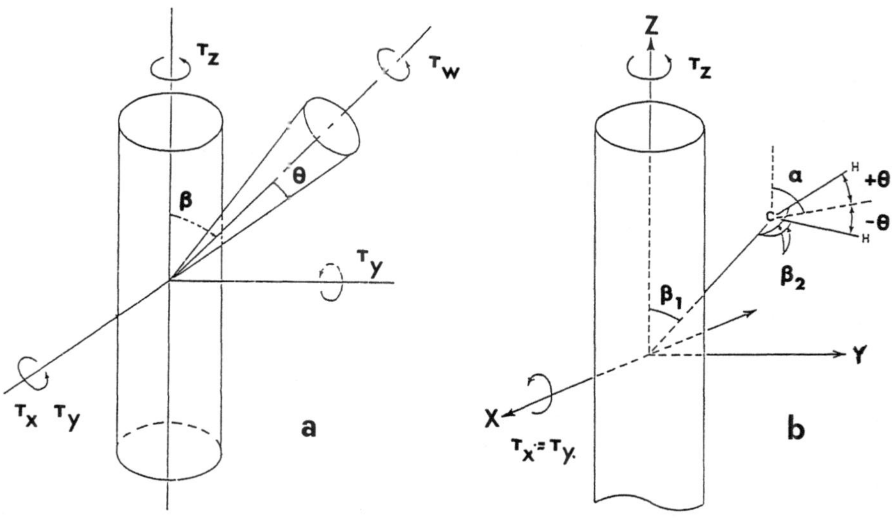

Fig. 1. (a) Diffusion in a cone model; (b) Bistable (2-state) jump model.

Another interesting finding of this study is revealed by comparison of the calculated fitting parameters τ_z and $\tau_x = \tau_y$ from the two models with the correlation times τ and τ obtained from hydrodynamic measurements (Eqs. 5, and 6), as a function of the number of monomer units per helical segment. These values are commensurate with a helical segment comprising, on the average, 40-50 monomer residues (wobbling in-a-cone model) and 20-60 monomer units (bistable jump model (Dais 1987).

Although the above treatment is successful in reproducing the relaxation data of amylose, it cannot be generalized to include other linear or branched polysaccharides having less pronounced helical segments in solution. Moreover, the complexity of the models used does not facilitate the description of the additional internal motion of the exocyclic hydroxymethyl group.

In a recent publication (Dais et al 1991) the backbone motion of amylose has been described by using a model developed by Dejean, Laupretre, and Monnerie (DLM) (Dejean et al 1988). The DLM model is a modification of an older model of Hall, Weber, and Helfand (HWH) (Hall et al 1982; Weber et al 1983), which takes into consideration an additional fast, anisotropic motion superimposed on the backbone rearrangement described by the HWH time-correlation function. This local motion is attributed to molecular librations of limited extent of the C-H vector inside a cone of half angle Θ, the axis of which is the rest position of the C-H bond.

The fitting parameters of this model are the angle Θ, the correlation time τ_2 for the librational motion, and the correlation times τ_0 and τ_1 describing the backbone motion in terms of conformational transitions in the HWH model. The DLM model was able to reproduce the multifield relaxation data for the ring carbons of amylose in dimethylsulfoxide (Dais et al 1991).

The different relaxation data for the anomeric carbon relative to those for the remaining ring carbons was explained on the basis of the simulated values of the half-angle Θ of the cone. This angle is 21° for C-1 and 26° on the average for the other endocyclic carbons, indicating that these two types of carbon sites do not experience the same local dynamics. The smaller Θ value for the anomeric carbon is indicative of greater steric hindrance to the librational motion of the corresponding equatorial C-H vector relative to that of the axial vectors in the remaining ring sites. Shortening of the C-1-O-1 and C-1-O-5 bonds relative to the remaining C-O bonds, which is observed in the crystal structure of amylose (Winter et al 1974) and other α-glucopyranosides (Sundaralingam 1968; Marchessault et al 1979), appears to be a major cause for restricting the amplitude of local libration at the anomeric carbon site.

Rotation of the hydroxymethyl group about the exocyclic bond is considered next. Free internal rotation about the C-5-C-6 bond superimposed on segmental motion as described by a new composite TCF (Dais et al 1991) based on the HWH model and the Woessner equations (Woessner et al 1969) for stochastic diffusion and jump processes were not able to reproduce the experimental data for the C-6 carbon.

Models incorporating restricted rotations were then considered. One is the internal two-state jump model (London et al 1981; London 1978) similar to that described before, and the other is the restricted diffusion model (London et al 1978), in which the O-6 bond moves continually between two limiting values of an angle χ (i.e. the amplitude of the restricted motion is 2χ). The TCF of internal motion, as described by these two models, can be combined with the HWH TCF to give two new composite TCFs[22].

A. HWH + Two-state jump TCF:

$$G(t) = (1-C)\exp(-t/\tau_0)\exp(-t/\tau_1)\, I_0(t/\tau_1) + C\exp(-t/\tau_0)\exp(-t/\tau_c)\exp(-t/\tau_1)\, I_0(t/\tau_1) \qquad (7)$$

$$\tau_c^{-1} = \tau_A^{-1} + \tau_B^{-1}$$

B. HWH + Restricted internal diffusion TCF:

$$G(t) = \sum_{a=-2}^{+2} \sum_{n=0} d_{ao}(\beta)^{2} E(a,n)^{2} \exp(-t/\tau_o) \cdot \exp(-t/\tau_n) \exp(-t/\tau_1) I_o(t/\tau_1) \qquad (8)$$

$$\tau_n^{-1} = \frac{n^2\pi^2}{24\tau_i\chi^2}$$

One of the imput parameter is the angle β, formed between the relaxation vector (C-H bond) and the internal axis of rotation (or jump axis), i.e. the C-5-C-6 bond. The others are correlation times τ_o and τ_1 of the HWH model, obtained from the fit of the data for the backbone carbons. The fitting parameters for the two state jump model are lifetimes τ_A and τ_B, and for the restricted diffusion model the correlation time τ_i for internal rotation. 2χ is the allowed range of motion (or the jump range) for both models. The parameter C is given (London 1978; Dais et al 1991) as a function of the angle β and the lifetimes τ_A, and τ_B.

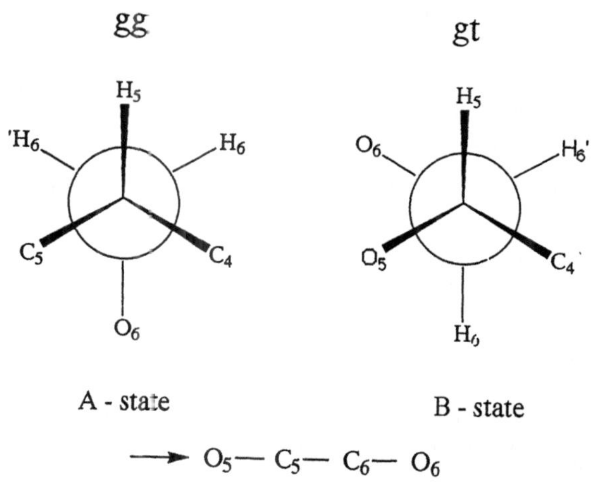

Fig. 2. Newman type projections showing the two states A and B corresponding to the gauche-gauche and gauche-trans conformations of the hydroxymethyl group of amylose.

Both models were able to reproduce the experimental data for the C-6 carbon of amylose (Dais et al 1991), although they differ somewhat in the time scale and range of restricted motion. Nevertheless, the restricted diffusion model is inadequate, since it does not respond properly at the limit of free rotation ($2\chi = 360^{o}$), which is an important criterion for its

applicability (London et al 1978). Indeed, a large discrepancy was found (Dais et al 1991) between the calculated relaxation parameters corresponding to a full range of motion, i.e. $2\chi = 360^{o}$, and parameters derived by using the free internal rotation model. Furthermore. the internal two-state jump model appears to be more realistic in describing the nature of the motion of the hydroxymethyl group in its preferred conformation in accord with crystallographic data (Sundaralingam 1968; Marchessault et al 1979).
The longer lifetime τ_A ($\tau_A = 3\tau_B$), derived by the two-state model, may be ascribed to the most stable gauche-gauche conformation (60%) as compared to the gauche-trans (40%) state for the gluco configuration. The Newman type projections of these two preferred conformations are depicted in Fig. 2.

REFERENCES

Benesi AJ and Brant DA (1985) Trends in molecular motion for a series of glucose oligomers and the corresponding polymer pullulan as measured by ^{13}C NMR relaxation. Macromolecules 18: 449-452.

Brant DA and Christ MD (1990). Realistic conformational modelling of carbohydrates. Applications and limitations in the context of carbohydrate-high polymer. In: French AD and Brady JW (eds). Computer modelling of carbohydrate molecules. ACS Symposium series No 430: Washington DC: 42-68

Buchanan CM, Hyatt JA, Kelley SS and Little JL (1990) a-D-cellooligosaccharide acetates: Physical and spectroscopic characterization and evaluation as models for cellulose triacetate. Macromolecules 18: 3747-3755

Dais P (1985) ^{13}C NMR relaxation study on amylose in dimethyl sulfoxide. Macromolecules 18: 1351-1354

Dais P (1987) Carbon-13 magnetic relaxation and local chain motion of amylose in dimethylsulfoxide. Carbohydr. Res. 160: 73-93

Dais P and Marchessault RH (1991) ^{13}C nuclear magnetic relaxation of amylose and dynamic behavior of the hydroxymethyl group. Macromolecules 24: 4611-4614

Dejean dela Battie R, Laupretre F and Monnerie L (1988) Carbon-13 NMR investigation of local dynamics in bulk polymers at temperatures well above the glass transition temperature. 1. Poly (vinyl methyl ether). Macromolecules 21: 2045-2052

Dill K and Allerhand A (1979) Studies of the carbohydrate residues of glycoproteins by natural abundance carbon-13 nuclear magnetic resonance spectroscopy. J. Biol. Chem. 254: 4524-4531

Gorin PAJ (1981) Carbon-13 Nuclear magnetic resonance of polysaccharides. Adv. Carbohydr. Chem. Biochem., 38: 13-104

Gorin PAJ and Mazurek M (1979) Structure-dependent, molecular-motion parameters of branched-chain polysaccharides. Carbohydr. Res. 72: C1-C5

Hall CK and Helfand E. (1982) Conformational state relaxation in polymers. Time-correlation functions. J. Chem. Phys. 77 3275-3282

Jones AA and Stockmayer WH (1977) Models for spin relaxation in dilute solutions of randomly coiled polymers. J. Polym. Sci. Polym. Phys. Ed. 15: 847-861

Kadkhodael M, Wu H and Brant DA (1991) Comparison of the conformational dynamics of the (1--- 4)- and (1--- 6)-linked α-D-glucans using ^{13}C NMR relaxation. Biopolymers 31: 1581-1592

Lipari G and Szabo A (1980) Effect of librational motion on fluorescence depolarization and nuclear magnetic resonance relaxation in macromolecules and membranes. Biophys. J. 30: 489-506

Lipari G and Szabo A (1981) Nuclear magnetic resonance relaxation in nucleic acid fragments: Models for internal motion. Biochemistry 20: 6250-6256

London RE and Phillipi MA (1981) A ^{13}C NMR study of solution dynamics of 1,3,5-triphenylbenzene; Analysis of motion about the phenyl-phenyl bond. J. Magn. Reson. 45: 476-489

London RE (1978) On the interpretation of ^{13}C spin-lattice relaxation resulting from ring puckering in proline. J. Am. Chem. Soc. 100: 2678-2685

London RE and Avitabile J (1978) Calculated ^{13}C NMR relaxation parameters for a restricted internal diffusion model. Application to methionine relaxation in dihydrofolate reductase. J. Am. Chem. Soc., 100: 7159-7165

Lyerla Jr JR and Levy GC (1974) Carbon-13 nuclear spin relaxation. In: Topics in carbon-13 NMR spectroscopy. Wiley & Sons New York. Vol. 1: 79-148

Matsuo K (1984) ^{13}C NMR relaxation of polysaccharides. dextran and amylose in dimethyl sylfoxide and water. Macromolecules 17: 449-452

Marchessault RH and Perez S (1979) Conformations of the hydroxymethyl group in crystalline aldohexopyranoses. Biopolymers 18: 2369-2374

Perlin AS and Casu B (1982) Spectroscopic methods. In: Aspinall GO (ed) The polysaccharides. Academic Press New York. Vol. 1. 133-193.

Saito H, Miyata E and Sasaki T (1978) A ^{13}C NMR study of gel-forming (1--- 3)-β-D-glucans: Molecular weight dependence of helical conformation and the presence of junction jones for association of primary molecules. Macromolecules 11: 1244-1251

Schaefer J (1973) Distribution of correlation times and the carbon-13 nuclear magnetic resonance spectra of polymers. Macromolecules 6: 882-888

Seymour FR and Knapp RD (1980) Structural analysis of α-D-glucans by ^{13}C-nuclear magnetic resonance, spin-lattice relaxation studies. Carbohydr. Res. 81: 67-103

Sundaralingam M (1968) Some aspects of stereochemistry and hydrogen bonding of carbohydrates related to polysaccharide conformation. Biopolymers 6: 189-213

Torchia DA, Hasson MA and Hascall VC (1977) Investigation of molecular motion of proteoglucans in cartilage by ^{13}C magnetic resonance. J. Biol. Chem. 252: 3617-3625

Vignon M, Michon F. and Joceleau JP (1983) Molecular motion of branched-chain polysaccharides studied by ^{13}C NMR spin-lattice relaxation rates. Macromolecules 16: 835-838

Weber TA and Helfand E (1983) Time correlation functions from computer simulation of polymers. J. Chem. Phys. 87: 2881-2889

Winter WT and SArko T (1974) Crystal structure and molecular structure of the amylose-DMSO complex. Biopolymers 13: 1461-1482

Woessner DE, Snowden BS and Meyer GH (1969) Nuclear spin-lattice relaxation in axially symmetric elipsoids with internal motion. J. Chem. Phys. 50: 719-721

^{13}C-^{1}H SPIN-COUPLING CONSTANTS IN FURANOSYL RINGS: APPLICATION TO OLIGONUCLEOTIDE STRUCTURE DETERMINATION

Anthony S. Serianni
Department of Chemistry and Biochemistry, University of Notre Dame, Notre Dame, Indiana 46556

INTRODUCTION

Conformational analysis of the five-membered (furanose) rings of oligonucleotides commonly relies on the use of three-bond (vicinal) ^{1}H-^{1}H spin-coupling constants ($^{3}J_{HH}$) obtained from an analysis of 2D NMR cross-peaks (Schmitz et al 1990; Emsley et al 1993; Majumdar and Hosur 1992). For a $H_aC_xC_yH_b$ fragment, the magnitude of $^{3}J_{Ha,Hb}$ depends on the C_x-C_y torsion angle, with maximal coupling observed when H_a and H_b are eclipsed or antiperiplanar, and minimal coupling observed when these atoms are approximately orthogonal (Karplus 1959); in the ensuing years since this discovery, more refined Karplus relationships have been proposed that account for substitution and/or electronegativity effects at C_x and C_y on $^{3}J_{Ha,Hb}$ (Haasnoot et al 1980).

The 2-deoxy-ß-D-*erythro*-pentofuranosyl (2-deoxyribofuranosyl) ring **1** (Scheme 1) found in DNA oligomers contains five $^{3}J_{HH}$ ($^{3}J_{H1',H2'R}$, $^{3}J_{H1',H2'S}$, $^{3}J_{H2'R,H3'}$, $^{3}J_{H2'S,H3'}$ and $^{3}J_{H3',H4'}$), whereas only three $^{3}J_{HH}$

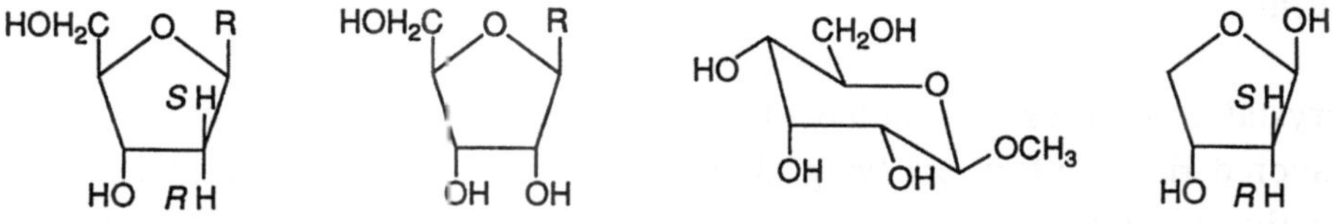

1 **2** **3** **4**

Scheme 1

are found within the ß-D-ribofuranosyl ring **2** (Scheme 1) of RNA oligomers ($^{3}J_{H1',H2'}$, $^{3}J_{H2',H3'}$ and $^{3}J_{H3',H4'}$). The structural interpretation

NATO ASI Series, Vol. H 87
NMR of Biological Macromolecules
Edited by C. I. Stassinopoulou

of these values is complicated by the fact that the energy differences between the non-planar forms of these rings, and furanose rings in general, are small (Levitt and Warshel 1978; Westhof et al 1983), leading to the possibility of considerable conformational heterogeneity in solution. Thus, NMR parameters obtained on these rings, including $^3J_{HH}$, will frequently be weighted averages of the parameters of the contributing conformers, as discussed in detail previously (Jardetzky 1980). Typically $^3J_{HH}$ values in **1** and **2** are treated by a two-state model which assumes conformational interconversion between representative north (N) and south (S) conformers of the pseudorotational itinerary (Altona and Sundaralingam 1972) ($^3E \rightleftharpoons {}^2E$; Fig. 1). The validity of this model is based on observations made on

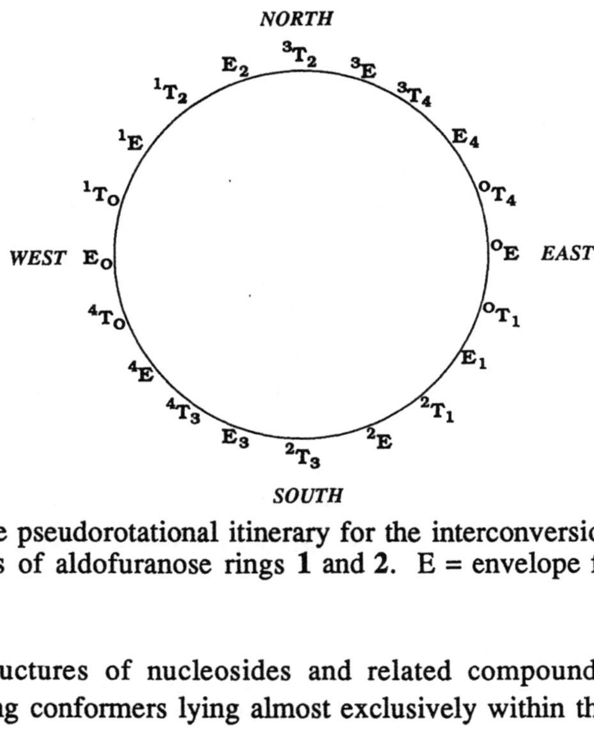

Fig 1. The pseudorotational itinerary for the interconversion of non-planar conformers of aldofuranose rings **1** and **2**. E = envelope forms; T = twist forms.

crystal structures of nucleosides and related compounds which reveal favored ring conformers lying almost exclusively within these two regions of the itinerary (de Leeuw et al 1980; Rinkel and Altona 1987), and on computational studies (Pearlman and Kim 1985). However, recent work by Reid and coworkers (Salazar et al 1993) has suggested that N/S interconversion may not always occur in oligonucleotides; in some cases furanose rings may assume one or more conformations lying outside the N/S regions. Given the flexibility of **1** and **2**, the relatively small number of $^3J_{HH}$ values available in these structures, while useful, nevertheless provides limited information about conformational preferences.

BACKGROUND

Recent advances in the ^{13}C-labeling of DNA (Lancelot et al 1993; Wu and Serianni 1993) and RNA (Nikonowicz et al 1992) promise to yield new information to assess, among other features, furanose structure in these biopolymers. The presence of one or more ^{13}C labels within **1** and **2** gives access to one- ($^{1}J_{CH}$), two- ($^{2}J_{CH}$), and three- ($^{3}J_{CH}$) bond ^{13}C-^{1}H spin couplings for structure elucidation. It has been argued previously (Cyr and Perlin 1979; Serianni and Barker 1984) that these couplings, as well as ^{13}C-^{13}C spin couplings, should provide valuable information with which to assess furanose conformation in solution. For example, in addition to the four $^{1}J_{CH}$ within the ring of **2**, eighteen $^{2}J_{CH}$ and $^{3}J_{CH}$ exist within this structure (Table 1); some ^{13}C-^{1}H couplings observed in [1'-^{13}C]uridine and [1'-^{13}C]2'-deoxyadenosine involving C-1' are shown in Fig. 2. In addition to the three intraring $^{1}J_{CC}$ values, three potentially useful longer range J_{CC} ($^{2}J_{C3,C5}$, $^{3}J_{C1,C5}$, $^{3}J_{C2,C5}$) are available in **2**.

Table 1. Two- and Three-Bond ^{13}C-^{1}H Coupling Pathways in **2**[a]

Intraring		Hydroxymethyl
C1 - H2	C3 - H2	C3 - H5*R*[b]
C1 - H3[b]	C3 - H4	C3 - H5*S*[b]
C1 - H4[b]	C4 - H1[b]	C4 - H5*R*
C2 - H1	C4 - H2[b]	C4 - H5*S*
C2 - H3	C4 - H3	C5 - H4
C2 - H4[b]	C5 - H3[b]	
C3 - H1[b]		

[a] Three additional C-H couplings are available in **1**
[b] Vicinal pathways

We confine the discussion to J_{CH} here. The application of these couplings to structure determination requires that (1) experimental methods are available to measure them reliably in simple and complex structures and (2) rules relating their magnitudes to molecular torsion angles are known. Considerable progress on (1) has been made in recent years (for example, see: Montelione et al 1989). In particular, homonuclear 2D NMR spectra of ^{13}C-labeled compounds can be used to extract precise J_{CH} values from crosspeak displacements (Fig. 3). Similar information can be obtained from 2D and 3D HMQC-TOCSY data (Hines et al 1993). Furthermore, an inspection of the relative displacement of TOCSY crosspeaks can establish the sign of $^2J_{CH}$, which can be either positive or negative in carbohydrates (Schwarcz et al 1975; Bock and Pedersen 1977) (Fig. 3); this method is based on the assumption that $^3J_{CCCH}$ is always positive in carbohydrates, as reported previously (Schwarcz and Perlin 1972).

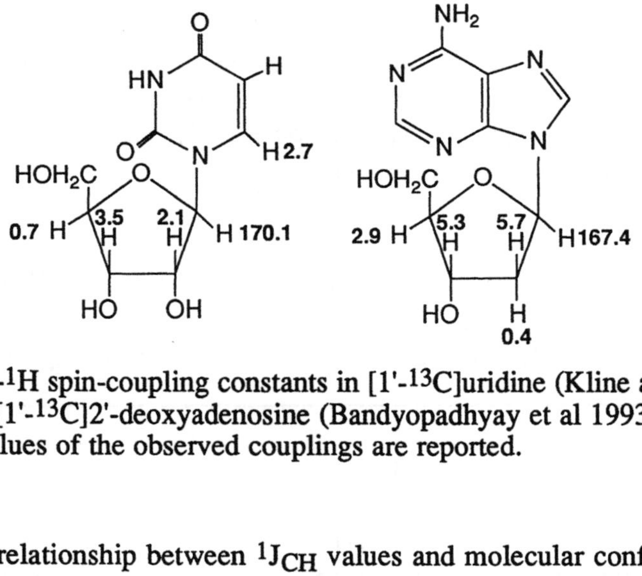

Fig 2. ^{13}C-^{1}H spin-coupling constants in [1'-^{13}C]uridine (Kline and Serianni 1990) and [1'-^{13}C]2'-deoxyadenosine (Bandyopadhyay et al 1993). Only the absolute values of the observed couplings are reported.

The relationship between $^1J_{CH}$ values and molecular conformation of **1** and **2** is at present not completely understood. Seminal work by Pederson and coworkers (Bock et al 1973; Bock and Pedersen 1975), however, provides a firm basis for further investigation. In pyranoses, $^1J_{CH}$ has been shown to depend on the orientation of the C-H bond. For example, $^1J_{C1,H1}$ in α-D-glucopyranose (169.5 Hz) is about 10 Hz larger than $^1J_{C1,H1}$ in ß-D-glucopyranose (160.0 Hz), supporting the general observation that $^1J_{CH}$ is larger for equatorial C-H bonds than for axial C-H bonds in pyranosyl rings.

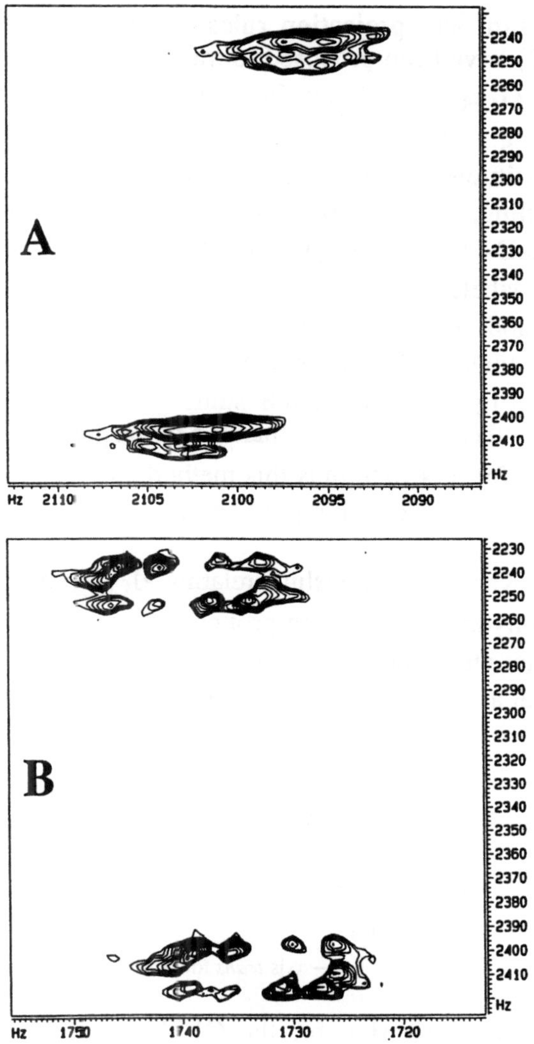

Fig 3. The partial 500 MHz TOCSY spectrum of methyl ß-D-[1-^{13}C]allopyranoside **3**. (A) The H1-H3 paired crosspeaks, showing the displacement from which $^3J_{C1,H3}$ was measured (6.0 Hz). The relative displacement of these signals (downfield signal displaced to the left) is correlated with a J_{CH} having a positive sign. (B) The H1-H2 paired crosspeaks, showing a displacement from which $^2J_{C1,H2}$ was determined (6.5 Hz). The relative displacement of these signals has the opposite sense of that observed in (A), indicating that $^2J_{C1,H2}$ has a negative sign.

Vector sum and projection rules (Schwarcz et al 1975; Bock and Pedersen 1977) have been proposed to interpret $^2J_{CH}$ in carbohydrates. The magnitude of $^2J_{CH}$ depends on the relative orientation of electronegative substituents (e.g., oxygen) on the coupling fragment. In general, electronegative substituents *anti* to the C-H bond contribute positively to the observed coupling. For example, consider the coupling between C-1 and H-2 of methyl ß-D-[1-^{13}C]allopyranoside **3**. A Newman projection viewed from C-1 (the labeled carbon) to C-2 is shown in Fig. 4. Using the projection sum method, an axis *trans* to the ^{12}C-H bond is visualized, and the cosines of the angle each oxygen substituent makes with respect to this axis are summed (Fig. 4). Projection sums greater than 0.5 correspond to positive $^2J_{CH}$, whereas sums less than 0.5 correspond to negative $^2J_{CH}$. The projection sum for **3** is -0.5; thus this method predicts the observed coupling of 6.5 Hz to be negative. This prediction is supported by experiment (Fig. 3).

Finally, Karplus relationships relating $^3J_{CCCH}$ and $^3J_{COCH}$ to dihedral angle in carbohydrates have been proposed (Schwarcz et al 1972; Hamer et al 1978), although refinement of these relationships is still under investigation (Cano et al 1987; Tvaroska et al 1989).

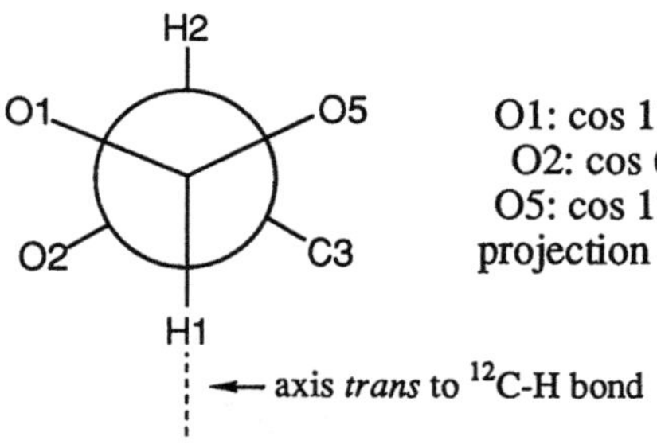

O1: cos 120° = -0.5
O2: cos 60° = 0.5
O5: cos 120° = -0.5
projection sum = -0.5

Fig 4. Newman projection for the C1-C2 bond in **3**. The dashed line identifies the axis *trans* to the ^{12}C-H bond which is used to compute the projection sum (see text).

NEW EXPERIMENTAL APPROACHES

We have been investigating the relationships between J_{CH} and furanose ring conformation in **1** and **2**. *Ab initio* molecular orbital calculations have been conducted on an intact furanose ring having a structure related to that found in these rings. Complete geometric optimization of the planar and ten envelope forms of 2-deoxy-ß-D-*glycero*-tetrofuranose **4** (Scheme 1) was conducted using the 3-21G (Garrett and Serianni 1990a), 6-31G* (Garrett and Serianni 1990b) and more recently the MP2/6-31G* basis sets (Serianni, unpublished result). In these calculations, only one endocyclic torsion angle was held constant (at 0°) in the non-planar forms, and two in the planar form. While a more thorough treatment of these data remains to be published, one important observation is worthy of discussion here. The calculations with each basis set show that C-H bonds in **4** vary systematically in length as the ring undergoes pseudorotation. More specifically, a given C-H bond within this ring is maximal in length when oriented quasi-axially, and minimal in length when quasi-equatorial (Fig. 5). This behavior is observed for all C-H bonds in **4**, although the magnitude of change varies. This result is significant in light of the above-noted relationship between $^{1}J_{CH}$ in pyranoses and C-H bond orientation. If we assume that the shorter equatorial C-H bonds have more s-character than the longer axial C-H bonds, then the former would be expected to exhibit larger $^{1}J_{CH}$, since $^{1}J_{CH}$ increases as s-character of the C-H bond increases (Muller and Pritchard 1959).

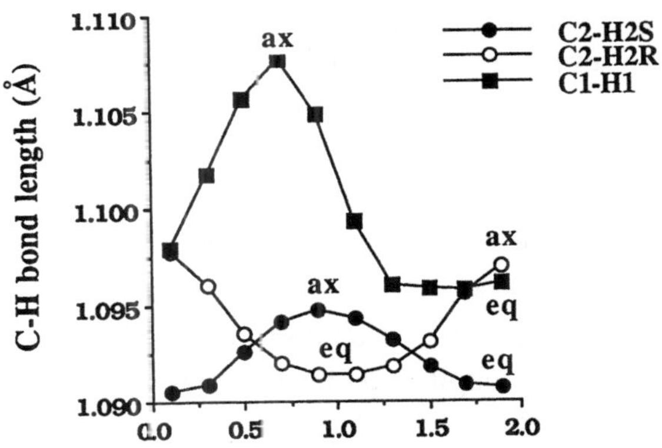

Fig 5. A plot of the effect of ring conformation on the C1-H1, C2-H2*R* and C2-H2*S* bond lengths in **4**. Data were obtained from MO calculations on **4** using the MP2/6-31G* basis set (see text). Ax = axial; Eq = equatorial. Only data for envelope forms are shown. $^{3}E = 0.1$ P/π radians.

This conclusion is consistent with observations made recently in RNA oligomers by Tinoco and coworkers (Varani and Tinoco 1991), who found that $^{1}J_{C2',H2'} > {}^{1}J_{C3',H3'}$ in ^{3}E conformers, while $^{1}J_{C3',H3'} > {}^{1}J_{C2',H2'}$ in ^{2}E forms. In the former, the C2'-H2' bond is quasi-equatorial, whereas in the latter it is quasi-axial. Using SCF methods, we have computed crude $^1J_{CH}$ values in two conformers of **4** using MP2/6-31G* optimized parameters (Fig. 6). While the absolute values of the computed couplings are smaller than those observed experimentally, we expect the difference between corresponding couplings in each form to be fairly accurate. In ^2E,

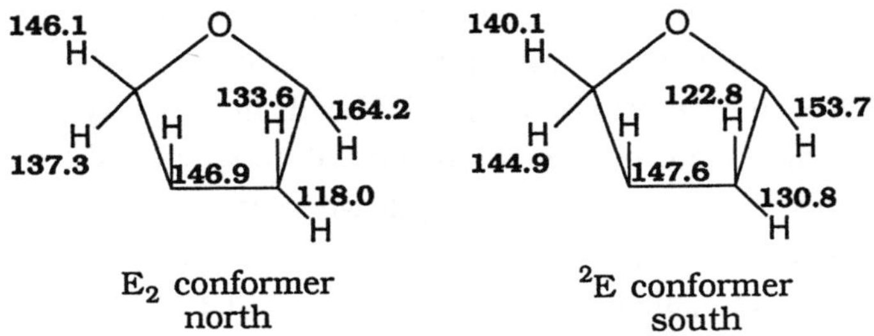

Fig 6. Crude $^1J_{CH}$ values computed at the SCF level for two conformers of **4**.

$^1J_{C2,H2R} > {}^1J_{C2,H2S}$, whereas in E_2, $^1J_{C2,H2S} > {}^1J_{C1,H2R}$. The C2-H2*R* bond is quasi-equatorial in the former conformer and quasi-axial in the latter. While these results are preliminary, they nevertheless suggest that $^1J_{CH}$ may be a useful probe of furanose conformation in oligonucleotides, especially in larger oligomers where the smaller $^2J_{CH}$ and $^3J_{CH}$ may be difficult to measure.

Geometric parameters obtained on **4** may be used to obtain a crude estimate of the effect of ring conformation on $^3J_{CH}$ in **2**. We assume that substitution of H2'*R* by an hydroxyl group does not appreciably affect torsion angles involving H2'*S*. Using crude Karplus relationships derived for COCH and CCCH coupling pathways, and appropriate torsion angles obtained from optimized envelope forms of **4**, a plot of $^3J_{CH}$ versus ring conformation was constructed (Fig. 7). This plot reveals six curves organized into three discrete groups ($^3J_{C1,H3}$ and $^3J_{C2,H4}$; $^3J_{C1,H4}$ and $^3J_{C4,H1}$; $^3J_{C3,H1}$ and $^3J_{C4,H2}$), showing maximal coupling at S, W and N ring forms, respectively, and minimum coupling at E forms.

As a complement to data in Fig. 6, and in pursuit of additional information on the effect of ring conformation on $^2J_{CH}$, a series of conformationally-fixed model compounds (methyl aldopyranosides) has been identified that contains two- and three-bond C-H coupling pathways found in E_2 (N) and ^2E (S) conformers of **2** (Table 2). For example, $^2J_{C1,H2}$ in methyl α-D-mannopyranoside should be similar in magnitude to $^2J_{C1,H2}$ in the E_2

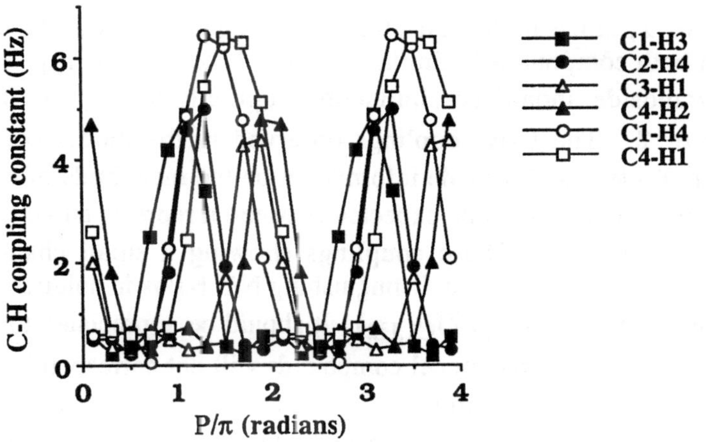

Fig 7. Predicted effect of ring conformation of **2** on the six intraring $^3J_{CH}$ values. Data are extrapolated from MO results obtained on **4** (MP2/6-31G* basis set). Two cycles of the pseudorotational itinerary are shown; ^3E = 0.1 P/π and 2.1 P/π.

Table 2. Modeling of C-H Coupling Fragments in **2** Using Methyl Aldopyranosides

coupled nuclei (in **2**)	north (E_2) conformer	south (^2E) conformer
C1-H2	α-*manno* (1.2 Hz)	ß-*allo* (-6.5 Hz)
C1-H3	α-*manno* (0 Hz)	ß-*allo* (6.0 Hz)
C1-H4 (NE,SW)[a]	ß-*gluco*	ß-*arabino*
C2-H1	α-*manno* (~1.8 Hz)	ß-*allo* (0 Hz)
C2-H3	α-*manno* (1.4 Hz)	ß-*allo* (-4.8 Hz)
C2-H4	ß-*altro (allo)* (~2.0 Hz)	α-*arabino*
C3-H1	α-*manno* (4.6 Hz)	ß-*allo* (0 Hz)
C3-H2	α-*manno* (-3.7 Hz)	ß-*allo* (1.3 Hz)
C3-H4	ß-*allo (gluco)*	α-*arabino*
C4-H1 (E,W)[a]	ß-*gluco* (~1.0 Hz)	ß-*arabino* (6.1 Hz)
C4-H2	ß-*altro (allo)*	α-*arabino*
C4-H3	ß-*allo (gluco)*	α-*arabino*
C5-H3 (NE,SW)[a]	ß-*allo (gluco)*	1,6-anhydro-*altro*

[a] The model compounds proposed are appropriate for the conformers in parenthesis.

conformer of **2**, since molecular structure along the C1-C2-H2 coupling pathway in both compounds is similar (Fig. 8). Likewise, $^2J_{C1,H2}$ in methyl ß-D-allopyranoside should be similar in magnitude to $^2J_{C1,H2}$ in the ^2E conformer of **2**. The C-H couplings observed in the model compounds identified in Table 2 will provide information on the magnitude of change in a specific coupling that is to be expected between a representative north and south conformer of **2**. Those couplings showing a small change (e.g., $^2J_{C2,H1}$) will not be as useful in distinguishing N/S ratios in solution as those showing larger changes (e.g., $^2J_{C1,H2}$). It should be appreciated that the C-H couplings observed in the model compounds will not correspond exactly to those expected in N and S forms of **2**, since the latter are not puckered maximally at 60°, ring strain (and its potential effect on J_{CH}) in furanosyl and pyranosyl rings differ, and substitution at C1 of **2** and the *O*-glycoside models are not the same. The model compounds in Table 2 are also inappropriate for a similar evaluation of J_{CH} in **1**, as deoxygenation at C2 is likely to affect the relationship between coupling magnitude and conformation (Bandyopadhyay et al 1993). Thus, a different series of (deoxy) model compounds must be chosen to provide related information.

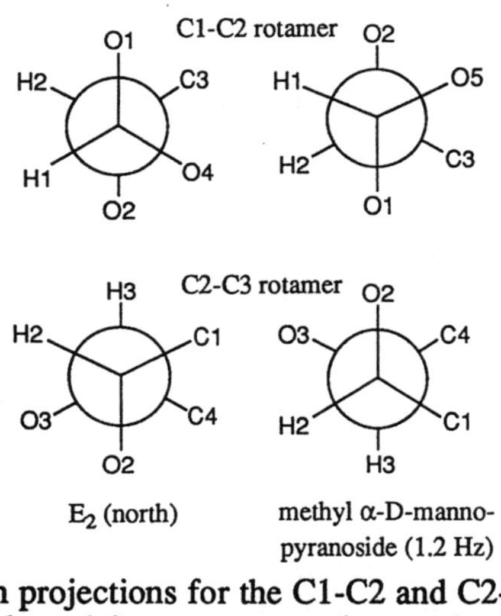

Fig 8. Newman projections for the C1-C2 and C2-C3 bonds of methyl α-D-mannopyranoside and for a representative north conformer (E_2) of **2**.

SUMMARY

In this article we have discussed the potential of ^{13}C-^{1}H spin-coupling constants as a means to provide structural information on furanosyl rings, in particular those found in nucleic acids (**1, 2**). Since these rings are conformationally flexible, it is desirable to have the largest, most diverse set of NMR parameters when assessing their solution structures. While $^{3}J_{HH}$ values have traditionally provided information on furanose conformation in DNA and RNA oligomers, use of the more abundant ^{13}C-^{1}H (and ^{13}C-^{13}C) couplings can provide valuable complementary information. These couplings will also complement other NMR parameters used to assess conformation (e.g., NOE, relaxation times).

With the development of methods to introduce ^{13}C and other stable isotopes into oligonucleotides, either site-specifically or uniformly, the measurement of J_{CH} values is likely to become standard procedure in future structural studies of these molecules. The preliminary results and experimental methods discussed in this paper represent an approach to determine how these parameters may be interpreted in terms of furanose conformation. However, additional work will be required before J_{CH} can be applied with confidence in structural studies of these biomolecules.

REFERENCES

Altona C and Sundaralingam M (1972) Conformational analysis of the sugar ring in nucleosides and nucleotides. A new description using the concept of pseudorotation. J Am Chem Soc 94: 8205-8212

Bandyopadhyay T, Wu J and Serianni AS (1993) [1'-^{13}C] 2'-deoxyribonucleosides: Structural and conformational insights derived from ^{13}C-^{1}H spin-coupling constants involving C1'. J Org Chem 58: 5513-5517

Bock K, Lundt I and Pedersen C (1973) Assignment of anomeric structure to carbohydrates through geminal ^{13}C-^{1}H coupling constants. Tet Letters: 1037-1040

Bock K and Pedersen C (1975) A study of ^{13}CH coupling constants in pentopyranoses and some of their derivatives. Acta Chem Scand Ser B B29: 258-264

Bock K and Pedersen C (1977) Two- and three-bond ^{13}C-^{1}H couplings in some carbohydrates. Acta Chem Scand Ser B B31: 354-358

Cano FH, Foces-Foces C, Jimenez-Barbero J, Alemany A, Bernabe M and Martin-Lomas M (1987) A conformational study of methyl ß-cellobioside-d_8 by ^{13}C nuclear magnetic resonance spectroscopy: Dihedral angle dependence of $^{3}J_{C\text{-}H}$ in ^{13}C-O-C-^{1}H arrays. J Org Chem 52: 3367-3372

Cyr N and Perlin AS (1979) The conformations of furanosides. A ^{13}C nuclear magnetic resonance study. Can J Chem 57: 2504-2511

de Leeuw HPM, Haasnoot CAG and Altona C (1980) Empirical correlations between conformational parameters in ß-D-furanoside fragments derived from a statistical survey of crystal structures of nucleic acid constituents. Isr J Chem 20: 108-126

Emsley L, Dwyer TJ, Spielmann HP and Wemmer DE (1993) Determination of DNA conformational features from selective two-dimensional NMR experiments. J Am Chem Soc 115: 7765-7771

Garrett EC and Serianni AS (1990a) Ab Initio molecular orbital calculations on carbohydrates: Conformational properties of deoxygenated furanose sugars, in Computer Modeling of Carbohydrate Molecules. French AD and Brady JW eds, ACS Symposium Series 430, American Chemical Society, Washington, DC pp 91-119

Garrett EC and Serianni AS (1990b) *Ab Initio* molecular orbital calculations on furanose sugars: A study with the 6-31G* basis set. Carbohydr Res 206: 183-191

Haasnoot CAG, de Leeuw FAAM and Altona C (1980) The relationship between proton-proton NMR coupling constants and substituent electronegativities I - An empirical generalization of the Karplus equation. Tetrahedron 36: 2783-2792

Hamer GK, Balza F, Cyr N and Perlin AS (1978) Experimental evidence of deviations from a Karplus-like relationship of vicinal carbon-proton coupling constants in some conformationally rigid carbohydrate derivatives. Can J Chem 56: 3109-3116

Hines JV, Varani G, Landry SM and Tinoco I Jr (1993) The stereospecific assignment of H5' and H5" in RNA using two-bond carbon-proton scalar couplings. J Am Chem Soc, submitted for publication

Jardetzky O (1980) On the nature of molecular conformations inferred from high-resolution NMR. Biochem Biophys Acta 621: 227-232

Karplus M (1959) Contact electron-spin coupling of nuclear magnetic moments. J Chem Phys 30: 11-15

Lancelot G, Chanteloup L, Beau J-M and Thuong NT (1993) Selectively ^{13}C-enriched DNA: ^{13}C and ^{1}H Assignments of the Lac operator by two-dimensional relayed HMQC experiments. J Am Chem Soc 115: 1599-1560

Levitt M and Warshel A (1978) Extreme conformational flexibility of the furanose ring in DNA and RNA. J Am Chem Soc 100: 2607-2613

Majumdar A and Hosur RV (1992) Simulation of 2D NMR spectra for determination of solution conformations of nucleic acids. Prog NMR Spec 24: 109-158

Montelione GT, Winkler ME, Rauenbuehler P and Wagner G (1989) Accurate measurements of long-range heteronuclear coupling constants from homonuclear 2D NMR spectra of isotope-enriched proteins. J Mag Reson 82: 198-204

Muller N and Pritchard DE (1959) C^{13} Splittings in proton magnetic resonance spectra I. Hydrocarbons J Chem Phys 31: 768-771

Nikonowicz EP, Sirr A, Legault P, Jucker FM, Baer LM and Pardi A (1992) Preparation of ^{13}C and ^{15}N labeled RNAs for heteronuclear multidimensional NMR studies. Nucl Acids Res 20: 4507-4513

Pearlman DA and Kim S-H (1985) Conformational studies of nucleic acids II. The conformational energetics of commonly occurring nucleosides. J Biomol Struct Dyn 3: 99-125

Rinkel LJ and Altona C (1987) Conformational analysis of the deoxyribofuranose ring in DNA by means of sums of proton-proton coupling constants: A graphical method. J Biomol Struct Dyn 4: 621-649

Salazar M, Federoff OY, Miller JM, Ribeiro NS and Reid BR (1993) The DNA strand in DNA•RNA hybrid duplexes is neither B-form nor A-form in solution. Biochemistry 32: 4207-4215

Schmitz U, Zon G and James TL (1990) Deoxyribose conformation in [d(GTATATAC]$_2$: Evaluation of sugar pucker by simulation of double-quantum-filtered COSY cross-peaks. Biochemistry 29: 2357-2368

Schwarcz JA, Cyr N and Perlin AS (1972) Orientational dependence of vicinal and geminal ^{13}C-^{1}H coupling. Can J Chem 50: 3667-3676

Schwarcz JA, Cyr N and Perlin AS (1975) Orientational effects and the sign of two-bond ^{13}C-^{1}H coupling. Can J Chem 53: 1872-1875

Serianni AS and Barker R (1984) [^{13}C]-Enriched tetroses and tetrofuranosides: An evaluation of the relationship between NMR parameters and furanosyl ring conformation. J Org Chem 49: 3292-3300

Tvaroska I, Hricovini M and Petrakova E (1989) An attempt to derive a new Karplus-type equation of vicinal proton-carbon coupling constants for C-O-C-H segments of bonded atoms. Carbohydr Res 189: 359-362

Varani G and Tinoco I Jr (1991) Carbon assignments and heteronuclear coupling constants for an RNA oligonucleotide from natural abundance ^{13}C-^{1}H correlated experiments. J Am Chem Soc 113: 9349-9354

Westhof E and Sundaralingam M (1983) A method for the analysis of puckering disorder in five-membered rings: The relative mobilities of furanose and proline rings and their effects on polynucleotide and polypeptide backbone flexibility. J Am Chem Soc 105: 970-976

Wu J and Serianni AS (1993) [^{13}C]-Labeled oligodeoxyribonucleotides: A solution study of a CCAAT-containing sequence at the NF-I recognition site of human adenovirus. Submitted for publication

Through Bond Sugar-Phosphate Backbone Assignment in ^{13}C labeled RNA by Triple Resonance ^{1}H, ^{13}C, and ^{31}P NMR Spectroscopy.

S.S.Wijmenga[#], H.A. Heus[&], F.J.M. van de Ven[&], and C.W. Hilbers[#&]

Nijmegen SON Research Centre for Molecular Design, Structure and Synthesis, [&]Laboratory of Biophysical Chemistry, [#]SON/NWO National HF-NMR Facility, Toernooiveld, 6525 ED Nijmegen, The Netherlands.

INTRODUCTION

Successful studies of the structure of biomolecules by means of NMR depend crucially on the ability to interpret their complicated spectra and to derive as many structural constraints from those spectra as possible. In protein NMR studies big strides forward have been made by the introduction of molecules uniformly enriched in ^{15}N and/or in ^{13}C isotopes (Clore and Gronenborn 1991). This allows, by means of multi-dimensional NMR spectroscopy, the spreading of spectral information along ^{13}C and/or ^{15}N frequency axes in addition to or in place of the traditional ^{1}H axes. This increases the spectral resolution appreciably and since in these experiments use is made of coherence transfer methods spectral assignment does not depend on NOE effects and thus not on assumptions about structure (Ikura et al. 1990).

In NMR studies of nucleic acids similar developments have lagged behind because procedures for uniform enrichment with ^{13}C and/or ^{15}N were not available. Yet, for these molecules uniform enrichment would be of extreme importance since their sugar proton resonances strongly overlap. This is particularly manifest for RNA where these resonances are confined to a 1 ppm region (except for the H_1'-resonances) between roughly 4 and 5 ppm. The distribution of the resonances of an RNA molecule is schematized in Fig 1. Fortunately, the situation has changed recently. The production of well-defined, uniformly enriched RNA fragments has now become feasible on a more or less routine basis

NATO ASI Series, Vol. H 87
NMR of Biological Macromolecules
Edited by C. I. Stassinopoulou

(Milligan et al. 1987). This opens up the way for the development and application of heteronuclear 3D and 4D NMR methods to structural studies of RNA as well. Some applications have already been reported (Nikonowicz and Pardi 1992a, 1992b, 1992c, 1993).

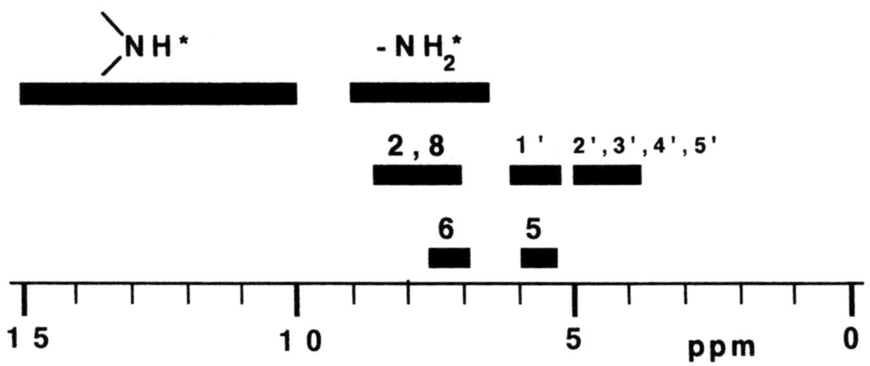

Figure 1. *Schematic overview of the ^{1}H chemical shift ranges in RNA fragments.*

Here we discuss a new experiment which makes use of uniformly enriched RNA and allows sequential assignment of the RNA spectrum using the sugar-phosphate backbone atoms. The nucleic acid sequence considered in these experiments is r(GGG-CCAAAGCCU) which forms a hairpin under the experimental conditions used in these studies. The -CAAA- tertranucleotide was chosen, because it represents one of the most frequently occuring loop sequences of the RNA genomes of the Coli-phages (J. van Duin, personal communication). While from phylogenetic comparison of ribosomal RNA the GNRA and UNCG (where N can be any nucleotide and R is a purine) tetranucleotide consensus sequences appeared to be the most common (Woese et al., 1983; Gutell et al., 1985), GNRA sequences do not appear that frequently in the RNA Coli-phages and the UNCG sequence is almost absent. Instead a preference is observed for YNNA sequences (where Y is a pyrimidine). Thus to extend our structural knowledge of tetranucleotide loops and to investigate the influence of a C rather than a U at the first position of the loop the -CAAA- sequence was chosen.

SYNTHESIS OF ENRICHED HAIRPIN r-(GGGC-CAAA-GCCU)

The synthesis of the 13C enriched hairpin involved enzymatic synthesis of 13C enriched 5-nucleoside triphosphates (NTPs), following a modification of the procedure described by Whitesides and coworkers (Simon et al., 1990), followed by enzymatic synthesis of the RNA hairpin using standard procedures (Milligan et al., 1987; Heus and Pardi, 1991). A number of reports dealing with enzymatic synthesis of enriched RNA, which all have there pro's and con's have been published (Nikonowicz et al., 1992a; Batey et al., 1992; Michnicka et al., 1993). The major advantage of the following procedure, which is now routinely used in our laboratory, is that it is very convenient and fast.

E.coli MC 4100 was grown on minimal medium (Sambrook et al., 1989), supplemented with 2g/L [13C]-glucose as the sole carbon source and 1g/L ammonium-sulphate as the sole nitrogen source. Cells were harvested at late log phase by centrifugation and resuspended in 20mM Tris-HCl, 10mM MgCl2, 2mM CaCl2, 1mM EDTA, 1.7mM 2-mercaptoethanol. Yields were 3-4 grams of cell paste per liter. The cells were broken by sonification and the cell debris was spun down at 30,000g for 30 minutes in a Beckman JA-20 rotor. Ribosomes from the supernatant of the low spin centrifugation were spun down at 100,000g in a 70.1Ti rotor for 3 hours. The ribosomes were dissolved in approximately 10ml 100mM sodium acetate pH 5.8 and ribosomal RNA was extracted by phenol extraction in the presence of 1% SDS, followed by chloroform extraction and the ribosomal RNA was concentrated by ethanol precipitation (Heus and van Knippenberg, 1988). The ribosomal RNA was dissolved in approximately 20mL 0.1 M sodium acetate, pH 5.8 and digested to 5'-nucleoside-monophosphates (NMPs) with 37ng (0.03 units) nuclease P1 (Pharmacia) per A260 unit (OD) rRNA at 37°C. Conversion into NMPs was followed by analytical FPLC chromatography on a Mono Q HR 5/5 column (Pharmacia). The reaction was complete after ~20 hours with typical yields of 18mg NMPs per gram cell paste or, in other words, 30mg NMPs per gram [13C]-glucose. NMPs were enzymatically converted into NTPs in a "one pot" reaction. After the nuclease P1 reaction the reaction mixture was briefly heated to 90°C (2 minutes), to inactivate P1 and used directly in the NTP synthesis. Typically the 20mL reaction mixture contained about 5 mM of each NMP, in 12.6mM Tris-HCl, 4mM MgCl2, 5mM KCl, 2.5mM DTT, 2.5 equivalents of PEP (60mM), and 2µM ATP. The reaction mixture was pH'd to 7, after

which the enzymes (Sigma) were added: 250 units myokinase, 0.5 units guanylate kinase and 250 units pyruvate kinase. Incubation was at 37°C, and initiallly proceeded rapidly as judged from a strong increase in pH, which was adjusted by adding aliquots of 1M HCl. The reaction was complete after about 24 hours with yields of >96% GTP, >98% ATP, UTP, CTP estimated from analytical FPLC chromatography on Mono Q. The NTPs were either used directly in the RNA synthesis or occasionally precipitated with three volumes of ethanol in the presence of 0.1M sodium acetate pH 4.8.
The [^{13}C] enriched -CAAA- hairpin was synthesized by transcription of partially duplex DNA template with T7-RNA polymerase. The 20mL reaction mixture contained 40mM Tris-HCl, 25mM MgCl2, 5mM DTT, 1mM spermidine, 400nM DNA template and 0.75mM [13C]-NTPs. The reaction duration was 6 hours at 37°C, after which the reaction was stopped by adding EDTA to a final concentration of 26mM. The reaction mixture was diluted twice with double distilled water and concentrated by ethanol precipitation. The spun down RNA was dissolved in 8ml 100mM Tris-HCl, 100mM sodium acetate, 1mM EDTA, pH 8.0. RNA oligomers were separated from NTPs by gel filtration on FPLC on Superdex 75 (Pharmacia) to recover the non-reacted NTPs for future purposes. The RNA oligomers were concentrated by ethanol precipitation and separated by preparative gel electrophoresis. Full length 12-mer RNA oligomers were detected by UV shadowing, excised from the gels and isolated by electroelution (LKB electroeluter). The purified RNA was dialyzed into 0.2 mM EDTA pH 6.8 using microconcentrators (Centricon), lyophilized and dissolved in 500ml D2O. Final yield was 71 OD units, or approximately 1.8mM using 1OD=50mg and a molecular weight of 3960 D for the 12-mer hairpin.

CONNECTIVITY PATHWAYS AND FEASIBILITY OF SEQUENTIAL ANALYSIS

So far, the sequential analysis of nucleic acid spectra has to a large extent depended on the interpretation of NOE crosspeaks between the H6/H8 base protons on one hand and the H1'/H2'/H2" resonances of the sugar of the particular base considered and those of the sugar of the 5'-neighboring residue on the other hand. The assignment procedure is based on the assumption that the residues considered are part of a backbone with an A- or B-type helix conformation. Although this approach will work in double helical regions its success is not garanteed in single stranded regions. Similarly to assignment procedures developed for NMR protein studies (Ikura et al. 1990) a way around this problem is provided by the connectivities generated through J-couplings existing between the backbone nuclei. The sequential assignment procedure proposed here depends on making correlations between ^{1}H, ^{13}C and ^{31}P (or ^{15}N) via through bond coherence transfer.

A series of pathways which may aid in the sequential analysis are depicted in Fig. 2. Fig. 2A represents a pathway in which, via C4'(n) → P(n + 1) → C4'(n + 1), connectivities sequential assignments are obtained. Results derived through this pathway can be supported by measuring connectivities available from other pathways. Thus, a sequential analysis can also be performed by considering the combination of the connectivities occurring between C5'(n+1) → P(n+1) (Fig. 2B) and P(n+1) → C3'(n) (Fig. 2C). Additionally, depending on the conformation of the backbone (vide infra) C2'(n) → P(n + 1) connectivities may provide supplementary information (Fig. 2D). The assigned backbone signals can be connected to the resonances of the sugar rings (Fig. 2E) and these in turn to the resonances of the bases (Fig. 2F). To connect these signals different types of experiments are necessary, namely a HCP experiment (Heus et al. 1994) to obtain the correlations depicted in Figs. 2A-D, a HCCH (Bax 1990, Nikonowicz and Pardi 1992b, 1993) or (H)CCHTOCSY (Heus et al. 1994) experiment to obtain the correlation depicted in Fig. 2E, and a HCN experiment (Sklenar et al. 1993) for the correlations depicted in Fig. 2F. Here we provide a more detailed analysis of the sequential assignment by through-bond coherence transfer via the sugar-phosphate backbone.

To estimate the feasibility of such an approach J-couplings between different atom pairs in the sugar-phosphate backbone are indicated in Figs. 3. Most of these J-couplings, i.e. the three-bond couplings are conformation dependent. The $^{3}J_{H3'P}$ and $^{3}J_{H5'/5''P}$ couplings depend on the torsion angles ε and β respectively. Since β is found overwhelmingly in the trans domain (e.g. ≈190° in A-RNA and ≈214° in B-DNA) $^{3}J_{H5'P}$ and $^{3}J_{H5''P}$ are normally found to be of the order of 1 and 4-9 Hz, respectively (Wijmenga et al. 1993) . Furthermore, $^{3}J_{H3'P}$ is 6-9 Hz for the nominal value of ε (≈ 210°) (Wijmenga et al. 1993). Also, $^{3}J_{C4'(n)P(n)}$ and $^{3}J_{C4'(n)P(n+1)}$ depend on the torsion angles β and ε,

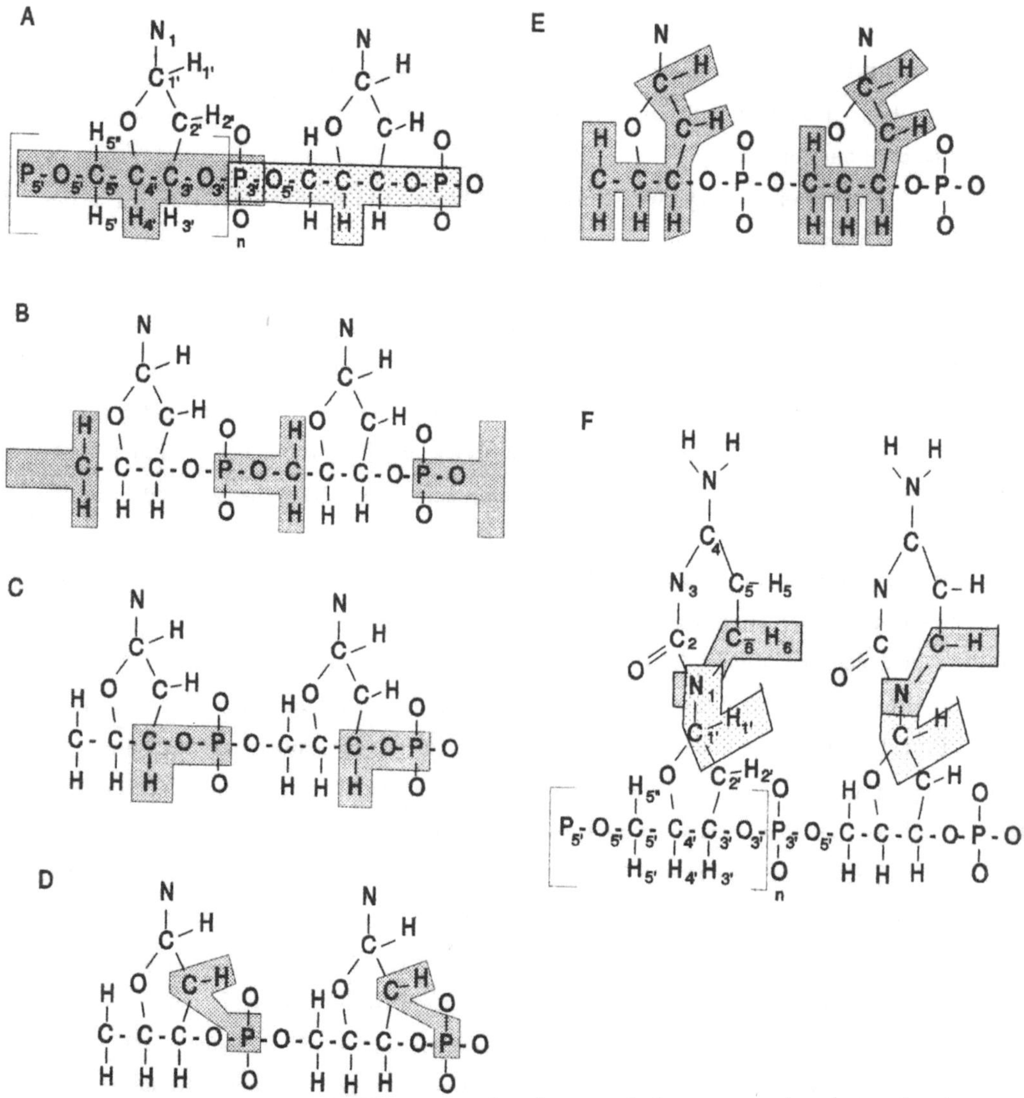

Figure 2. *Schematics of RNA sugar-phosphate and base-sugar-phosphate showing pathways for sequential assignment via through-bond coherence transfer (shaded regions). A-E. connectivities observed in HCP experiment (see text); (E) connectivities in HCCH and (H)CCHTOCSY experiment (see text); (F) connectivities observed in HCN experiment (see text).*

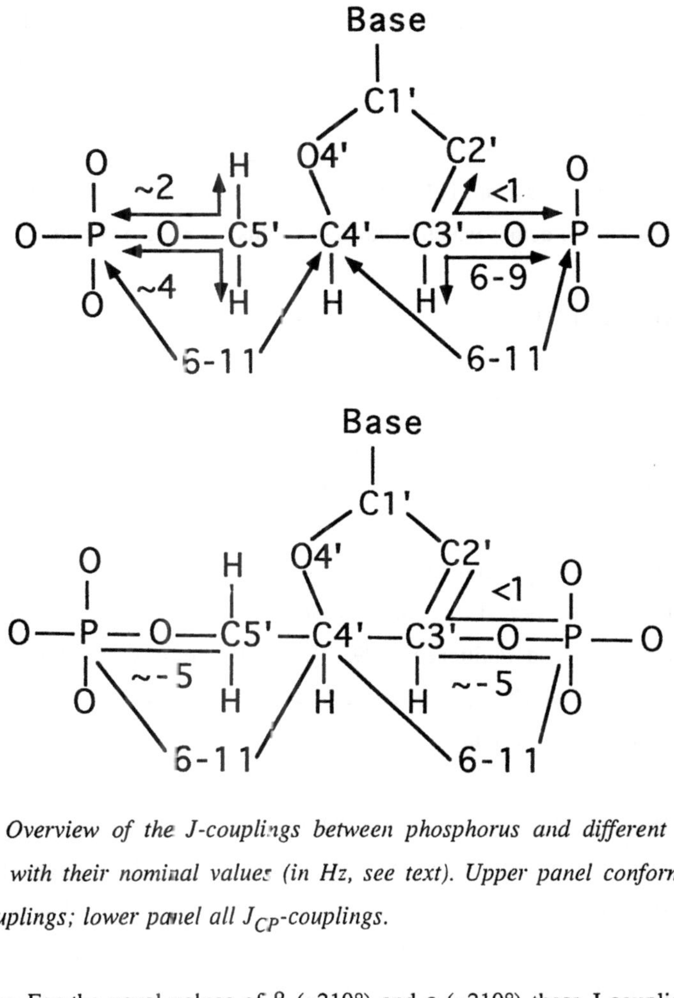

Figure 3. *Overview of the J-couplings between phosphorus and different atoms of the sugar ring with their nominal values (in Hz, see text). Upper panel conformation dependent ^3J-couplings; lower panel all J_{CP}-couplings.*

respectively. For the usual values of β (≈210°) and ε (≈210°) these J-couplings are of the order of 11 Hz. Depending on the backbone conformation these torsion angles may deviate from these values appreciably. The torsion angle ε may roughly vary between 180° and 270°, the allowed range for β is larger, i.e. 120° to 240° (Saenger 1984, Mooren 1993). This implies that $^3J_{C4'(n)P(n+1)}$ may vary between 11 and 0.7 Hz, depending on ε, and $^3J_{C4'(n)P(n)}$ between 11 and 6 Hz depending on β.

It can be concluded from this evaluation that a sequential assignment based on through-bond connectivities is feasible when the backbone has an A- or B-type helical conformation. Difficulties may arise when ε shifts to the g^--domain because then the small value of $^3J_{C4'(n)P(n+1)}$ prohibits sufficient coherence transfer. Fortunately, under these conditions $^3J_{C2'(n)P(n+1)}$ becomes large, i.e. when $^3J_{C4'(n)P(n+1)}$ decreases from 9 to 0.7 Hz, $^3J_{C2'(n)P(n+1)}$ increases from 0.7 to 9 Hz (Fig. 5C) and the sequential analysis can still be performed via the detour which makes use of this coupling constant.

THE HCP EXPERIMENT

The pulse sequence of this 3D-NMR triple resonance experiment which correlates ^1H, ^{13}C, and ^{31}P (HCP) is shown in Fig. 4. It closely resembles the pulse sequence of the CT-HNCO experiment (Grzesiek and Bax 1992). The ^1H H_y coherence, obtained after the first ^1H-$\pi/2$ pulse, is transferred via an INEPT sequence into H_zC_y anti-phase coherence, which is not affected by J_{HH} couplings. The subsequent ^{13}C evolution (t1) takes place within the constant time period, T, during which the J_{CC}- and the J_{CH}-couplings are removed (Van de Ven and Phillippens 1992, Grzesiek and Bax 1992) by a judicious choice of the times at which π-pulses are applied to the ^1H and ^{13}C spins. A further advantage of this constant evolution period is the posibility to enhance resolution by reflecting the FIDs around the origin and the subsequent application of linear prediction methods to extend the FIDs (Zhu and Bax 1990). During the period T the J_{CH} and J_{CP} couplings are active and depend on the delay times τ' and δ, respectively, so that coherence transfer can take place to ^{31}P. Thus, during the time T the following coherence pathway is traversed: $H_zC_y \Rightarrow C_x \Rightarrow C_xP_z$. The subsequent simultaneous 90° pulse on ^{13}C and ^{31}P transfers C_xP_z into C_zP_y coherence. Then during t2, while ^{13}C and ^1H are simultaneously decoupled from ^{31}P, ^{31}P evolution takes place. Finally, via the reverse route, this coherence is transferred back into ^1H-coherence, which is detected during t3, while ^{13}C and ^{31}P are simultaneously decoupled. In the HCP 3D-NMR triple resonance experiment the following correlations can thus be found, C4'(n) ⇒ P3'(n + 1) ⇒ H4'(n), C4'(n) ⇒ P5'(n) ⇒ H4'(n), C5'(n) ⇒ P5'(n) ⇒ H5'/H5"(n), C3'(n) ⇒ P3'(n + 1) ⇒ H3'(n). This allows a sequential walk along the backbone resonances using the

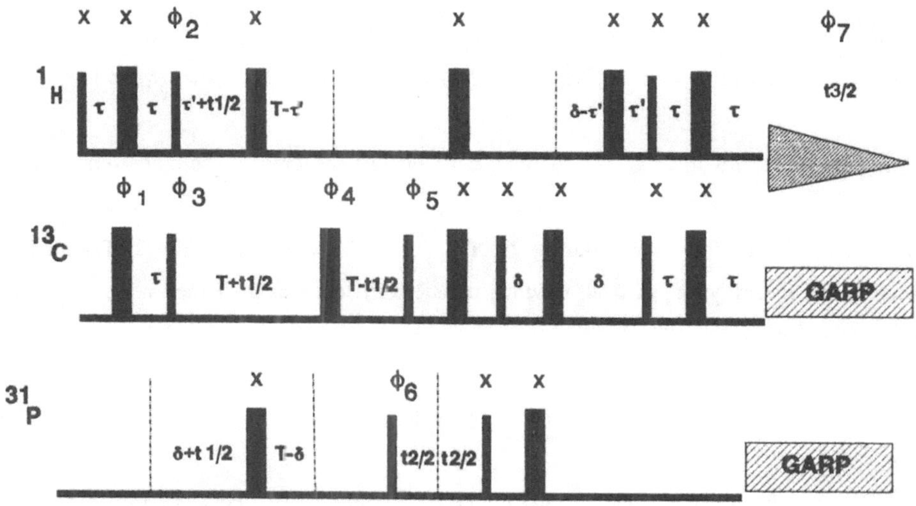

Figure 4. *Pulse sequence of the HCP experiment. The narrow and wide bars indicate 90° and 180° pulses, respectively. Constant time* ^{13}C *evolution applied (van de Ven and Philippens 1992, Grzesiek and Bax 1992) and phase cycling as follows:* $\phi_1 = x\ {-x}$; $\phi_2 = x\ {-x}$; $\phi_3 = x$ *(+TPPI (t1));* $\phi_4 = 4x\ 4y$; $\phi 5 = x\ {-x}$; $\phi_6 = x\ x\ {-x}\ {-x}$ *(+TPPI(t2));* $\phi_7 = x\ x\ {-x}\ {-x}\ {-x}\ {-x}\ x\ x$. *TPPI, time proportional phase incrementation (Marion and Wuthrich 1983).*

C4'(n) ⇒ P(n + 1) ⇒ C4'(n + 1) correlations which are indicated in the boxes in Fig. 2. In case the torsion angle ε is shifted to the g^--domain the pathway C4'(n) ⇒ P(n + 1) ⇒ C4'(n) is blocked and the pathway C2'(n) ⇒ P(n + 1) ⇒ C2'(n) should be considered.

As mentioned above, during the evolution time, T, the J_{CC}- and J_{CH}-couplings are removed, while the J_{CP}-couplings remain active. Thus, one has to carefully calculate the coherence transfer functions for the transfer from H ⇒ C ⇒ P to be able to make optimal adjustments for the delay times in this period. For the different carbon atoms the transfer functions (Trnf) are given by:

$$\mathrm{TrnfC2'} = \cos(2\pi J_{CC}T)^2 * \sin(2\pi J_{C2'(n)P5'(n+1)}\delta) * \sin(2\pi J_{C2'H2'}\tau')$$
$$\mathrm{TrnfC3'} = \cos(2\pi J_{CC}T)^2 * \sin(2\pi J_{C3'(n)P5'(n+1)}\delta) * \sin(2\pi J_{C3'H3'}\tau')$$

$TrnfC4' = \cos(2\pi J_{CC}T)^2 * \sin(2\pi J_{C4'(n)P5'(n+1)}\delta) * \sin(2\pi J_{C4'(n)P5'(n)}\delta) * \sin(2\pi J_{C4'H4'}\tau')$

$TrnfC5' = \cos(2\pi J_{CC}T) * \sin(2\pi J_{C5'(n)P5'(n)}\delta) * \sin(2\pi J_{C5'H5'}\tau') * \cos(2\pi J_{C5'H5''}\tau')$

Here, J_{CC} represents the ^{13}C-^{13}C coupling constant, the meaning of the other subscripts is self evident. In considering the coherence transfer also the influence of relaxation has to be accounted for. This is T_{2H} during 2τ, T_{2C} during 2T and T_{2P} during t2. Taking into account the reverse route as well, the total transfer efficiency of the experiment, Trnft, can be shown to be:

$$TrnftCx = (TrnfCx * \exp(-\tau/T_{2H}) * \exp(-T/T_{2C}) * \exp(-t2/T_{2P}))^2$$

The J_{CC} couplings are ≈40 Hz and the direct J_{CH} couplings vary between 140 to 150 Hz. Thus, since J_{CC} and J_{CH} are virtually constant the transfer efficiency of the experiment is determined by the J_{CP} couplings.

In Fig. 5A, the transfer functions TrnfC4', TrnfC3' and TrnfC5' are depicted as a

Figure 5. *Overview of the C ⇒ P coherence transfer efficiencies and relevant J_{CP}-couplings. C ⇒ P coherence transfer efficiencies are calculated with J_{CC} = 40 Hz, J_{HC} = 145 Hz; for transfers involving C4', C2' and C3' τ' = 1.7 ms, while for transfers involving C5' τ' = 0.9 ms. (A) TrnfC4' = TrnfC4'(n)P(n+1) = TrnfC4'(n)P(n), TrnfC3' = TrnfC3'(n)P(n+1), and TrnfC5' = TrnfC5'(n)P(n) calculated as a function of δ with $J_{C4'(n)P(n+1)} = J_{C4'(n)P(n)}$ = 11 Hz, $J_{C3'(n)P(n+1)}$ = 5 Hz, and $J_{C5'(n)P(n)}$ = 5 Hz, respectively; (B) As in A with TrnfCx' multiplied with $\exp(-2\delta/T_{C2})$ with T_{C2} = 4 Hz; (C) $J_{C4'(n)P(n+1)}$ and $J_{C2'(n)P(n+1)}$ as a function of ε, for the J_{CP} couplings the most recent parametrization of the Karplus equation was used (Mooren et al. 1993, Wijmenga et al. 1993); (D) $J_{C4'(n)P(n)}$ as a function of β. (E) TrnsfC4' = TrsnfC4'(n)P(n) calculated as a function of the directly related torsion angle β for two extreme values of ε; (F) TrnfC4' = TrnfC4'(n)P(n+1) calculated as a function of the directly related torsion angle ε for the nominal value of β = 180°, and two extreme values 150° and 240°; TrnfC2' = TrnfC2'(n)P(n+1) as a function of ε.*

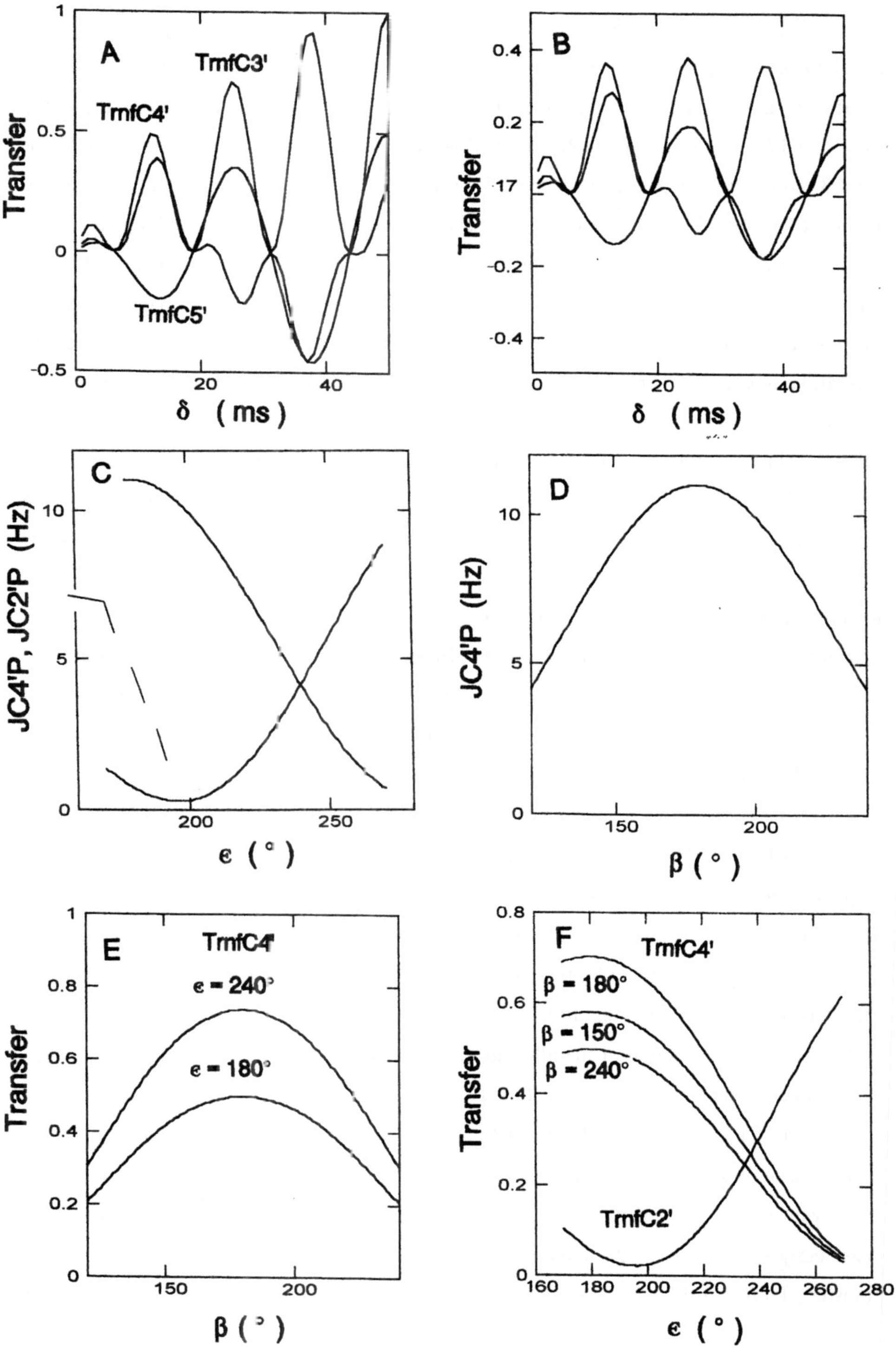
A
TmfC3'
TmfC4'
TmfC5'
Transfer
δ (ms)
B
Transfer
δ (ms)
C
JC4'P, JC2'P (Hz)
ε (°)
D
JC4'P (Hz)
β (°)
E
TmfC4'
ε = 240°
ε = 180°
Transfer
β (°)
F
TmfC4'
β = 180°
β = 150°
β = 240°
TmfC2'
Transfer
ε (°)

function of the delay time δ, which in the experiment discussed in this paper was chosen equal to the delay time T, the maximum evolution time. In the calculations the nominal values for the J-couplings were used. The first maxima (minima) are seen at the delay time $\delta \approx 12$ ms, corresponding for the C4'-correlation to a transfer efficiency of 0.5, for the C3'-correlation of 0.35 and for the C5'-correlation of -0.2. Other maxima (minima) are found at 25 and 37 ms, however, for these longer delay periods additional signal loss is observed due to T_2 relaxation of the carbon spins. A value for the line width of 4 Hz (corresponds to a T_2 of 80 ms as found for the molecule studied in this paper), leads to the transfer shown in Fig. 5B. This demonstrates that the loss of coherence due to T_2 relaxation has only a limited effect on the transfer efficiency at the first optimum. It reduces the transfer efficiency at the other maxima more significantly. In addition, at these other maxima the transfer values are more strongly influenced by small variations in for example J_{CC} couplings. The best choice for δ is therefore 12 ms, i.e. the first maximum (minimum). At the 12 ms maximum the total transfer for $C_{4'}$ is then $0.37^2 = 0.14$. Since such an efficiency corresponds to a strong NOE in a 2D NOE experiment, the 3D HCP experiment is predicted to be quite sensitive, since the signal to noise ratio (S/N) is proportional to the total measurement time (Griesinger and Ernst, 1988). Furthermore, the sensitivy remains acceptable for larger molecules. For a linewidth of 16.5 Hz, expected for a molecule of approximately 50 nucleotides, the efficiency of the P $\Rightarrow$ C transfer step is reduced by only a factor of 4 at the first maximum.

A feature that needs further consideration is the conformational dependence of the transfer functions. The important variables are the torsion angle dependent J-couplings, $J_{C4'(n)P(n)}$ and $J_{C4'(n)P(n+1)}$ plotted for the allowed ranges of β and ε in Fig. 5C and 5D, respectively. The first maximum in the transfer function is always at 12 ms independent of the values of the corresponding torsion angles β and ε, since its location is determined by the value of the J_{CC} couplings. In Fig. 5E we have therefore plotted TrnfC4'(n)P(n) (=P5') at $\delta = 12$ ms as a function of the allowed values of β for values of ε for which $J_{C4'(n)P(n+1)}$ acquires its most extreme values. In Fig. 5F the same is done for TrnfC4'(n)P(n + 1) as a function of ε for values of β for which $J_{C4'(n)P(n)}$ acquires its most extreme values. As can be seen from Fig. 5E TrnfC4'(n)P(n) increases for values of $\varepsilon > 180°$, because for these values of ε the C4'(n) $\Rightarrow$ P(n+1) transfer becomes less efficient. The transfer stays reasonably high even for the most extreme values of β (lowest transfer 0.2). For ε the transfer, TrnfC4'(n)P(n+1), is not much influenced the value of β (Fig. 5F). However, when ε becomes g^- (270°) the transfer efficiency, TrnfC4'(n)P(n + 1), drops to very low values, of the order of 0.05 (Fig. 5F), so that the corresponding cross peak is expected to be absent in the spectrum. However, such a change in ε is accompanied by a concomitant increase in TrnfC2'(n)P(n + 1) (due to the increased $J_{C2'(n)P(n + 1)}$) from the very low values 0.02 at ε = 180°, to 0.6 at ε = 270°. The disappearance of the cross peak, C4'(n) $\Rightarrow$ P(n + 1), will thus be accompanied by a concomitant appearance of a C2'(n) $\Rightarrow$ P(n + 1) cross peak (Fig. 5F).

APPLICATION TO r(GGGCCAAAGCCU)

The HCP spectrum of a hairpin formed by r(5'GGGCCAAAGCCU3') (1.8 mM strand concentration) was recorded on a Bruker AM400 spectrometer modified to handle the three frequencies and equipped with a triple resonance probe (1H, 13C and a broadband channel adjusted to ^{31}P). The proton coil of the probe was doubly tuned to ^{13}C and ^{1}H. The processing of the data was performed with NMRi software.

The theoretically calculated sensitivity, which is of the order of 12-14%, could not be achieved in practice. A careful check of the factors determining the extra losses found in practice turned out to be related to rf inhomogeneity of the coils. The main loss is found in the INEPT transfer step for which theoretically the transfer efficiency is predicted to be of the order of 70 % when T2 relaxation times are included, while in practice it turns out to be approximately 30-40 %, as evidenced by a comparison of the S/N of (^{13}C, ^{1}H) CT-HSQC spectrum (Van de Ven and Phillippens 1992) with a 1D spectrum (S/N(CT-HSQC) : S/N(1D) = 10-15%). The transfer from ^{13}C to ^{31}P is roughly 60% of the theoretical transfer which indicates a rather limited loss due to rf inhomogeneities effects. Nevertheless, the HCP spectrum shown in Fig. 6 was recorded in 20 hours, illustrating that in practice still a reasonable sensitivity is obtained. In Fig. 6 cross sections through the 3D-HCP spectrum are shown, i.e. the ^{31}P, ^{1}H planes taken at the different C4' resonance frequencies along the ^{13}C-axis. These planes are arranged in such a way that a sequential walk can be performed along the C4' and P atoms of the backbone. In Fig. 6A we start with the plane through the C4'-resonance of the 5'-terminal residue (G1). The corresponding H4'-resonance is connected to two phosporous resonances, P(1) and P(2), of the 5'-terminal and the second residue of the chain, respectively. The plane to be linked to Fig. 6A is the one in which the P(2) resonance is connected to the H4'(2) resonance (of the

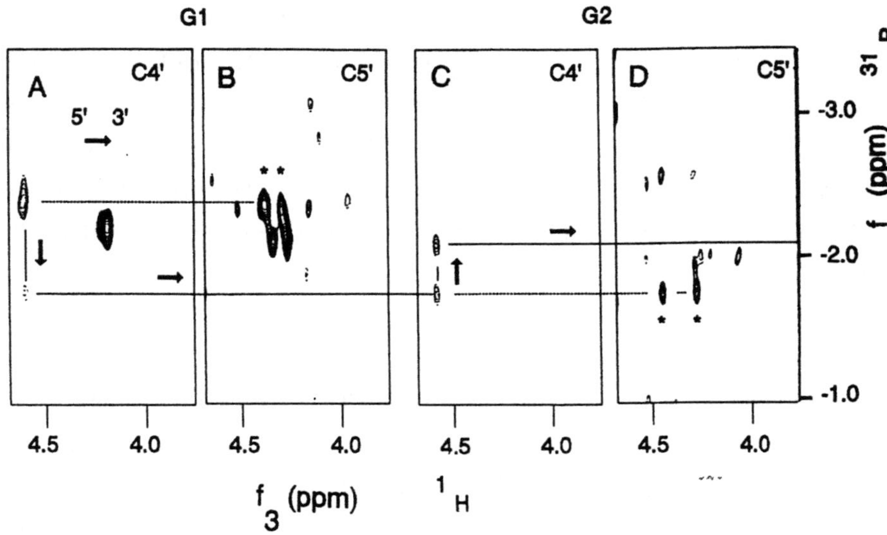

Figure 6. *Sequentially arranged ^{31}P, ^{1}H planes from the HCP spectrum, taken at the different C4' and C5' resonance frequencies along the ^{13}C axis. The planes at the different C4' frequencies show the sequential walk through the sugar-phosphate backbone for the first two residues, G1 (5'-end) and G2, as indicated by the arrows and drawn lines; the two cross peaks indicated by stars in the plane at the C5' frequency of residue i represent the H5'/5" resonances positions of residue i; the dotted line connects these H5'/5" resonances with the P5' cross peak in the plane at the C4' frequency of residue i. The HCP was recorded in approximately 20 hours with the following acquisition settings: 16 scans for each FID of 1024 data points (t3), 128 t1 values ($t1^{max}/2 = T = \delta = 12.5$ ms), 28 t2 values, $\tau = \tau' = 1.5$ ms ($\approx 1/4J_{CH}$), $T\text{-}\delta = 5\ \mu s$, 1 s relaxation delay with solvent presaturation, low power (3 Watt) GARP decoupling of ^{31}P and ^{13}C, spectral width, 2941 Hz, 400 Hz, and 2941 Hz and carrier position, 73.6 ppm, -2.23 ppm, and 4.75 ppm for ^{13}C, ^{31}P, and ^{1}H, respectively. Typical processing parameters: zero-filling twice in t1, t2 (preceeded by zeropadding to 32 points), and t3; applying a sin^2 window-function shifted by $\pi/5$, $\pi/3$, and $\pi/2$ in t1, t2, and t3, respectively. The spectrum was calibrated relative to TSP as described by Ikura et al 1990.*

second residue) which is in turn connected to P(3) of the third residue, as shown in Fig 6C. The 5' $\Rightarrow$ 3' direction in a plane through the C4'-resonance of residue i can be inferred from the plane through the C5'-resonance of the same residue (Figs. 6B and D). Continuation of the analysis in this manner allowed us to assign the backbone resonances of the molecule.

Acknowledgements

This work was supported by the Netherlands Foundation for Chemical Research (SON) with financial aid from the Netherlands Organization for Advanced Research (NWO). We wish to thank J.W.M. van Os, J.W.G. Janssen, and J.J. Joordens for excellent technical assistance. H.A. Heus is supported by a grant from the Royal Netherlands Academy of Arts and Sciences.

References

Batey, R.T; Inada, M; Kujawinski, E; Puglisi, J.D; Williamson, J.R. *Nucleic Acids Res.* **20**, 4515-4523 (1993).

Clore, G.M.; Gronenborn, A.M. *Science* **252**, 1390-1399 (1991).

Grzesiek, S.; Bax A. *J. Magn. Res.* **96**, 432-440 (1992).

Gutell R.R.; Weiser B.; Woese C.R.; Noller HF *Prog. Nucleic Acids Res. and Mol. Biol.* **32**, 155-216 (1985).

Heus, H.A.; Van Knippenberg, P.H. in *Methods in Enzymology* **164**, Dahlberg, J.E.; Abelson, J.N. (eds.) Academic Press, Inc. San Diego, pp. 188-200 (1988).

Heus, H.A.; Pardi, A. J. Mol. Biol. 217, 113-124 (1991).

Ikura, M.; Kay, L.E.; Bax, A. *Biochemistry* **29**, 4659-4667 (1990).

Marion, D.; Wuthrich, K. *Biochem. Biophys. Res. Comm.* **113**, 967-974 (1983).

Michnicka, M.J.; Harper, J.W.; King, G.C. *Biochemistry* **32**, 395-400 (1993).

Milligan, J.F; Groebe, D.R; Witherell, G.W.; Uhlenbeck, O.C. *Nucleic Acids Res.*, 15, 8783-8798 (1987).

Mooren, M.M.W. *thesis*; University of Nijmegen, The Netherlands, 1993.

Nickonowitz, E.P.; Sirr, A.; Legault, P.; Jucker, F.M; Baer, L.M.; Pardi, A. *Nucleic Acids*

Res. **20**, 4507-4513 (1992a).

Nikonowicz, E.P.; Pardi, A. *J. Am. Chem. Soc.* **114**, 1082-1083 (1992b).

Nikonowicz, E.P.; Pardi, A. *Nature* **355**, 184-186 (1992c).

Nikonowicz, E.P.; Pardi, A. *J. Mol. Biol.* **232**, 1141-1156 (1993).

Saenger, W. *Principles of nucleic acid structure.* Springer Verlag, Heidelberg (1984).

Sambrook, J.; Fritsch, E.F.; Maniatis, T. *Molecular cloning: A Laboratory Manual.* Cold Spring Harbor Laboratory Press, Cold Spring Harbor, New York (1989).

Simon, E.S.; Grabowski, S.; Whitesides, G.M. *J. Org. Chem.* **55**, 1834-1847 (1990).

Sklenar, V.; Peterson, D.R.; Rejante, M.R.; Feigon, J. *J. Biomol. NMR* **3**, 721-727 (1993).

Woese C.R.; Gutell R.R.; Gupta R.; Noller HF *Microbiol. Rev.* **447**, 621-669 (1983).

Van de Ven, F.J.M.; Phillippens, M.E.P. *J. Magn. Reson.*, **97**, 637-644 (1992).

Wijmenga, S.S.; Mooren M.M.W.; Hilbers, C.W. *NMR of Nucleic Acids; from spectrum to structure*; in *NMR of Macromolecules*, ed. Roberts, G.C.K.; Oxford University Press: Oxford, pp 217-288 (1993).

Heus, H.A.; Wijmenga, S.S.; Van de Ven F.J.M.; Hilbers, C.W. *J. Am. Chem. Soc.* to be published 1994.

Zhu, G.; Bax, A. *J. Magn. Reson.* **90**, 405-410 (1990).

NMR STUDIES OF PROTEIN-NUCLEIC ACID INTERACTION: THE LAC OPERON

R. Kaptein, R. Boelens, V.P. Chuprina and J.A.C. Rullmann
Bijvoet Center for Biomolecular Research, Utrecht University
Padualaan 8, 3584 CH Utrecht
The Netherlands

INTRODUCTION

Many fundamendal cellular processes rely on the interaction of proteins with specific DNA base sequences. Therefore, the problem of protein-DNA recognition constitutes an active area of research in molecular biology. The approaches to this problem range from biochemical and genetic studies (mutagenesis of DNA-binding proteins) to the application of the methods of structural biology (X-ray crystallography and NMR spectroscopy). In the early '80s the first crystal structures became available for a number of bacterial proteins such as λ and cro repressors and CAP (for recent reviews see Pabo and Sauer, 1992; Steitz, 1990; Travers 1993). NMR spectroscopy started to contribute around 1985 with the structure elucidation of the *lac* repressor headpiece (Kaptein *et al.* 1985) and a low resolution structure of the headpiece-operator complex in 1987 (Boelens *et al.*, 1987). While the first structures all contained a helix-turn-helix motif as the essential DNA-binding subdomain, later a plethora of DNA-binding motifs was characterized including zinc-fingers, leucine zipers, helix-loop-helix proteins and even β-sheet DNA-binding proteins. NMR has signicifantly contributed to this, since often DNA-binding domains of proteins are relatively small independently folded domains that can be expressed and studied separately. Thus, in particular for the various subclasses of zinc-fingers the first structural information came from NMR. This was the case for the TFIIIA-type zinc fingers (Lee *et al.*, 1989; Klevit *et al.*, 1990) for the nucleocapsid proteins (Summers *et al.*, 1990) and for the nuclear hormone receptors (Härd *et al.*, 1990; Schwabe *et al.*, 1990).

Often a combination of various biochemical and genetic methods with structural work is most fruitful for understanding protein-DNA recognition on a molecular level. X-ray or NMR structures of protein-DNA complexes allow an interpretation of the results of mutagenesis, which would be very difficult in the absence of structural data. Conversely, binding affinities of mutant proteins or DNA variants give insight in the importance of the various interactions seen in X-ray or NMR structures in terms of the free energy of binding.

The *lac* repressor is a case in point. For this system over the years an enormous body of biochemical and genetic data has been generated, which can now be interpreted in structural

NATO ASI Series, Vol. H 87
NMR of Biological Macromolecules
Edited by C. I. Stassinopoulou

1 5 10 15
Met Lys Pro Val Thr Leu Tyr Asp Val Ala Glu Tyr Ala Gly Val
helix I

16 20 25 30
Ser Tyr Gln Thr Val Ser Arg Val Val Asn Gln Ala Ser His Val
helix II

31 35 40 45
Ser Ala Lys Thr Arg Glu Lys Val Glu Ala Ala Met Ala Glu Leu
helix III

46 50 55 60
Asn Tyr Ile Pro Asn Arg ¦ Val Ala Gln Gln Leu ¦ Ala Gly Lys ¦ Gln
HP 51 ¦ HP 56 ¦ HP 59 ¦

Fig. 1. Amino acid sequence of the N-terminal part of lac repressor. The α-helical regions as determined by NMR are indicated. Cleavage sites are shown of the enzymes clostripain, chymotrypsin and trypsin, yielding headpiece fragments of 51, 56 and 59 amino acid residues, respectively.

terms. Mostly by the work of Miller (1984) and Müller-Hill (Lehming *et al.*, 1987) and their coworkers a staggering number (thousands) of mutants has been characterized. In particular mutant proteins with altered DNA-binding specificity are informative in this respect. With recent structure refinement of the *lac* headpiece-operator complex by NMR (Chuprina *et al.*, 1993), this work can be put on a firm structural footing. In this chapter we shall discuss the *lac* repressor-operator system as an example to illustrate the NMR approach to protein-DNA recognition.

LAC REPRESSOR

The *lac* repressor of *E. coli* is a tetrameric protein of molecular weight 154 000. The native repressor is too large for high-resolution NMR studies. However, each subunit has a separate DNA-binding domain (headpiece) that can be cleaved off by proteolytic enzymes (Geisler and Weber, 1977). The amino acid sequence of the N-terminal region is shown in Figure 1. Depending on the proteolytic enzyme used, headpieces can be prepared containing 51, 56 or 59 amino acid residues (HP 51, HP 56 or HP 59). These headpieces retain their original three-dimensional structure and their ability to recognize the *lac* operator specifically (Ogata and Gilbert, 1979). The trypsin-resistant core is involved in the subunit interaction and contains the inducer binding site. The sequence of the natural *lac* operator reveals an approxiate two-fold symmetry (Gilbert and Maxam, 1973) and in agreement with that, two subunits of *lac* repressor suffice to recognize lac operator (Kania and Brown, 1976). Therefore *lac* repressor should bind with two headpieces to each half of the operator.

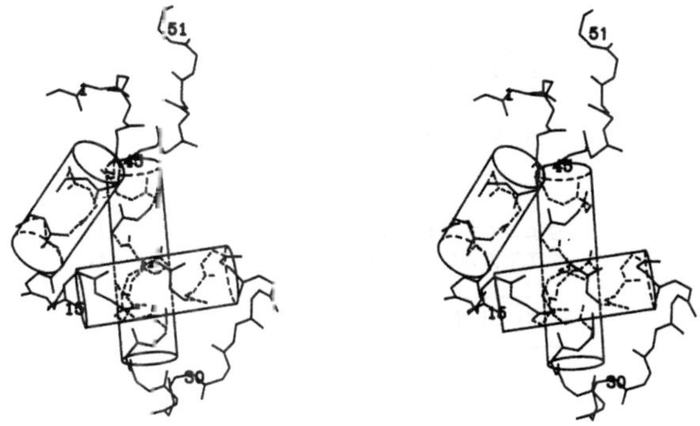

Fig. 2. Stereo diagram of the backbone conformation of headpiece 51. Cylinders represent the three α-helices. The structure was determined from a set of 169 proton-proton distance constraints from NOEs, using a restrained molecular dynamics procedure (De Vlieg *et al.*, 1986).

Lac headpiece (HP 51) was one of the first proteins for which the three-dimensional structure was determined by NMR (Kaptein *et al.*, 1985; De Vlieg *et al.*, 1986). The structure determination was accomplished using the restrained molecular dynamics (RMD) method. In this method one adds pseudo-energy terms to the molecular dynamics force field, that represent distance restraints obtained from 2D NOE data and restraints on dihedral angles from J-couplings. The RMD method has good sampling and minimization properties and is still often used for structure refinement both in NMR and X-ray crystallography. The resulting headpiece structure is shown in Figure 2. The helix-turn-helix region consisting of the helices I and II of the headpiece can be clearly seen with the third helix packing against the first two forming a hydrophobic core. The RMD run also indicated that the three-helical core of the protein is rather rigid, whereas the N-terminal and C-teminal region and also the loop between helices I and III showed higher mobility.

LAC HEADPIECE-OPERATOR COMPLEX

Lac operator of E. *coli* is defined genetically as the control region in the *lac* operon, where operator constitutive mutants occur. The region protected by *lac* repressor is 20-25 bp long,

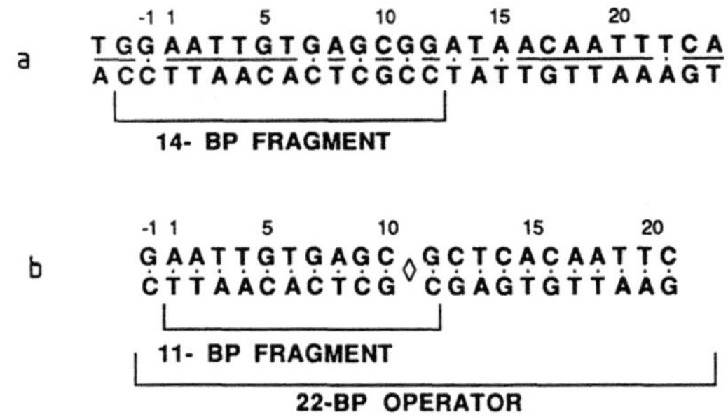

Fig. 3. Sequences of native *lac* operator (a) (Gilbert and Maxam, 1973) and 'ideal' symmetric *lac* operator (b) (Sadler *et al.*, 1983; Simons *et al.*, 1984). Synthetic operator fragments of 11, 14 and 22 bp used in the NMR studies are indicated.

with a pseudo-dyad axis going through GC 11 (Gilbert and Maxam, 1973). It was found by Sadler *et al.* (1983) and Simons *et al.* (1984) that symmetrical *lac* operators lacking the central GC base-pair bind *lac* repressor up to an order of magnitude stronger than the native one. The sequences of the operators and the fragments discussed here are shown in Figure 3. The initial NMR studies were made with the 14 bp fragment, which turned out to be the stronger binding half-operator as it also occurs in the symmetrical operator (Fig. 3b). The binding affinity of the isolated headpiece with a half-operator is not extremely high ($K_D \approx 10^{-6}$M), so that the free and DNA-bound forms are in fast exchange on the NMR time-scale. This greatly helps in assigning the resonances in the NMR spectra of the protein-DNA complex, since the resonances of the free species can be followed in titrations, either by adding, for instance, protein to DNA or by titrating the complex with increasing amounts of salt, by which it will gradually dissociate. Figure 4 shows one of the most readily accessible regions of a 2D NOE spectrum of the complex of HP 56 with the 14 bp operator. It contains a window, where only intra-DNA cross-peaks occur (H6/H8-H1' and cytosine H5-H6). These cross-peaks provided a start for the assignment of the DNA resonances, as is shown in Figure 4 for one strand by the lines connecting intra- and internucleotide cross-peaks. In this way, assignments were obtained for all non-exchangeable protons of the DNA in the complex, except for some of the H5' and H5" protons. The general pattern of intra-DNA NOEs is still that of a B-DNA type conformation. Also, most of the ^{1}H resonance positions show small shifts upon complex formation, with a maximum of 0.2 p.p.m. for the H8 proton of G5 and the H1' proton of G7. These results are also consistent with the idea that small adjustments of the DNA conformation occur. Similarly, a large number of ^{1}H assignments have been made for the protein part of the HP 56-14 bp operator complex. For this, a combination of 2D NOE spectra and homonuclear Hartman-Hahn (HOHAHA) spectra was used.

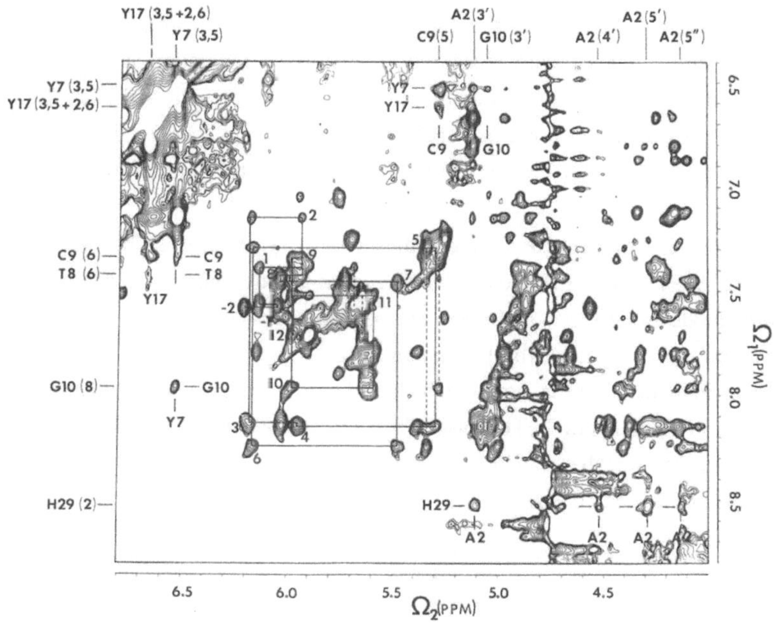

Fig. 4. Part of the 500 MHz 2D NOE spectrum of the HP 56-14 bp operator complex. Sequential NOEs of one of the DNA strands are indicated by the connecting lines. The protein-DNA NOE cross-peaks present in this part of the spectrum are also indicated.

The early low-resolution structure of the HP 56-14 bp operator complex was obtained by docking the protein to DNA in standard B conformation guided by 11 NOEs observed between protein and DNA (Boelens *et al.*, 1987). This model was similar to other complexes of helix-turn-helix proteins in that helix II of the headpiece, the "recognition helix" of the helix-turn-helix motif, is inserted in the major groove of DNA and makes the majority of the protein-DNA interactions. However, a surprising result was that the orientation of the recognition helix was opposite to that found in all other known complexes. This means that the first helix points towards the dyad axis of the operator, while for other proteins, such as λ and cro repressors and CAP, it points away from it. Since in 1987 NMR was not yet as well established as a structural tool as it is now, this result was not easily accepted by the scientific community. And of course, there was the question whether the whole *lac* repressor would bind the same way as the headpiece does. That this is indeed the case was shown in genetic experiments by Lehming *et al.* (1987). They constructed a *lac* repressor mutant with the first two amino acids of the recognition helix replaced by those of *gal* repressor (Tyr 17 → Val, Gln 18 → Ala). This mutant repressor had high affinity for the gal operator, which differs from lac operator at positions 7 and 9. Although this already gives some clue as to the orientation of the recognition helix, a more definitive result was their finding of a repressor mutant with Arg 22 replaced by Asn, which now had specificity for a *lac* operator with GC5 replaced by TA. This fixes unambiguously the orientation of the recognition helix as the opposite of that of cro and λ repressors.

Studies of a complex of two HP 56 molecules with a symmetric 22 bp *lac* operator (Lamerichs *et al.*, 1989) showed that the two headpieces bind independently and in essentially the same binding mode as they do in half-operator complexes. Therefore, a more detailed study was made of a complex with an 11-bp half-operator (cf. Fig. 3b) (Lamerichs *et al.*, 1990). Based on a larger number of NOEs, among which 39 betwen protein and DNA, a restrained molecular dynamics refinement of the headpiece operator complex was carried out (Chuprina *et al.*, 1993). Some statistics of the calculation are shown in Table 1. We note that the protein-DNA complex was put in a box containing over 3 000 water molecules to which salt ions were

Table 1

Restrained Molecular Dynamics Refinement of Lac Headpiece-11 bp operator complex

Start:	optimized vacuum structure			
Size:	989 atoms complex, 28 Na^+, 10 Cl-3346 waters			
Constraints:	intra protein	260		
	intra DNA	241		
	protein-DNA	39		
Forcefield:	GROMOS			
Trajectory	60 ps equilibration			
	85 ps analysis			
Annealing:	(6x)	1000 K →	300 K	
		5 ps		

added, and periodic boundery conditions were applied. Inclusion of solvent was felt necessary for a reliable simulation and also because water molecules may play an important role in the interface between protein and DNA. A fairly long equilibration of 60 ps was necessary primarily to equilibrate the ion distribution around the complex. Then a 85 ps trajectory was used for analysis. In addition an annealing procedure was undertaken in order to assess the precision of the structure determination. This consisted of heating the complex to 1000 K and letting it cool down to 300 K in a 5 ps RMD calculation followed by energy minimization. Six structures were obtained by this procedure using different snapshots from the trajectory as starting points.

The RMD runs yielded a satisfactory structure for the complex with a low total energy (- 175 x 10^3 $kJmol^{-1}$) while the restraint energy did not exceed the average thermal energy. There were on average 42 bounds (out of 980 total), which were violated by 0.5 Å or more, mostly within the DNA (none involving protein-DNA contacts).

Among a set of 15 structures, 9 from the trajectory and 6 from the annealing procedure, the r.m.s.d. for backbone atoms for both DNA and protein (residues 4-48 for the headpiece) with respect to the mean was found to be 0.9 Å. This value can be considered as a measure of the precision of the structure determination. From the analysis of the 85 ps trajectory a picture emerged of the interactions by which *lac* operator is recognized by the headpiece. Apart from electrostatic interactions, which are probably non-specific, a number of hydrogen bonds are observed between protein and DNA, both to the sugar-phosphate backbone and to the bases. In addition an extensive network of apolar interactions is observed involving the side-chains of Tyr 17, Gln 18 and Ser 21, and the methyl groups of Thy 6 and Thy 8. These interactions are schematically shown in Figure 5. Two direct hydrogen bonds between amino acid side-chains and bases are seen, which appear to be essential for specific recognition. These are the ones between Gln 18 and Cyt 7 and between Arg 22 and Gua 5. Important anchoring hydrogen bonds with DNA phosphates include those of the amide NH of Leu 6 and the side-chain amide of Asn 25. These are conserved among the whole family of helix-turn-helix proteins. Histidine 29, present in the loop between helices II and III is involved in a multitude of interactions: electrostatic interactions because of its positive charge, a hydrogen bond with a phosphate group and van der Waals interactions with the methyl group and sugar ring of Thy 3. For a more complete description of these interactions we refer to the paper by Chuprina *et al.* (1993).

An important observation was that some of these hydrogen bonds are not rigid but fluctuate in time. Thus, while anchoring H-bonds involving Leu 6 and Asn 25 are formed close to 100% of the time, the one between Arg 22 and Gua 5, for instance, breaks up during the trajectory and is replaced by an intra-protein H-bond with Glu 26. Apparently, alternative hydrogen bonding schemes are possible with very similar energies. We believe that indeed protein-DNA complexes are rather dynamic in nature and that the current more rigid picture of these complexes needs correction.

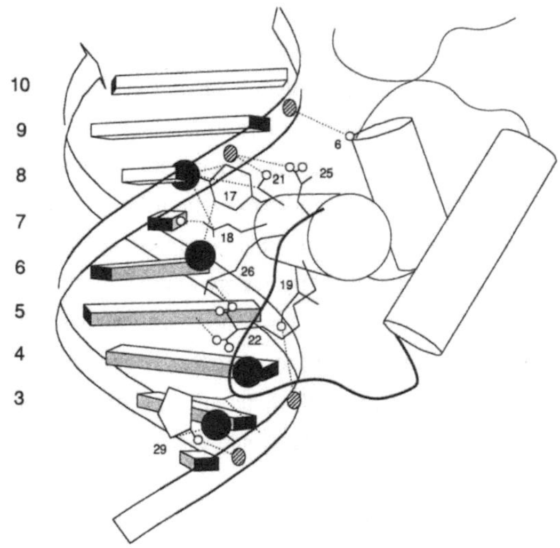

Fig. 5. Schematic view of the interactions between *lac* headpiece and *lac* operator. Shown are the hydrogen bonds between protein and DNA and the apolar interactions in the hydrophobic cluster formed by side-chains of Tyr 17, Ser 21, Gln 18 and the methyl groups of Thy 6 and Thy 8. Thymine methyl groups are shown as black balls, phosphate groups are striped circles and protons in H-bonds as small circles.

COMPARISON WITH GENETIC DATA

Ebright (1985) has demonstrated that wild-type *lac* repressor with Gln 18 substituted by Gly, Ser and Leu loses the ability to distinguish between the base pairs GC, TA and AT at position 7. In contrast, all three substituted variants are able to discriminate at position 5, 6, 8, 9, and 10. Consequently Ebright proposed that Gln 18 contacts base pair 7. Based on the affinity of repressor variants to operators with particular base pair substitutions Müller-Hill and co-workers (Lehming *et al.*, 1988; Sartorius *et al.*, 1989) suggested a direct contact between Arg 22 and base pair GC 5 in addition to the Gln 18-GC 7 contact. Both contacts are found in our structural model.

From the work of Caruthers (1980) a hydrophobic interaction involving the methyl group of Thy 8 is known to be essential for binding of *lac* repressor. We observe a stable interaction between this methyl group and a hydrophobic pocket formed on the protein by Tyr 17 and Ser 21. The carbonyl Cδ of Gln 18 is also close to the Thy 8 methyl. These results nicely correlate with the mutant studies by Sartorius *et al.* (1991), indicating that residues in the fifth position of the recognition helix (Ser 21 in wild type) contribute to the recognition by residues 1 and 2 (Tyr 17 and Gln 18 in wild type). Although no firm evidence was found for a direct specific contact of Ser 21 to any base pair, the authors suggested that Ser 21 is involved in the recognition of bp 7 (our numbering). This is not corroborated by our results.

Sartorius *et al.* (1989) concluded that the specific binding of *lac* repressor mutants with amino acid exchanges at positions 1 and 2 of the recognition helix is mainly directed towards bp 7 and to a lesser extent towards bp 6 of *lac* operator (our numbering). This is not incompatible with our results, that show interactions between Gln 18 and bp 7, and between Tyr 17 and bp 6, 8 and 9. Interactions with bp 8 and 9 have not been observed in the genetic studies by Sartorius *et al.* (1989, 1991), which is probably due to the fact that no repressor mutants were identified with a non-negligible affinity for any *lac* operator variant substituted in bp 8. Apparently Thy 8 is a crucial element in repressor-operator recognition. Furthermore, substitutions at positions 1 and 2 of the recognition helix apparently lead to rearrangements of side chains, giving rise to amino acid - base pair correlations in complexes of mutant operators with mutant repressors that are somewhat different from the correlations in the wild type complex.

In our structure Asn 25 makes a hydrogen bond contact with the Thy 8 phosphate, as has been proposed by Sartorius *et al.* (1991) based on genetic data. Thr 19 and Ser 21 are also important for repression, since their substitution almost invariably leads to an I$^-$ mutant (Kleina and Miller, 1990). Our data suggest that the observed loss of repression upon replacement of these amino acids may be due to loss of the operator contacts formed by these residues. Thus Thr 19 appears to be important not only for the protein tertiary structure, as suggested by

Boelens *et al.* (1987), but also for DNA binding. Ser 21 is not only invoved in a base specific contact to Thy 8, as discussed above, but in a phosphate contact as well.

We can conclude that our structure of the lac headpiece-operator complex agrees very well with the genetic data and indeed provides a basis for a detailed interpretation of these data in structural terms.

REFERENCES

Boelens R, Scheek RM, van Boom JH and Kaptein R (1987) Complex of *lac* Repressor Headpiece wth a 14 Base-pair *lac* Operator Fragment Studied by Two-dimensional Nuclear Magnetic Resonance. J. Mol. Biol. 193: 213-216

Caruthers MH (1980) Deciphering the protein-DNA recognition code. Acc. Chem. Res. 134: 155-160

Chuprina VP, Rullmann JAC, Lamerichs RMJN, van Boom JH, Boelens R and Kaptein R (1993) Structure of the Complex of *lac* Repressor Headpiece and an 11 Base Pair Half-Operator determined by NMR Spectroscopy and Restrained Molecular Dynamics. J. Mol. Biol. 234: 446-462

De Vlieg J, Boelens R, Scheek RM, Kaptein R and van Gunsteren WF (1986) Restrained molecular dynamics procedure for protein tertiary structure determination from NMR data: A *lac* repressor headpiece structure based on information on J-coupling and from presence and absence of NOE's. Israel J. Chem. 27: 181-188

Ebright RH (1985) Use of "loss-of-contact" substitutions to identify residues involved in an amino acid-base pair contact: effect of substitution of Gln18 of *lac* repressor by Gly, Ser and Leu. J. Biomol. Struct. Dyn. 3: 281-297

Geisler N and Weber K (1977) Isolation of the Amino-Terminal Fragment of Lactose Repressor Necessary for DNA Binding. Biochemistry 16: 938-943

Gilbert W and Maxam A (1973) The Nucleotide Sequence of the *lac* Operator. Proc. Natl. Acad. Sci. USA 70: 3581-3584

Härd T, Kellenbach E, Boelens R, Kaptein R, Dahlman K, Carlstedt-Duke J, Freedman LP, Maler BA, Hyde EI, Gustafsson J-Å, Yamamoto KR (1990) ^{1}H NMR Studies of the Glucocorticoid Receptor DNA-Binding Domain: Sequential Assignments and Identification of Secondary Structure Elements.Biochemistry 29: 9015-9023

Kania J and Brown DT (1976) The functional repressor parts of a tetrameric *lac* repressor-β-galactosidase chimaera are organized as dimers. Proc. Natl. Acad. Sci. USA 73: 3529-3533

Kaptein R, Zuiderweg ERP, Scheek RM, Boelens R and van Gunsteren WF (1985) A Protein Structure from Nuclear Magnetic Resonance Data: *lac* Repressor Headpiece. J. Mol. Biol. 182: 179-182

Kleina LG and Miller JH (1990) Genetic Studies of the *lac* Repressor XIII. Extensive Amino Acid Replacements Generated by the Use of Natural and Synthetic Nonsense Suppressors. J. Mol. Biol. 212: 295-318

Klevit RE, Herriott JR, Horvath SJ (1990) Solution Structure of a Zinc Finger Domain of Yeast ADRI. Proteins 7: 215-226

Lamerichs RMJN, Boelens R, van der Marel GA, van Boom JH, Kaptein R, Buck F, Fera B and Rüterjans H (1989) ^{1}H NMR Study of a Complex between the *lac* Repressor Headpiece and a 22 Base Pair Symmetric *lac* Operator. Biochemistry 28: 2985-2991

Lamerichs RMJN, Boelens R, van der Marel GA, van Boom JH, and Kaptein R (1990) Assignment of the ^{1}H-NMR spectrum of a *lac* repressor headpiece-operator complex in H_2O and identification of NOEs. Consequences for protein-DNA interaction. Eur. J. Biochem. 194: 629-637

Lee MS, Gippert GP, Soman KV, Case DA, Wright PE (1989) Three-Dimensional Solution Structure of a Single Zinc Finger DNA-Binding Domain. Science 245: 635-637

Lehming N, Sartorius J, Niemöller M, Genenger G, von Wilcken-Bergmann B and Müller-Hill B (1987) The interaction of the recognition helix of *lac* repressor with *lac* operator. EMBO J 6: 3145-3153

Lehming N, Sartorius J, Oehler S, von Wilcken-Bergmann B, and Müller-Hill B (1988) Recognition Helix of *lac* and λ Repressor are Oriented in Opposite Directions and Recognize Similar DNA Sequences. Proc. Natl. Acad. Sci. USA 85: 7947-7951

Miller JH (1984) Genetic Studies of the *lac* Repressor. XII. Amino Acid Replacements in the DNA Binding Domain of the *Escherichia coli lac* Repressor. J. Mol. Biol. 180: 205-212

Ogata RT and Gilbert W (1979) DNA-binding site of *lac* repressor probed by dimethylsulfate methylation of *lac* operator. J. Mol. Biol 132: 709-728

Pabo CO and Sauer RT (1992) Transcription factors: Structural families and principles of DNA recognition. Annu. Rev. Biochem. 61: 1053-1095

Sadler JR, Sasmor H and Betz JL (1993) A perfectly symmetric *lac* operator binds the *lac* repressor very tightly. Proc. Natl. Acad. Sci. USA 80: 6785-6789

Sartorius J, Lehming N, Kisters B, von Wilcken-Bergmann B and Müller-Hill B (1989). *Lac* repressor mutants with double or triple exchanges in the recognition helix bind specifically to *lac* operator variants with multiple exchanges. EMBO J. 8: 1265-1270

Sartorius J, Lehming N, Kisters-Woike B, von Wilcken-Bergmann B and Müller-Hill B (1991) The roles of residues 5 and 9 of the recognition helix of *lac* repressor in lac operator binding. J. Mol. Biol. 218: 313-321

Schwabe JWR, Neuhaus D and Rhodes D (1990) Solution Structure of the DNA-binding Domain of the Oestrogen Receptor. Nature 238: 458-461

Simons A, Tils D, von Wilcken-Bergmann B and Müller-Hill B (1984) Possible ideal *lac* operator: *Escherichia coli lac* operator-like sequences from eukaryotic genomes lack the central G.C pair. Proc. Natl. Acad. Sci. USA 81: 1624-1628

Steitz TA (1990) Structural studies of protein-nucleic acid interaction: the sources of sequence-specific binding. Quart. Rev. Biophys. 23: 205-280

Summers MF, South TL, Kim B, Hare DR (1990) High-Resolution Structure of an HIV Zinc Finger-like Domain Via a New NMR-Based Distance Geometry Approach. Biochemistry 29: 329-340

Travers A (1993) DNA-Protein Interactions. Chapman & Hall, London

DETERMINATION OF HIGH-RESOLUTION, SEQUENCE-DEPENDENT DNA DUPLEX STRUCTURES IN SOLUTION[1]

Thomas L. James[2], Karl D. Bishop, Forrest Blocker, Carlos Gonzalez, Anil Kumar, He Liu, Anwer Mujeeb, Uli Schmitz, Nikolai B. Ulyanov, Yate-Ching Yuan and Klaus Weisz

Department of Pharmaceutical Chemistry, University of California, San Francisco, CA 94143-0446, U.S.A.

INTRODUCTION

There can be rather subtle structural variations in the DNA double helix which are sequence-dependent; but these subtle structural differences may direct protein, mutagen or drug recognition. These subtle variations demand detailed knowledge of the structure. This has not been easy to achieve, but the ability to determine an accurate, high-resolution structure of nearly any DNA double helix of length less than 15 base pairs (bp) is now possible if sufficient care and effort are expended. The structure of any molecule can be determined with a sufficient number of structural restraints, e.g., internuclear distances and bond torsion angles, in conjunction with holonomic constraints of bond lengths, bond angles, and steric limitations. NMR, in conjunction with appropriate computational algorithms, has become the method of choice for determination of the high-resolution solution structure of proteins, nucleic acids and complexes. Multidimensional NMR has the capability of yielding interproton distances and bond torsion angles as experimental structural restraints (James and Basus, 1991; Oppenheimer and James, 1989; Wagner et al, 1992). These structural restraints *per se* do not constitute a structure. However, use of algorithms, such as distance geometry (DG) and restrained molecular dynamics (rMD), which search conformational space to define structures consistent with the experimental restraints will provide a "structure" or envelope of closely related structures. Structure determination via NMR is distinctly different from the

---

[1] This work was supported by the National Institutes of Health via grants GM39247, GM41639, and RR01695.

[2] Author to whom correspondence should be addressed.

NATO ASI Series, Vol. H 87
NMR of Biological Macromolecules
Edited by C. I. Stassinopoulou

situation with x-ray crystallographic determination of structure, where Fourier transformation of the diffraction pattern basically yields an atomic array (assuming the phase problem has been solved). With x-ray crystallography, refinement largely entails minimization of the residual index (R-factor), yielding a single most favorable structure. In contrast, with NMR, an effort is made to generate an array of structures which are consistent with the available NMR data in an effort to define the limits of the structure. Indeed, it is desirable to make the array of structures large enough to map out the conformational space which will accommodate the available experimental data. The ability to determine solution structures of biomolecules by NMR spectroscopy is limited by the quality and quantity of distance and torsion angle restraints that can be extracted from the NMR data. A higher resolution structure can be achieved with more structural restraints and with more accurate structural restraints. Techniques used to establish these restraints are constantly evolving and improving.

EXPERIMENTAL RESTRAINTS AND BOUNDS

Distance Restraints

We have established that complete relaxation matrix analysis of proton homonuclear 2D NOE spectra enables numerous accurate interproton distances to be calculated (Borgias and James, 1988; Borgias and James, 1990; James, 1991; Keepers and James, 1984; Kumar et al, 1992; Liu et al, 1992). The most efficient techniques for calculating accurate distances entail an iterative approach (Boelens et al, 1989; Borgias and James, 1988; Borgias and James, 1989; Borgias and James, 1990). In particular, the MARDIGRAS algorithm permits the determination of a large number of accurate distance restraints and aids in individually setting bounds for those distances. For the present studies, we have benefited from recent improvements in MARDIGRAS accounting for the presence of molecular motions(Liu et al, 1992) and incorporation of exchange with bulk water for exchangeable protons (Liu et al, 1993).

The effect of cross-relaxation between two neighboring protons during the mixing time period τ_m of the 2D NOE experiment is to transfer magnetization between them. The efficiency of this transfer depends on the distance between the two protons and upon the rate of molecular motion. With transfer of magnetization, the cross-peak intensities in the spectrum will be modified. Consequently, the cross-peak intensities have structural information, i.e., distances, embedded in them. In interesting molecules, the two protons giving rise to a particular cross-peak are not the only protons in the molecule. Rather, they belong to an array of all protons in the molecular structure which, in principle, experience dipole-dipole interactions with all the others. So cross-relaxation between the two protons is part of a coupled relaxation network. To be rigorous, the whole relaxation network should be considered. However, distances are typically estimated from 2D NOE cross-peak intensities by applying the two-spin or Isolated Spin Pair Approximation (ISPA):

$$r_{ij} = r_{ref}\left[\frac{a_{ref}}{a_{ij}}\right]^{1/6} \tag{1}$$

where r_{ij} is the interproton distance to be estimated and a_{ij} is the corresponding 2D NOE cross-peak intensity, while r_{ref} and a_{ref} are, respectively, a known interproton distance and its cross-peak intensity. This equation derives from truncation after the linear term of the Taylor series expansion of the complete rate expression accounting for all proton dipole-dipole interactions (Macura and Ernst, 1980):

$$\mathbf{a}(\tau_m) = e^{-\mathbf{R}\tau_m} \approx \mathbf{1} - \mathbf{R}\tau_m + \tfrac{1}{2}\mathbf{R}^2\tau_m^2 - \cdots + \frac{(-1)^n}{n!}\mathbf{R}^n\tau_m^n + \cdots \tag{2}$$

where **a** is the matrix of 2D NOE intensities, **R** is the matrix describing the complete dipole-dipole relaxation network, and τ_m is the mixing time for the experiment. The truncation is valid in the limit of short mixing time τ_m. The elements of the relaxation matrix **R** depend on spin state transition probabilities, written below as zero-, single- and double-quantum transition probabilities W_m^{ij}, as follows:

$$R_{ii} = 2(n_i - 1)(W_1^{ii} + W_2^{ii}) + \sum_{j \neq i} n_j(W_0^{ij} + 2W_1^{ij} + W_2^{ij}) + R_{1i}; \tag{3a}$$

$$R_{ij} = n_i(W_2^{ij} - W_0^{ij}). \tag{3b}$$

Here n_i is the number of equivalent spins in a group such as a methyl rotor. For a rigid molecule undergoing isotropic random reorientation with correlation time τ_c, the following equations for transition probabilities hold:

$$W_0^{ij} = \frac{q\tau_c}{r_{ij}^6}; \quad W_1^{ij} = 1.5\,\frac{q\tau_c}{r_{ij}^6}\,\frac{1}{1+(\omega\tau_c)^2}; \quad W_2^{ij} = 6\,\frac{q\tau_c}{r_{ij}^6}\,\frac{1}{1+4(\omega\tau_c)^2} \tag{4}$$

where ω is the Larmor frequency of the protons, $q = 0.1\gamma^4\hbar^2$ and γ is the proton gyromagnetic ratio. The term R_{1i} represents external sources of relaxation such as paramagnetic impurities. Truncation of the series expansion of equation 2 to the first term linear in **R** is valid if τ_m is sufficiently short, which is equivalent to assuming that each cross-peak intensity depends only on the cross-relaxation rate between the two pertaining protons, i.e., ISPA. The general method of assessing whether or not τ_m is sufficiently short is to obtain build-up curves of

2D NOE cross-peak intensities, i.e., the intensities are measured as a function of mixing time. If the build-up is approximately linear, it is assumed that the nonlinear terms in equation 2 can be ignored. However, this is not a good assumption (Borgias and James, 1988; Borgias and James, 1989; Post et al, 1990). For example, for mixing times generally accepted as being sufficiently short (i.e., 50 to 100 ms) and not including internal motions, ISPA can result in systematic errors of 45 - 80% in distances over 3.5 Å, a range which is quite important in defining molecular structure.

An effective approach to calculating 2D NOE intensities from a model structure, taking into account all dipole interactions, is to utilize linear algebra and the simplifications which arise from working with the characteristic eigenvalues and eigenvectors of a matrix (Keepers and James, 1984). The rate matrix **R** can be represented by a product of matrices:

$$\mathbf{R} = \boldsymbol{\chi}\boldsymbol{\lambda}\boldsymbol{\chi}^{\mathrm{T}} \tag{5}$$

where $\boldsymbol{\chi}$ is the unitary matrix of orthonormal eigenvectors ($\boldsymbol{\chi}^{-1} = \boldsymbol{\chi}^{\mathrm{T}}$), and $\boldsymbol{\lambda}$ is the diagonal matrix of eigenvalues. Since $\boldsymbol{\lambda}$ is diagonal,

$$\mathbf{a} = \boldsymbol{\chi} e^{-\boldsymbol{\lambda}\tau_m} \boldsymbol{\chi}^{-1} \tag{6}$$

We have developed a program, CORMA (an acronym for COmplete Relaxation Matrix Analysis), for performing this calculation (Borgias and James, 1988; Keepers and James, 1984). Calculation of 2D NOE peak intensities (intensity matrix **a**) from a model structure is often called "back-calculation".

Rather than simply testing model structures, we will generally be interested in determining structure, so we must have structural restraints. There are basically two ways of accounting for multi-spin effects using the relaxation matrix approach in generating solution structures. One is the direct determination of distances without making the isolated spin pair approximation – essentially using the reverse of CORMA. The other is to automatically refine the structure based on the 2D NOE intensities. The direct method, i.e., reverse of CORMA, is clearly the ideal method when it can be used. The matrix of experimental intensities at time τ_m, however, will generally be incomplete for biopolymers due to experimental limitations such as resolution (i.e., peak overlap) and (realistic) signal-to-noise. As a consequence, the corresponding relaxation matrix does not represent the spatial arrangement of protons leading to the intensities; a direct back-calculation of **R** will therefore fail to yield accurate, unbiased distances (Borgias and James, 1988; Post et al, 1990). It is possible to overcome this lack of NOE intensities by supplementing experimental intensities with intensities calculated for a model structure to form a hybrid intensity matrix (Boelens et al, 1989; Borgias and James, 1989; Borgias and James, 1990). The algorithm we have developed for these purposes is called MARDIGRAS. Work with MARDIGRAS has shown that the accuracy of the distances obtained increases with (a) fraction of experimental cross-peaks observed, (b) improved signal-to-noise, and (c) accuracy of initial model structure; but, more importantly, fairly accurate distances can still be obtained with a poor initial model (Borgias and James, 1989; Borgias and James, 1990; Thomas et al, 1991). Recent improvements in MARDIGRAS account for the presence of the internal molecular motions of methyl rotors and methylene or aromatic ring flips (Liu et al, 1992); unresolved aromatic ring or methylene proton resonances are similarly treated. If information about order parameters has either been estimated from molecular dynamics calculations or determined from relaxation time measurements, that additional knowledge can also be incorporated (Kumar et al, 1992). MARDIGRAS also now incorporates exchange effects, which permits us to account for exchange with bulk water for exchangeable protons (Liu et al, 1993). In summary, MARDIGRAS permits the determination of a large number of accurate distance restraints and aids in individually setting bounds for those distances (*vide infra*).

To properly assess the extent of conformational space consistent with our experimental data, we need some estimate of the accuracy of our structural restraints. Explicitly, they are needed for setting bounds in distance geometry or flat-well size in restrained molecular dynamics calculations. Most structural studies have utilized ISPA to estimate interproton distances, but estimates of the error, reflected by the upper and lower bounds assigned, vary widely throughout the literature. Tighter distance bounds (smaller error bars) lead to a higher resolution structure. But distance bounds made tighter than warranted by experimental accuracy mislead to a highly defined (small atomic RMSD) but incorrect structure (Thomas et al, 1991). Thus we need to make the bounds as tight as possible but not so tight that the real distance can lie outside the bounds.

More accurate distances available via iterative complete relaxation matrix methods minimize the possibility of an estimated distance lying outside the bounds. Also, MARDIGRAS can be used to aid our choice of bounds individually for each proton pair in a logical fashion: (a) The "error bar" for the distance between nuclei i and j can be estimated from the fit of the ij element of the converged MARDIGRAS matrix with the experimental cross-peak intensities, using as a minimum the error dictated by the experimental noise level (Borgias and James, 1990). (b) For MARDIGRAS, distances are determined from a single 2D NOE spectrum, rather than acquiring a set of spectra at several mixing times and measuring build-up curves. However, depending on the internuclear distance and on the proton environment, different mixing times will be optimal for different proton pairs. For this reason and for comparing results, it is still desirable to obtain spectra at a few different mixing times. This enables distances determined independently from spectra acquired at different mixing times to be compared. Of course, the spread in the distance measurements for any given proton pair can aid in the choice of bounds. (c) We have found MARDIGRAS to be surprisingly robust, generating a largely correct set of distances independent of the starting model used in the iterative process (Borgias and James, 1989; Borgias and James, 1990). However, there are some distance variations with starting structure; these also serve as input for our choice of bounds.

(d) Quite importantly, MARDIGRAS calculates bounds for distances to individual protons undergoing motional or overlap averaging, i.e., methyl, methylene and aromatic protons (Liu et al, 1992). As an iterative technique, MARDIGRAS converges to an internally consistent relaxation rate matrix (and equivalently, a 2D NOE intensity matrix) in good agreement with the experimental data. Each element of this converged relaxation rate matrix that corresponds to an experimental cross-peak is then used to calculate an interproton distance, generally assuming a single isotropic correlation time. However, distances entailing protons averaged by motion or spectral overlap, e.g., methyl, methylene and aromatic protons, may be in serious error if the averaging is ignored. The cross-relaxation rates in these cases will depend on orientation as well as distance. MARDIGRAS does a second level of iteration, varying the orientation and distances of all dipole-dipole interactions to find the best fit (Liu et al, 1992). Perhaps more importantly, however, MARDIGRAS lists the distances corresponding to the worst-case geometries, enabling upper and lower bounds to be set for distances entailing protons averaged by overlap or internal motions.

(e) Chemical exchange can alter cross-peak intensities (Choe et al, 1991; Jeener et al, 1979; Landy and Nagaswara Rao, 1989), leading to inaccurate distances. In particular, exchangeable protons in biopolymers (e.g., imino, amino, amide) depend on the exchange rate with bulk water. Exchange effects have now been included in the MARDIGRAS algorithm, with exchange involving bulk water emphasized (Liu et al, 1993). To be significant, exchange rates with water must be >1 sec^{-1}. Ignoring these effects will lead to overestimated distances. Even if the exchange rate is not known, it may be possible to place an upper limit on the rate of exchange; incorporation of this exchange rate upper limit into the calculation can lead to an estimation of the lower distance bound (Liu et al, 1993; Mujeeb et al, 1993; Weisz et al, 1993). More recently, we have begun measuring imino proton exchange rates, so the effects of that exchange could be explicitly known and more accurate distances involving imino protons calculated (Bishop et al, 1994).

All of the above considerations enable a selection of bounds separately for each proton pair. With a list of such bounds, we can employ any of the various methods of structure refinement.

Torsion Angle Restraints

Bond torsion angles can be determined using various correlation spectroscopic techniques, e.g., E.COSY, PCOSY and double-quantum-filtered COSY (2QF-COSY) (Griesinger et al, 1985; Marion and Bax, 1988; Piantini et al, 1982). Broad lines have prevented direct analysis of all coupling constants in DNA oligomers greater than ~ 8 bp in length, but fitting of simulated to experimental cross-peaks enables extraction of vicinal coupling constants and, subsequently, torsion angle restraints (Celda et al, 1989; Schmitz et al, 1990; Widmer and Wüthrich, 1987); the torsion angles for the deoxyribose rings are determined using a parameterization of the relationship between torsion angles and coupling constants established by Rinkel and Altona (Rinkel and Altona, 1987; Weisz et al, 1992). The major difficulty with this approach is in establishing the correct linewidth to be employed (Schmitz et al, 1990). However, the choice of linewidth can usually be constrained such that limits (bounds) on the torsion angles describing sugar pucker can be made.

In our studies of about ten DNA duplexes, we have found, as we examine all the scalar coupling-based cross-peaks, that a single conformer is inadequate to account for all the measured coupling constants, but reasonable fits were obtained by using a two-state model representing a rapid interconversion between S- and N-type sugar puckers (Altona and Sundaralingam, 1972). Unrestrained molecular dynamics calculations in our lab and elsewhere suggest the possibility that more than two sugar pucker states may exist, and recent molecular dynamics calculations using exponentially-weighted, time-averaged restraints also indicate that sugar puckers other than standard N- and S-states are significantly populated (Schmitz et al, 1992; Schmitz et al, 1993). However, Occam's Razor dictates that we utilize the simplest model to account for the data. So, studies in our lab in the past few years have utilized the two-state model. In general, for nonterminal nucleotides, we find that one conformer (S-state) dominates, being populated 75-95% of the time.

STRUCTURE GENERATION USING RESTRAINT BOUNDS

Methods entailing systematic searches of accessible conformational space have been advocated, but they are computationally very expensive at present for molecules above 1 kilodalton. Consequently, an intelligent search of conformational space is recommended at present. Restrained molecular dynamics (also called simulated annealing) attempt to reconcile the experimental structural restraints with energetic considerations (van Gunsteren, 1993). Many popular MD programs can accommodate NMR restraints. These programs are quite similar in that the potential energy is calculated for a set of atomic coordinates using a force field:

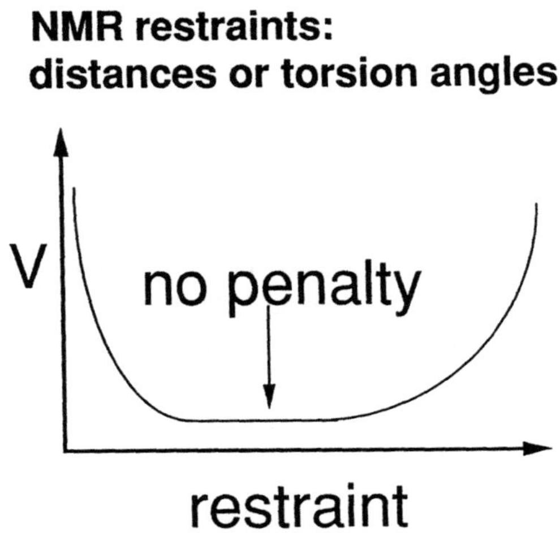

Fig. 1. Penalty function for NMR restraints (either distance or torsion angle) used in rMD calculations. The size of the flat portion of the well is determined by the values determined for the restraint bounds.

$$V_{total} = V_{bondlengths} + V_{bondangles} + V_{dihedrals} + V_{electrostatics} + V_{NOE} + V_{Jcoupling} \quad (7)$$

The first four terms on the right monitor the classical potential energy of the molecule. The final two pseudo-energy terms serve as penalty terms for the NOE-derived distance restraints and scalar coupling-derived torsion angle restraints. We utilize a flat-well potential for our penalty terms, e.g.,

$$V_{NOE} = \sum_{\text{all NOEs}} \begin{cases} k_2(r-r_l)^2 & \text{when } r < r_l \\ 0 & \text{when } r_l \le r \le r_u \\ k_3(r-r_u)^2 & \text{when } r_u < r \\ 4k_2(r-r_u) & \text{when } r > r_u + 2.0 \end{cases} \quad (8)$$

where the lower (r_l) and the upper (r_u) bounds can be derived from MARDIGRAS (*vide supra*) to define the size of the flat-well width individually for each proton pair. The form of equation 8 is depicted in Figure 1. We use the AMBER 4.0 force field (Pearlman et al, 1991). Many studies have not incorporated torsion angle restraints, but all have utilized distance restraints.

As an rMD calculation is able to pass over energy barriers of amplitude ~kT, a global minimum in the energy expressed by equation 7 is sought with a balance between maintenance of the "classical" energy terms and the experimental restraints' penalty functions. Figure 2 illustrates the approach. Starting structures are selected from different parts of

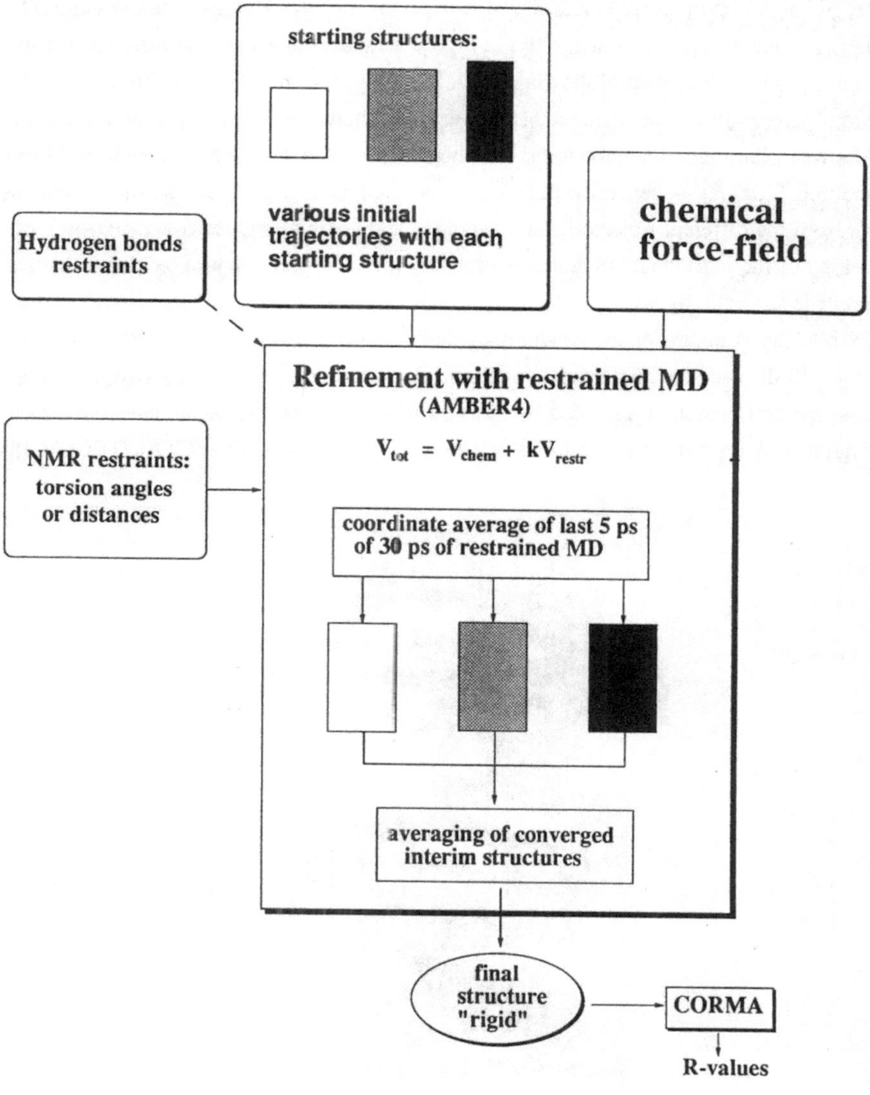

Fig. 2. Structure generation via restrained molecular dynamics. See text for explanation.

conformational space. In figure 2, this is represented by the differently shaded and shaped boxes at the top. Furthermore, different random initial trajectories should be utilized for each starting structure. Newton's equations of motion are solved with the forces generated by taking the derivative of the potential, equation 7, with respect to the coordinates (van Gunsteren, 1993). If all goes well, a global minimum is achieved. In fact, the same global minimum should be achieved with all the different starting structures and different initial trajectories. This is suggested in Figure 2 by the convergence of the differently shaded boxes to the same shape following rMD. In reality, the structures resulting from different simulations will not be *identical*. However, they should be similar with small atomic RMSD between the individual structures little bigger than atomic displacements from librational motions (~0.5 Å) — generally <1.0 Å is considered adequate. Typically, the structures resulting from the different trajectories are averaged and a "final" structure reported. In fact, the envelope of the individual structures is probably a better representation of the structure. An example is shown in figure 3.

While the structure of the trisdecamer duplex shown in figure 3 is fairly well restrained by ~10 distance restraints and 5 torsion angle restraints per nucleotide (Mujeeb et al, 1993), the spectral resolution afforded by other structures studied in our lab have enabled a greater number of restraints to be established. For [d(GATGCAAATG)·d(CTACGTTTAC)],

Fig. 3. Superimposed structures of [d(AGCTTGCCTTGAG)·d(CTCAAGGCAAGCT)], a completely conserved sequence from the HIV-1 genome, resulting from ten rMD simulations, each of 30-ps duration, five using A-DNA and five using B-DNA as starting structure, but each with a different initial trajectory. Convergence to an atomic RMSD of ~0.9 Å is achieved, compared with RMSD ~6.5 Å between the two different starting models, A-DNA and B-DNA. A total of 267 distance restraints (~10 per nucleotide) and 130 torsion angle restraints (5 per nucleotide) were utilized.

which contains the octamer motif found in the promoter and enhancer region of immunoglobulin genes which is recognized by the POU family of protein domains, ~20 distance restraints and 5 torsion angle restraints per nucleotide were obtained (Weisz et al, 1992). Several rMD simulations resulted in a well-defined structure. With the largess of restraints available to us, it was possible to examine the effect on the structure derived of randomly eliminating some of the distance restraints. As seen in figure 4, randomly reducing the distance restraints to 15 per residue has little effect on the final structure derived. However, there was a significant diminution in our ability to define the structure as we diminished the distance restraints to <12 per residue; of course, it must be kept in mind that even with 50% of the restraints eliminated, the experimental restraints are still largely determining the structure — keep in mind the difference of more than 5 Å RMSD between energy-minimized A- and B-DNA structures.

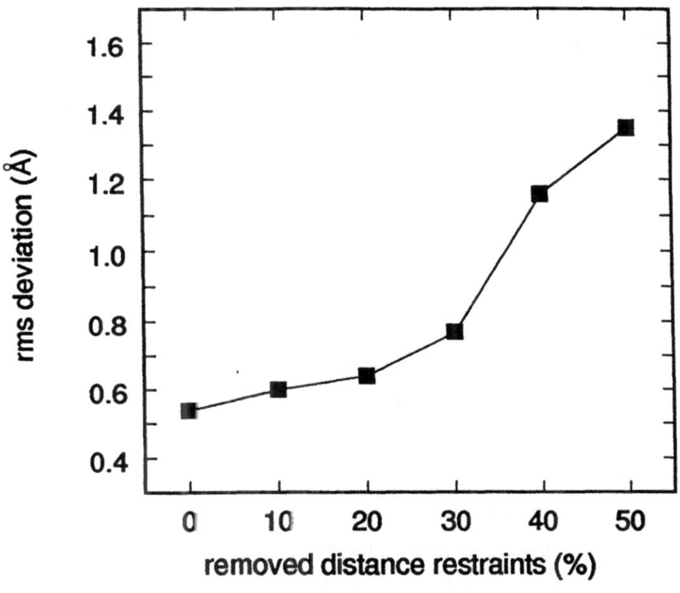

Fig. 4. Influence of the fraction of distance restraints utilized in rMD refinement on the structure of [d(GATGCAAATG)·d(CTACGTTTAC)]. The atomic RMS difference between structures refined against full and partial data sets is plotted as a function of the fraction of experimentally determined distance restraints randomly deleted from the restraint list. The starting structure was energy-minimized B-DNA, and 0% of the distances removed reflects the difference between two independently refined structures with different initial velocities.

CRITERIA FOR ASSESSING STRUCTURE QUALITY

Having determined a structure from NMR restraints, it is pertinent to evaluate the quality of the structure. While this is rarely done with protein structures which have been reported, it should be done. Among criteria which could be used, we can consider how well the final structure or structures determined compare with the experimental NMR data. In x-ray crystallography, an evaluation of the fit of a derived structure with the original electron densities via a residual index or R factor is required. Likewise, we can quantitatively compare calculated 2D NOE spectral intensities for any proposed molecular structure with the experimental intensities. We could use a residual index analogous to the crystallographic R factors:

$$R_1 = \frac{\sum_i | a_o^i - a_c^i |}{\sum_i (a_o^i)} \tag{9}$$

R_1 gives equal weighting to all deviations between intensities a_o and a_c, so stronger peaks (and shorter distances, i.e., <2.5 Å) will generally dominate. Consequently, we prefer a sixth-root residual index which scales roughly with atomic distances (James, 1991; Thomas et al, 1991):

$$R_1^x = \frac{\sum_i | a_o^{1/6}(i) - a_c^{1/6}(i) |}{\sum_i a_o^{1/6}(i)} \tag{10}$$

R_1^x permits longer interactions, e.g., ~4 Å, to have a role in the calculated R factor. Longer distance restraints are most important in structure determination, so it would be good to have their input into assessing structure quality.

Consistency of the final structure with scalar-coupling-based multidimensional NMR spectra may be assessed as well. For example, we could calculate all the spectral cross-peaks to be expected for any given structural model using SPHINX-LINSHA (Widmer and Wüthrich, 1986), and compare with the experimental cross-peaks. This, of course, assumes that a well-parameterized Karplus relationship has been established between the dihedral angles and three-bond coupling constants (Rinkel and Altona, 1987). Alternatively, the root mean square deviation between experimental and theoretical coupling constants can be calculated according to $J_{rms} = (1/N)\sqrt{\sum (J_{exp} - J_{theor})^2}$, the summation being carried out over N, which constitutes all or any subset of the coupling constants (Ulyanov et al, 1993).

Of course, any structure we generate should be consistent with any other experimental data available on the oligomer in solution. In principle, we can compare our structure with any available crystal structure. However, in the case of a DNA duplex, it is highly unlikely that a good diffracting crystal in anything other than A-form, i.e., not comparable to the solution structure, can be obtained. Furthermore, one must be cautious in interpreting any B-form crystal structure because of the influence of crystal packing forces.

In summary, it is now possible to determine the time-averaged structure of a DNA helix at high-resolution and to assess the quality of the structure. To do so, however, requires a large number of accurate structural restraints – especially distances – and a careful refinement of the structure using those restraints. At this stage, it is time to examine the dynamic aspects of the structure; such an investigation has been initiated (Schmitz et al, 1992; Schmitz et al, 1993).

References

Altona, C. and Sundaralingam, M. (1972) J. Am. Chem. Soc. *94*: 8205-8212.

Bishop, K. D., Blocker, F., Egan, W. and James, T. L. (1994) Biochemistry *33*: in press.

Boelens, R., Koning, T. M. G., van der Marel, G. A., van Boom, J. H. and Kaptein, R. (1989) J. Magn. Reson. *82*: 290-308.

Borgias, B. A. and James, T. L. (1988) J. Magn. Reson. *79*: 493-512.

Borgias, B. A. and James, T. L. (1989) In Methods in Enzymology, Nuclear Magnetic Resonance, Part A: Spectral Techniques and Dynamics (Oppenheimer, N. J. and James, T. L., eds), vol. 176, pp. 169-183, Academic Press, New York.

Borgias, B. A. and James, T. L. (1990) J. Magn. Reson. *87*: 475-487.

Celda, B., Widmer, H., Leupin, W., Chazin, W. J., Denny, W. A. and Wüthrich, K. (1989) Biochemistry *28*: 1462-1470.

Choe, B., Cook, G. W. and Rama Krishna, N. (1991) J. Magn. Reson. *94*: 387-393.

Griesinger, C., Sørensen, O. W. and Ernst, R. R. (1985) J. Am. Chem. Soc. *107*: 6394-6396.

James, T. L. (1991) Curr. Opin. Struct. Biol. *1*: 1042-1053.

James, T. L. and Basus, V. J. (1991) Ann. Rev. Phys. Chem. *42*: 501-542.

Jeener, J., Meier, B. H., Bachmann, P. and Ernst, R. R. (1979) J Chem Phys *71*: 4546-4553.

Keepers, J. W. and James, T. L. (1984) J. Magn. Reson. *57*: 404-426.

Kumar, A., James, T. L. and Levy, G. C. (1992) Israel J. Chem. *32*: 257-261.

Landy, S. B. and Nagaswara Rao, B. D. (1989) J. Magn. Reson. *81*: 371-377.

Liu, H., Thomas, P. D. and James, T. L. (1992) J. Magn. Reson. *98*: 163-175.

Liu, H., Kumar, A., Weisz, K., Schmitz, U., Bishop, K. D. and James, T. L. (1993) J. Am. Chem. Soc. *115*: 1590-1591.

Macura, S. and Ernst, R. R. (1980) Mol. Phys. *41*: 95-117.

Marion, D. and Bax, A. (1988) J. Magn. Reson. *80*: 528-533.

Mujeeb, A., Kerwin, S. M., Egan, W., Kenyon, G. L. and James, T. L. (1993) Biochemistry *32*: in press.

Oppenheimer, N. J. and James, T. L. (1989) Methods in Enzymology, Nuclear Magnetic Resonance, Part A: Spectral Techniques and Dynamics, vol. 176, Academic Press, New York.

Pearlman, D. A., Case, D. A., Caldwell, J., Seibel, G. L., Singh, U. C., Weiner, P. K. and Kollman, P. A. (1991) AMBER 4.0 (UCSF), University of California, San Francisco.

Piantini, U., Sørensen, O. W. and Ernst, R. R. (1982) J. Am. Chem. Soc. *104*: 6800-6801.

Post, C. B., Meadows, R. P. and Gorenstein, D. G. (1990) J. Am. Chem. Soc. *112*: 6796-6803.

Rinkel, L. J. and Altona, C. (1987) J. Biomol. Struct. Dyn. *4*: 621-649.

Schmitz, U., Zon, G. and James, T. L. (1990) Biochemistry *29*: 2357-2368.

Schmitz, U., Kumar, A. and James, T. L. (1992) J. Am. Chem. Soc. *114*: 10654-10656.

Schmitz, U., Ulyanov, N. B., Kumar, A. and James, T. L. (1993) J. Mol. Biol. *233*: in press.

Thomas, P. D., Basus, V. J. and James, T. L. (1991) Proc. Nat. Acad. Sci. USA *88*: 1237-1241.

Ulyanov, N. B., Schmitz, U. and James, T. L. (1993) J. Biomolec. NMR *3*: 547-568.

van Gunsteren, W. F. (1993) Curr. Opin. Struc. Biol. *3*: 277-281.

Wagner, G., Hyberts, S. G. and Havel, T. F. (1992) Ann. Rev. Biophys. Biomol. Struc. *21*: 167-198.

Weisz, K., Shafer, R. H., Egan, W. and James, T. L. (1992) Biochemistry *31*: 7477-7487.

Weisz, K., Shafer, R. H., Egan, W. and James, T. L. (1993) Biochemistry ·: submitted.

Widmer, H. and Wüthrich, K. (1986) J. Magn. Reson. *70*: 270-279.

Widmer, H. and Wüthrich, K. (1987) J. Magn. Reson. *74*: 316-336.

LIST OF PARTICIPANTS

Director

Stassinopoulou Ioannidou Chariklia
NCSR "Demokritos", Institute of Biology,
153 10 Ag.Paraskevi, Athens, GREECE

ORGANIZING COMMITTEE

Dais Photis , Univ. of Crete, Dept. of Chemistry, Heraklion, GREECE.

Feeney James , National Inst. for Med. Research, The Ridgeway Mill Hill London, NW7 1AA, UK.

Gerothanassis Ioannis , Univ. of Ioannina, Dept. of Chemistry, Ioannina, GREECE.

Wagner Gerhard , Harvard Medical School, Dept. of Biol.Chemistry and Mol. Pharmacology, 240 Longwood Ave, Boston, MA 02115, USA.

Wuthrich Kurt, Inst. fur Molekularbiologie und Biophysik, ETH Honggerberg, CH-8093 Zurich, SWITZERLAND.

LECTURERS

Bertini Ivano , Lab. di Chimicha Inorganica e Bioinorganica, Via G.Capponi 7 50121 Firenze, ITALY.

Billeter M. , Inst. fur Molekularbiologie und Biophysik, ETH Honggerberg, CH-8093 Zurich, SWITZERLAND.

Dais Photis , Univ. of Crete, Dept. of Chemistry, Heraklion, GREECE.

Feeney J. , National Inst. for Med. Research, The Ridgeway Mill Hill London, NW7 1AA, UK.

Gerothanassis J. , Univ. of Ioannina, Dept. of Chemistry, Ioannina, GREECE.

Hilbers C.W. , Lab. of Biophys. Chemistry, Univ. of Nijmegen, Toernooiveld, 6525 Nijmegen, NETHERLANDS.

Hounsell Elizabeth , Glycoconjugates Sect., Clin. Res. Centre, Harrow, HA1 3UJ, UK.

James Th.L. , Dept. of Pharmaceutical Chem., Univ. of California, San Francisco, CA 94/43-0446, USA.

Kaptein Robert , Bijvoet Center, Univ. of Utrecht, Padulalaan 8, 3584 CH Utrecht, NETHERLANDS.

Kessler Horst , Organisch-Chemisches Institut, Technische Univ. Muenchen, D-8046 Garching, GERMANY.

Laue E.D. , Dept. of Biochemistry, Tennis Court Road, Cambridge CB2 1QW, UK.

Neidig Klaus-Peter , Bruker Analytische Messtechnik GmbH, Silberstreifen, D-7512 Rheinstetten 4, GERMANY.

Patel Dinshaw, Memorial Sloan-Kettering Cancer Center, 1275 York Avenue, New York, NY 1002, USA.
Perico Angelo , CEMAG, Corso Europa 30, 16132 Genova, ITALY.
Reid Brian , Dept. of Biochemistry, University of Washington, Seattle, Washington 98195, USA.
Serianni A.S. , Dept. of Chemistry and Biochemistry, Univ. of Notre Dame, Notre Dame, Indiana 46556, USA.
Smallcombe Steve , VARIAN Associates, 3120 Hansen way, Bldg 4, Palo Alto, CA 94304,-1030, USA.
Stassinopoulou Chariklia , NCSR "Demokritos", Institute of Biology, 153 10 Ag.Paraskevi, Athens, GREECE.
Wagner Gerhard , Harvard Medical School, Dept. of Biol.Chemistry and Mol. Pharmacology, 240 Longwood Ave, Boston, MA 02115, USA.

PARTICIPANTS

Aabloo Alvo , Univ. of Tartu, Dept. of Physics, Tahe 4 Street, Tartu EE2400, ESTONIA.
Albert S. Jeffrey , Univ. of Pittsburgh, Dept. of Chemistry, Pittsburgh, MA 15260, USA.
Andre Joao , Universidade do Minho, Dept. of Chemistry, 4700 Braga PORTUGAL.
Arnosti Carol , Massachusetts Inst. of Technology, Woods Hole Oceanographic Inst., Dept. of Chemistry, Eye 115, Woods Hole MA 02543, USA.
Arsan T.Bengi , Hacettepe Univ., Dept. of Food Engineering, Beytepe, 06532 Ankara, TURKEY.
Bagci Eyup , Firat University, Dept. of Biology, 23100 Elazig, TURKEY.
Barlos Kleomenis, Univ. of Patras, Dept. of Chemistry, 26110 Patras, GREECE.
Baydanoff Stephan Ivanoff , Univ. School of Medicine, Dept. of Biology-Immunology, 1. St.Kliment Ohridsky str., 5800 Pleven, BULGARIA.
Benaki Dimitra , University of Athens, Dept. of Biology, Panepistimiopolis, 157 01 Athens, GREECE.
Birlirakis Nikolaos , Lab. de RNM Haute Resolution, et de Chimie Bio-organique, Service de Chimie Moleculaire, Centre D'Etudes Nucleaires de Saclay, F-91191 Gif sur Yvette Cedex, FRANCE.
Botsi Antigone , NCSR "Demokritos", Institute of Phys.Chemistry, 153 10 Ag.Paraskevi, Athens, GREECE.
Campos Ramon , Consejo Superior de Investigaciones Cientificas , NMR Dept., Serrano 119, 28006 Madrid, SPAIN.
Chifotides Helen , NCSR "Demokritos", Inst. of Phys.Chemistry, 153 10 Ag.Paraskevi Athens, GREECE.
Cowburn David , The Rockefeller Univ., 1230 York Ave., 10021 New York, USA.
Cutts Rosalind , Surrey Univ., Dept. of Chemistry, Guildford, Surrey, GU2 5XN, UK.
Dafnomili Dimitra, Univ. of Crete, Dept. of Chemistry, 71409 Heraklion, GREECE.

Dervakos George , UMIST, Dept. of Chemical Engineering, P.O. Box 88, Manchester, M60 1QD, UK.
Detsikas Evangelos , Univ. of Ioannina, Dept. of Chemistry, 451 10 Ioannina, GREECE.
Ertosun Suheda , Istanbul Univ., Dept. of Pathology, Istanbul Medical Faculty, 343 90 Istanbul, TURKEY.
Felli Isabella , Univ. of Florence, Dept. of Chemistry, V.Giano Della Bella 24, 501 24 Florence, ITALY.
Gunes Mehasin , Bogazici University, Dept. of Biomedical Engineering, Bebek, 80815 Istanbul, TURKEY.
Hadjiliadis Nick , Univ. of Ioannina, Dept. of Chemistry, Lab. of Inorg.Chemistry, 451 10 Ioannina, GREECE .
Hamodrakas Stavros , Univ. of Athens, Dept. of Biology, Panepistimiopolis, 157 01 Athens, GREECE .
Hansen Svend Hoime , Novo Nordisk A/S, Heparin Research Lab., Niels Steensens Vej 1, Gentofte 2820, DENMARK .
Hard Karl , Utrecht University, Dept. of NMR Spectroscopy, Padualaan 8, 3584 CH Utrecht, The NETHERLANDS.
Harris James , Queen Mary & Westfield College, Dept. of Chemistry, Mile End Rd.,E14NS London, UK.
Hinton James , Arkansas Univ., Dept. of Chemistry/Biochemistry, Fayetteville Arkansas 72701, USA.
Hofmann Hans-Peter , Univ. Bayreuth, Dept.of Biochemie, Postfach 101251, 95440 Bayreuth,GERMANY.
Kalatzis George, Univ. of Athens, Dept.of Inorganic Chemistry, Panepistimiopolis, Kouponia, 157 01 Athens, GREECE.
Kodess Michail , Dept. of Fine Organic Synthesis of Ural, Division of Russian Academy of Sci., ul S.Kovalevskoj 20, Ekaterinburg 620219, RUSSIA.
Korkmaz Mustafa , Hacettepe Univ., Dept. of Physics Engineering, Beytepe, 06532 Ankara, TURKEY.
Kovala-Demertzi Dimitra , Univ. of Ioannina, Dept. of Chemical, Eleusa, Ioannina, GREECE.
Kukreti Shrikant , Jawaharlal Nehru University, Dept. of Biotechnology, N.Mehrauli Road, New Delhi 110 067, INDIA.
Kupka Teobald , Univ. of Silesia, Dept. of Chemistry, Szkolna 9 Street, Katowice 40-006, POLAND.
Leondiadis Leondios , Univ. of Crete, Dept. of Chemistry, GREECE.
Legault Pascale , University of Colorado, Boulder, Dept. of Chemistry and Biochemistry, Campus Box 215, Boulder Co, USA.
Liptaz Tibor , Slovak Technical Univ., Central Laboratory, Radlinskeho 9, Bratislava 812 37, SLOVAKIA.
Madsen Jens Christian , Carlsberg Lab., Dept. of Chemistry, Gamle Carlsberg Vej 10 Valby, Cph., DK-2500, DENMARK.
Marques M.Paula , Univ. of Coimbra, Dept. of Biochemistry, Apartado 3126 3049 Coimbra, PORTUGAL.
Martorell Gabriel , Nation Institute for Medical Res., The Ridgeway, Mill Hill NW7 1AA London, UK.
Matsoukas Ioannis, Univ. of Patras, Dept of Chemistry, 26110 Patras, GREECE.

Mavromoustakos Thomas , National Hellenic Research Foundation, Inst. Organic and Pharmaceutical Chemistry, 48 Vas.Constantinou Ave., 115 36 Athens, GREECE.

Meacham Kenneth , Univ. of Southampton, Dept. of Chemistry, Highfield, Southampton, SO9 5NH, UK.

Mikros Emmanuel , Univ. of Athens, Dept. of Pharmacy, Panepistimiopolis, 157 71 Zografou, GREECE.

Nilges Michael , EMBL, Dept. of Structures, Meyerhofstr 1, 6900 Heidelberg, GERMANY.

Pardi Arthur , Univ. of Colorado, Boulder, Dept. of Chemistry and Biochemistry Campus Box 215, Boulder 80309, USA.

Pelecanou Maria , Univ. of Patras, Dept. of Pharmaceutical Chemistry University of Patras, Patras, GREECE.

Petrou Athinoula , Univ. of Athens, Dept. of Inorganic Chemistry, Panepistimiopolis, Kouponia, 157 01 Athens, GREECE.

Pierattelli Roberta , Univ. of Florence, Dept. of Chemistry, V.G.Capponi 7, 90121 Florence, ITALY.

Pissas Dimitrios , Univ. of Crete, Dept. of Chemistry, 714 09 Heraklio, GREECE

Pistolis George , NCSR "Demokritos", Inst. of Phys.Chemistry, 153 10 Ag.Paraskevi Athens, GREECE.

Politou Anastassia, EMBL, Structures Dept., Meyerhof str. 1, 69117 Heidelberg, GERMANY.

Pook Karl-Heinz , Boehringer Ingelheim KG, Dept. of Analyt.Research, Binger str. Ingelheim, D-6507, GERMANY.

Reif Werner Bernd , University of Bayreuth, Lab. of Biochemistry, Universitatsstrasse 30, 6580 Bayreuth, GERMANY.

Sidorczyk Zygmunt , University of Lodz, Dept. of General Microbiology, Banacha Str. 12/16, POLAND.

Sijbesma Wilbert F.H. , I.T.Q.B., Oeiras, NMR Dept., Rua da Quinta Grande 6, ap.127, 2780 Oeiras, PORTUGAL.

Simorre Jean-Pierre , Inst. de Biologie Structurale, 41 Avenue des Martyrs, Grenoble, FRANCE.

Spithouris George , Agricultural Univ. of Athens, Dept. of General Chemistry, 75 Iera Odos, 118 55 Athens, GREECE.

Spyros Apostolos , Univ. of Crete, Dept. of Chemistry, 71409 Heraklion,GREECE.

Tato Marco , Farmitalia Carlo Erba, Computational Chem., Via Giovanni XXIII 23, 200 14 Nerviano - Milan, ITALY .

Teeaar Raivo , Estonian Academy of Sci., Inst. of Chemical Physics & Biophysics, Dept. of Chemical Physics, Ravala Pst 10, Tallinn EE0100, ESTONIA.

Troganis Anastassios , Ohio State Univ., Dept. of Chemistry, 120W 18th Ave. Columbus 43210, USA.

Trouard Ted , Univ. of Umea, Dept of Physical Chemistry, Umea, S-901 87, SWEDEN.

Tsikaris Vassilios , Univ. of Ioannina, Dept. of Chemistry, 451 10 Ioannina, GREECE.

Tsiveriotis Panagiotis , Univ. of Ioannina, Chemical Dept., 451 10 Ioannina, GREECE.

Tzougraki Chryssa , Univ. of Athens, Dept. of Chemistry, Panepistimiopolis 157 71 Athens, GREECE.
Vakka Constantina , Univ. of Ioannina, Dept. of Chemistry, 451 10 Ioannina GREECE.
Van Tilborg Marc , Utrecht Univ., Dept. of NMR-Spectroscopy, Padualaan 8 3584 CH Utrecht, The NETHERLANDS.
Vis Hans , Utrecht University, Dept. of NMR Spectroscopy, Padualaan 8, 3584 CH Utrecht, The NETHERLANDS.
Vorstenbosch Erik L.H. , Leiden Univ., Dept. of Biochemistry, P.O. Box 9502, 2300 RA Leiden, The NETHERLANDS.
Yannakopoulou Dina , NCSR "Demokritos", Institute of Phys.Chemistry, 153 10 Ag.Paraskevi, Athens, GREECE.

INDEX

NATO ASI Series H

Vol. 1: **Biology and Molecular Biology of Plant-Pathogen Interactions.** Edited by J.A. Bailey. 415 pages. 1986.

Vol. 2: **Glial-Neuronal Communication in Development and Regeneration.** Edited by H.H. Althaus and W. Seifert. 865 pages. 1987.

Vol. 3: **Nicotinic Acetylcholine Receptor: Structure and Function.** Edited by A. Maelicke. 489 pages. 1986.

Vol. 4: **Recognition in Microbe-Plant Symbiotic and Pathogenic Interactions.** Edited by B. Lugtenberg. 449 pages. 1986.

Vol. 5: **Mesenchymal-Epithelial Interactions in Neural Development.** Edited by J. R. Wolff, J. Sievers, and M. Berry. 428 pages. 1987.

Vol. 6: **Molecular Mechanisms of Desensitization to Signal Molecules.** Edited by T M. Konijn, P J. M. Van Haastert, H. Van der Starre, H. Van der Wel, and M.D. Houslay. 336 pages. 1987.

Vol. 7: **Gangliosides and Modulation of Neuronal Functions.** Edited by H. Rahmann. 647 pages. 1987.

Vol. 8: **Molecular and Cellular Aspects of Erythropoietin and Erythropoiesis.** Edited by I.N. Rich. 460 pages. 1987.

Vol. 9: **Modification of Cell to Cell Signals During Normal and Pathological Aging.** Edited by S. Govoni and F. Battaini. 297 pages. 1987.

Vol. 10: **Plant Hormone Receptors.** Edited by D. Klämbt. 319 pages. 1987.

Vol. 11: **Host-Parasite Cellular and Molecular Interactions in Protozoal Infections.** Edited by K.-P. Chang and D. Snary. 425 pages. 1987.

Vol. 12: **The Cell Surface in Signal Transduction.** Edited by E. Wagner, H. Greppin, and B. Millet. 243 pages. 1987.

Vol. 13: **Toxicology of Pesticides: Experimental, Clinical and Regulatory Perspectives.** Edited by L.G. Costa, C.L. Galli, and S.D. Murphy. 320 pages. 1987.

Vol. 14: **Genetics of Translation. New Approaches.** Edited by M.F. Tuite, M. Picard, and M. Bolotin-Fukuhara. 524 pages. 1988.

Vol. 15: **Photosensitisation. Molecular, Cellular and Medical Aspects.** Edited by G. Moreno, R. H. Pottier, and T. G. Truscott. 521 pages. 1988.

Vol. 16: **Membrane Biogenesis.** Edited by J.A.F Op den Kamp. 477 pages. 1988.

Vol. 17: **Cell to Cell Signals in Plant, Animal and Microbial Symbiosis.** Edited by S. Scannerini, D. Smith, P. Bonfante-Fasolo, and V. Gianinazzi-Pearson. 414 pages. 1988.

Vol. 18: **Plant Cell Biotechnology.** Edited by M.S.S. Pais, F. Mavituna, and J. M. Novais. 500 pages. 1988.

Vol. 19: **Modulation of Synaptic Transmission and Plasticity in Nervous Systems.** Edited by G. Hertting and H.-C. Spatz. 457 pages. 1988.

Vol. 20: **Amino Acid Availability and Brain Function in Health and Disease.** Edited by G. Huether. 487 pages. 1988.

NATO ASI Series H

Vol. 21: **Cellular and Molecular Basis of Synaptic Transmission.**
Edited by H. Zimmermann. 547 pages. 1988.

Vol. 22: **Neural Development and Regeneration. Cellular and Molecular Aspects.**
Edited by A. Gorio, J. R. Perez-Polo, J. de Vellis, and B. Haber. 711 pages. 1988.

Vol. 23: **The Semiotics of Cellular Communication in the Immune System.**
Edited by E.E. Sercarz, F. Celada, N.A. Mitchison, and T. Tada. 326 pages. 1988.

Vol. 24: **Bacteria, Complement and the Phagocytic Cell.**
Edited by F. C. Cabello und C. Pruzzo. 372 pages. 1988.

Vol. 25: **Nicotinic Acetylcholine Receptors in the Nervous System.**
Edited by F. Clementi, C. Gotti, and E. Sher. 424 pages. 1988.

Vol. 26: **Cell to Cell Signals in Mammalian Development.**
Edited by S.W. de Laat, J.G. Bluemink, and C.L. Mummery. 322 pages. 1989.

Vol. 27: **Phytotoxins and Plant Pathogenesis.**
Edited by A. Graniti, R. D. Durbin, and A. Ballio. 508 pages. 1989.

Vol. 28: **Vascular Wilt Diseases of Plants. Basic Studies and Control.**
Edited by E. C. Tjamos and C. H. Beckman. 590 pages. 1989.

Vol. 29: **Receptors, Membrane Transport and Signal Transduction.**
Edited by A. E. Evangelopoulos, J. P. Changeux, L. Packer, T. G. Sotiroudis, and K.W.A. Wirtz. 387 pages. 1989.

Vol. 30: **Effects of Mineral Dusts on Cells.**
Edited by B.T. Mossman and R.O. Begin. 470 pages. 1989.

Vol. 31: **Neurobiology of the Inner Retina.**
Edited by R. Weiler and N.N. Osborne. 529 pages. 1989.

Vol. 32: **Molecular Biology of Neuroreceptors and Ion Channels.**
Edited by A. Maelicke. 675 pages. 1989.

Vol. 33: **Regulatory Mechanisms of Neuron to Vessel Communication in Brain.**
Edited by F. Battaini, S. Govoni, M.S. Magnoni, and M. Trabucchi. 416 pages. 1989.

Vol. 34: **Vectors asTools for the Study of Normal and Abnormal Growth and Differentiation.**
Edited by H. Lother, R. Dernick, and W. Ostertag. 477 pages. 1989.

Vol. 35: **Cell Separation in Plants: Physiology, Biochemistry and Molecular Biology.** Edited by D. J. Osborne and M. B. Jackson. 449 pages. 1989.

Vol. 36: **Signal Molecules in Plants and Plant-Microbe Interactions.**
Edited by B.J.J. Lugtenberg. 425 pages. 1989.

Vol. 37: **Tin-Based Antitumour Drugs.** Edited by M. Gielen. 226 pages. 1990.

Vol. 38: **The Molecular Biology of Autoimmune Disease.**
Edited by A.G. Demaine, J-P. Banga, and A.M. McGregor. 404 pages. 1990.

NATO ASI Series H

Vol. 39: **Chemosensory Information Processing.**
Edited by D. Schild. 403 pages. 1990.

Vol. 40: **Dynamics and Biogenesis of Membranes.**
Edited by J. A. F. Op den Kamp. 367 pages. 1990.

Vol. 41: **Recognition and Response in Plant-Virus Interactions.**
Edited by R. S. S. Fraser. 467 pages. 1990.

Vol. 42: **Biomechanics of Active Movement and Deformation of Cells.**
Edited by N. Akkas. 524 pages. 1990.

Vol. 43: **Cellular and Molecular Biology of Myelination.**
Edited by G. Jeserich, H. H. Althaus, and T. V. Waehneldt. 565 pages. 1990.

Vol. 44: **Activation and Desensitization of Transducing Pathways.**
Edited by T. M. Konijn, M. D. Houslay, and P. J. M. Van Haastert. 336 pages. 1990.

Vol. 45: **Mechanism of Fertilization: Plants to Humans.**
Edited by B. Dale. 710 pages. 1990.

Vol .46: **Parallels in Cell to Cell Junctions in Plants and Animals.**
Edited by A. W Robards, W. J . Lucas, J . D. Pitts, H . J . Jongsma, and D. C. Spray. 296 pages. 1990.

Vol. 47: **Signal Perception and Transduction in Higher Plants.**
Edited by R. Ranjeva and A. M. Boudet. 357 pages. 1990.

Vol. 48: **Calcium Transport and Intracellular Calcium Homeostasis.**
Edited by D. Pansu and F. Bronner. 456 pages. 1990.

Vol. 49: **Post-Transcriptional Control of Gene Expression.**
Edited by J. E. G. McCarthy and M. F. Tuite. 671 pages. 1990.

Vol. 50: **Phytochrome Properties and Biological Action.**
Edited by B. Thomas and C. B. Johnson. 337 pages. 1991.

Vol. 51: **Cell to Cell Signals in Plants and Animals.**
Edited by V. Neuhoff and J. Friend. 404 pages. 1991.

Vol. 52: **Biological Signal Transduction.**
Edited by E. M . Ross and K . W. A. Wirtz. 560 pages. 1991.

Vol. 53: **Fungal Cell Wall and Immune Response.**
Edited by J. P. Latge and D. Boucias. 472 pages. 1991.

Vol. 54: **The Early Effects of Radiation on DNA.**
Edited by E. M Fielden and P. O'Neill. 448 pages. 1991.

Vol. 55: **The Translational Apparatus of Photosynthetic Organelles.**
Edited by R. Mache, E. Stutz, and A. R. Subramanian. 260 pages. 1991.

Vol. 56: **Cellular Regulation by Protein Phosphorylation.**
Edited by L. M. G. Heilmeyer, Jr. 520 pages. 1991.

NATO ASI Series H

NATO ASI Series H

Springer-Verlag and the Environment

We at Springer-Verlag firmly believe that an international science publisher has a special obligation to the environment, and our corporate policies consistently reflect this conviction.

We also expect our business partners – paper mills, printers, packaging manufacturers, etc. – to commit themselves to using environmentally friendly materials and production processes.

The paper in this book is made from low- or no-chlorine pulp and is acid free, in conformance with international standards for paper permanency.

MIX
Papier aus verantwortungsvollen Quellen
Paper from responsible sources
FSC® C105338

If you have any concerns about our products,
you can contact us on
ProductSafety@springernature.com

In case Publisher is established outside the EU,
the EU authorized representative is:
Springer Nature Customer Service Center GmbH
Europaplatz 3, 69115 Heidelberg, Germany

Printed by Libri Plureos GmbH
in Hamburg, Germany